Praise for *Mountain Gazette*:

"...This is not the usual jock-type sports magazine..."
—*Mother Earth News*

"...*Mountain Gazette* is, by turns, in-groupish, funny, maudlin, eclectic, eccentric, imaginative, serious, ribald (not often), monthly and enjoyable. Mostly eccentric, monthly and enjoyable."
—*Friends of the Earth*

"Verbose, amusing, opinionated and cantankerous..."
—*The Denver Post*

" A sometimes quirky, sometimes startling monthly journal. Not mass-society journalism; this is ascetic hedonists illuminating their times with harshness, misanthropy, condescension, humor, integrity and originality."
—*Aspen Times*

"...a connoisseur's magazine."
—*The Village Voice*

"What's most appealing about the *Gazette* is its refusal to pander to the Mountain Dew generation, with Xtreme sports mentality that shouts, 'Get out of my way!'"
—*Brill's Content*

" Best Alternative Regional Publication"
—*Utne Reader*

"...If more of the general public would read *Mountain Gazette*, there would be a wider understanding of man's involvement with nature and less temptation to ask that eternal query, 'Why do you climb mountains?'"
—*The American Alpine Journal*

"Among its glossy newsstand rivals, the resurrected *Gazette* looms like a cornice both physically and in raw literary potential."
—*High Country News*

WHEN IN DOUBT
GO HIGHER

SELECTIONS FROM:

Edward Abbey   Bruce Berger   Charles Bowden
Geoff Childs   Barry Corbet   Dick Dorworth   Katie Lee
John Nichols   Royal Robbins   David Roberts
Galen Rowell   Lacey Story
Lito Tejada-Flores

WITH A FOREWORD BY: Tim Cahill

Mountain GAZETTE
ANTHOLOGY

Edited by M. John Fayhee

MOUNTAIN SPORTS PRESS

Boulder, Colorado USA

When in Doubt, Go Higher: Mountain Gazette Anthology
© 2002 M. John Fayhee

All rights reserved: No part of this book may be reproduced, stored in a retrieval system, or transmitted, in any form or by any means, electronic or photocopy or otherwise, without the prior written permission of the publisher.

Printed in the United States of America.

ISBN 0-9676747-9-4

Library of Congress Cataloging-in-Publication data applied for.

First printing, May 2002

Front cover photograph: Cirque of the Unclimbables, Logan Mountains, Northwest Territories, Canada © 2001 Galen Rowell

Associate Publisher: Alan Stark
Editor in Chief: Bill Grout
Art Director: Michelle Klammer Schrantz
Associate Art Director/Designer: Scott Kronberg
Managing Editor: Chris Salt
Illustrator: Rob Pudim
Account Manager: Andy Hawk

929 Pearl Street, Suite 200
Boulder, CO 80302
303-448-7617

Dedicated to George Stranahan

Table of Contents

Foreword by Tim Cahill I

Introduction by M. John Fayhee V

Where's Tonto? by Edward Abbey 1

The Monkey Wrench Gang: A Review by George Sibley 11

Ascentuality by Michael Charles Tobias 19

A Dream of White Horses by Royal Robbins 27

Mountain Towns by Ted Kerasote 33

The Ride by Katie Lee 37

The Guardian of Sleep by Jeremy Bernstein 45

Hallucinations by Barry Corbet 51

Slouching Toward Simpletopia by George Sibley 63

N.E.D. 77

Hanging Around by David Roberts 83

Crooked Road to the Far North by Lito Tejada-Flores 91

Alaska: Journey by Land by Galen Rowell 109

Growing Up High by Randy LaChapelle 117

Fear by David Roberts 125

Flat Mountain by Charles Bowden 133

The Impsons, Ed & Ma'am by John Peters, M.D. 139

Lady with a Baby by Gilbert Preston .145

On the Frontier by Steve Wishart 155

Where the Trees Walk by Harvey Manning 167

**Breaking Free from the Human
    Potential Movement** by Mike Moore 173

Wild Red Dharma Pickup Truck by Lacey Story 191

Obituary: Lucette K. Car by B. Frank 197

Crossings by M. John Fayhee 203

There Was a River by Bruce Berger 215

For the Sport of It? by Gaylord Guenin 225

Confessions of a Butterfly Chaser by Rob Pudim 237

The South Side of the New England Soul by John Skow 249

Lobster Fishing in America by Geoffrey Childs 257

Confessions of a Sauna Junkie by Jack Aley 269

Gone Fishin' by John Nichols 279

Obituary: Scott Fly Rod by Michael Holzmeister 287

Coyote Song by Dick Dorworth 291

Climbing the Walls in Berkeley by Karen Recknagel 303

The Mogul Problem by Tad Hall 309

Bitches in Heats by Cindy Kleh 313

My Friend Ed by Doug Peacock 321

Appendix A: Letters to the Editor 335

Appendix B: About the Writers 350

Subscription Page for Mountain Gazette

FRANK DAVIDSON

FOREWORD

Dirty Little Secrets

BY
TIM CAHILL

There is a dirty little secret that resides at the core of what has got to be considered a great triumph. It is a matter that I have kept hidden away for over a quarter of a century. Why talk about it out in the open anyway? Nobody cares. So why would I "fess up"?

And then when I was least expecting it—when I figured my friends and I pulled this one off—I got a call from a guy named John Fayhee who asked if I had ever read a publication called *Mountain Gazette*.

"Used to be around in the 70s," I said. "Good magazine. I miss it."

"Well, we're resuscitating it," John said. He was also putting together an anthology of articles from *Mountain Gazette*, both the old version and new. Would I care to write the foreword?

"Sure," I said, and then months passed and I didn't write the foreword because, I realized, there was no way I could do it without telling my dirty little secret. Right now I'm late with it, Fayhee is screaming and I might as well go ahead and spill the proverbial beans.

It was sometime back in 1975, in the *Rolling Stone* offices, back when they were located in the warehouse district in San Francisco. Jann Wenner, the publisher, assigned editors Michael Rogers and Harriet Fier and me to come up with a new magazine idea, one that could be used as a vehicle to sell ads to all these little companies springing up all over that were making sleeping bags and such. Companies with strange names like "Patagonia," and "The North Face." Harriet objected to the concept outright: "You want us to invent a magazine designed to sell ads? That is just not classy."

"Then," Jann said, "you make it classy."

We did. We put together *Outside*, a glossy outdoor magazine of acknowledged literary and artistic merit. It was, from the first, a writer's magazine, and writers liked working for us, liked being in the company of other authors they admired.

But in the early days, we were roundly ridiculed for our efforts. People who went outdoors, media pundits declared, did not read. Outdoorsy folk were knuckle dragging mouth breathers and aesthetic imbeciles, as evidenced by *the very fact that they went outdoors*.

As an editor and writer for that magazine, I sometimes found myself defending it on radio talk shows or TV programs of the *Good Morning Cleveland* variety. I argued, to a succession of hosts that American literature—from *Moby Dick* through *Huckleberry Finn* through Faulkner and Hemingway—was about the outdoors. We weren't doing something new.

Well, today, the idea seems like a slam dunk, and *Outside* is the only magazine to have ever won the National Magazine Award for General Excellence three times in a row. We, all of us, succeeded in our intention of putting out a literate magazine about the outdoors. We've won all sorts of awards.

But here's the dirty little secret: *Mountain Gazette* was doing the same thing—publishing literate writing about the outdoors—and they were doing it years before we at *Outside* ever published our first issue.

I can tell you that we were acutely aware of *Mountain Gazette* from the first. In the initial phases of creating the magazine that *Outside* would become, Harriet, Michael and I spent several months reading every outdoor magazine then on the market. Most of what we read was service oriented, which is to say, if a magazine purported to be about canoeing, it told the reader how to buy and paddle a canoe twelve times a year. The articles were informational rather than inspiring.

But there was one magazine we admired: this dingbat effort out of Denver called *Mountain Gazette*. It was swollen with attitude, arrogant as all hell, and a complete delight to read. They had all these fantastic writers, Ed Abbey among them, producing stuff that we, the creators of *Outside*, would have been proud to publish. In fact, we did publish those writers later on. Abbey wrote for us, as did Lito Tejada-Flores, Gordon Wiltsie, David Roberts and Doug Robinson. Indeed, it was Doug who taught me to telemark (not his fault I'm so poor at it), and who took me rock climbing in Yosemite. He taught me to see and feel the wilderness in a way I had not imagined before, and I owe him deeply for that, but think I may be able to get away with buying him a few beers.

Later, *Mountain Gazette*'s founding editor came to *Outside*. Mike Moore was a classy guy and what we call in the biz "a tasty editor." What I learned from Mike is invaluable, and I can never repay him, but he probably wouldn't turn down a couple of beers either.

As the years ground on, *Mountain Gazette* began to flounder—I have no idea why—and it died a quiet death in 1979, after seven mostly glorious years. *Outside* was struggling at the time, but it published the work of many *Gazette* alumni. By the late 80s and early 90s, *Outside* was unstoppable: a financial and critical triumph. We were congratulated in the media for our foresight: imagine, a literary outdoor magazine. Who'd a thunk it?

Well, *Mountain Gazette* did, and four or five years before we did.

OK. There. I said it. That's my dirty little secret. I feel much better now.

INTRODUCTION

Guilty as Charged

BY
M. JOHN FAYHEE

My good friend Tom Jones Jr.—author of the official guidebook to the Colorado section of the Continental Divide Trail (if you've used his book and become disoriented as a result, I have Tom's home phone number, and I will gladly share it with you)—calls them "forehead-slappers." I wish I could lay claim to that term (a forehead-slapper in and of itself), but I live too close to Tom to do that. He would find out and lord my phrase-stealing over me till the end of time. Anyhow, the forehead-slapper-in-question took the form of a man who we'll call Bear phoning me up out of the blue one otherwise fine and beer-filled High Country autumn day to suggest that we put together some sort of *Mountain Gazette* book. Had I been sober enough to process a decent battle plan, I would have instantly stolen the idea and run like the wind, forevermore denying that I had ever even heard of this wicked profligate named Bear.

As it was, I was decidedly not on my toes and, the next thing I knew, I was ass-deep in this book project that I can't for the life of me believe: (1) I didn't think of it myself and (2) didn't steal it so fast Bear's head would still be spinning like a gyroscope on speed. The only reason I haven't ripped my eyes out in forehead-slapping shame is that I have in my possession a large stack of seriously incriminating photos of Bear taken at P.T.'s Emporium in Denver during the tag-team bikini jello wrestling finals, and I know, one day, the idea for this book will magically become mine not only legally, but historically.

When Bear proposed this idea, which, as we all know, he blatantly stole from me, I said this will be one serious piece of cake; all we would need to do is bind-up all 80-some-odd issues of the *Gazette*, slap a huge price tag upon the volume and proceed to rake in vast quantities of praise and money. The realization that we were then talking about more than 3,000 pages of tabloid-sized text sort of nixed that notion, so yours truly, who as usual sort of neglected to eyeball the fine print of the legal documents Bear and his team of corporate heavies made me sign (proving once again what Tom Waits said: "The large print giveth, the small print taketh away"), was compelled to read through all of those 3,000-plus pages and to pick and choose which stories would be included in this book.

Though this was difficult in the one sense (there are a lot of pieces that I love that are not included; more on that in a moment), it was pure joy in another sense, the most important sense, as it were. For decades, much to the consternation of my spouse—who does not share her husband's opinion that stacks of musty magazines piled on the coffee table make for cutting-edge decor—I have kept every issue of *Mountain Gazette* within arm's reach in my living room. I have thumbed through, digested, read and re-read and pondered every single issue, story, photo, illustration, cartoon and advertisement. This magazine has long been, if not my life, then a large part of my life. For three years, I ran a weekly publication called *Summit Outdoors* that was conceptually based upon (read that: "conceptually plagiarized" from) *Mountain Gazette*. Long ago I adopted a writing persona that was stylistically and attitudinally based upon *Gazette* articles penned by the likes of Edward Abbey and George Sibley (though, needless to say, the scale of comparison is slightly out-of-kilter).

Some background is necessary here, before I get too far ahead of myself. From 1966–72, a man named Mike Moore ran a publication called *Skier's Gazette*, which he was about to close down for financial reasons. Bob Craig, the ex-president of the American Alpine Club who authored the astounding book, *Storm and Sorrow in the High Pamir,* hooked Moore up with George Stranahan. The two men hobnobbed and, in mid-1972, with Stranahan's financial backing, *Skier's Gazette* became *Mountain Gazette*. Moore conceived and edited it for five years, before handing the reins over to Gaylord Guenin.

For a number of reasons, *Mountain Gazette*, despite its ample and justifiable critical acclaim, could not turn the financial corner into the greed-filled 80s. With much ceremony, in 1979, it was closed down with issue 78.

Fast-forward to 1999. I had just ended a ten-year stint with the company that owns the *Summit Daily News*. That company had allowed me to run *Summit Outdoors*, despite the fact that it lost money every second it existed. Finally, upper management, in an effort to reroute the flow of red ink into their personal bonus packages, told me that I was going to have to lay off a staff person. In an employment maneuver that is still known in this part of Colorado as "pulling a Fayhee," I did as they asked: I laid me off. Once I hid all the knives and guns from my betrothed, I had to decide how I was going to replace the (by my dirtbag standards) reasonable amount of regular money I had grown accustomed to spending on things like mortgage payments, food and carbonated beverages. The obvious choice was to return to freelance writing, which I did with decent success for years. But I hate going back to things, especially things that require grovelling, shit-eating and a feigned willingness to pen gear and destination stories for glossy magazines run by sales and marketing people.

While thumbing through my musty stack of *Mountain Gazettes* one day, an idea began to germinate. Several days later, I was literally getting ready to begin the process of hunting down the legal status of the name *Mountain Gazette*, when the phone rang. It was a hideous ne'er-do-well named Curtis Robinson, who, if you ever meet him, you should post haste smear garlic on the back of your neck and make the sign of the cross. Or else just hand him a pint of Guinness and hope for the best.

"You ever heard of something called *Mountain Gazette*?" Robinson queried, as my stomach tightened. Ended up that the aforementioned Stranahan, with whom Curtis was acquainted by way of a place called the Woody Creek Tavern, had decided he would like to relaunch the publication. Curtis and I, knowing the karmic ramifications of side-stepping a direct message from God (in our humble cosmos, that would be Stranahan), signed on and, in November 1999, we relaunched the *Gazette*, to much simultaneous applause and head-shaking. (Please understand that, in our relaunch issue, #78, Curtis and I took more than 3,000 words to describe all this; this horrible idea-stealing man named Bear has only given me 1,700—OK, 2,000—words for this entire intro.)

From the get-go, Curtis and I understood that we were treading upon some sacred editorial and conceptual turf. If we screwed this up—and there were so many ways we could do so we stopped counting at "get sent to prison for tax fraud"—we knew the ghost of Cactus Ed would follow us around like a rabid dog till we just laid down and died. We have been heartened at how supportive just about everyone has been. Many of what we call the "Gazette alumni"—including George Sibley, Dick Dorworth, Royal Robbins, Bob Chamberlain, Karen Chamberlain, Bruce Berger et al.—have been gracious enough to send words and images our way, and a whole slew of other notable writers and photographers—among them John Nichols, Charles Bowden and Katie Lee—have jumped aboard. And others—including many old *Gazette* writers and photographers whose work appears in this volume—have given us what support they could, including not talking too badly about us behind our backs when we make horrible mistakes.

Our relaunch effort was less than a year old when Bear called. After a quick head-clearing road trip up to the Northwest Territories, I jumped into the process, which was tough, because it sort of required me to interface with the thought processes of both Mike Moore and Gaylord Guenin, two people who are so much more creative, intelligent and perceptive than I am that my stomach quivers at the mere thought of having my name uttered in the same breath as theirs, which, so far, no one has done anyhow. So I guess I can relax.

When I first started working on this book, I glibly referred to it as, "*Mountain Gazette's Greatest Hits*." I stopped doing that early on because that's not necessarily what this is. It may be *MG's Greatest Hits* as John Fayhee perceives things, but I've had my hand on the *Gazette* tiller for a far shorter time than either Moore or Guenin. It is from their watch that most of these stories come and it is my guess that, were they sitting in my chair, they likely would have chosen a different line-up. So, we've gone with the "Anthology" appellation, which gives me cache to say pretty much "these are my favorites." There are going to be people out there who have had as long a running relationship with *MG* as I have (including Messrs. Moore and Guenin) who are going to wonder aloud why such-and-such masterpiece was not included. I have only my personal taste and a desire to be representative as a defense. It is my hope that this book only serves as Volume 1, and that subsequent *Gazette* anthologies follow, with stories picked by other people.

    Some necessary notes on structure. This man Bear, idea poacher that he is, being a professional book-producer for most of his 55 years, was inclined to actually organize this tome. Our organizational options ran the gamut from straight-up in-order chronological to subject-based (skiing, climbing, drinking, wild bestiality, etc.). But neither of those options felt right, and, here in our corporate offices in the Colorado High Country, feel is criteria enough for making any and all decisions. We then entertained the idea of having one section (section pi) reserved for the stories that make sense, and one section (section infinity) reserved for the stories that do not. Problem with that was Bear and I could not agree on which ones fall into which of those categories. So, we ended up shuffling the manuscripts like a deck of cards and, well, here we are.

    Before wishing you godspeed in your attempt to make your way through this book, I need to point out that some of the stories come from *Gazettes* that were published on my watch—maybe even a disproportionate number. I guess I feel like Joe Torre when he was managing the American League All-Stars and he picked 12 Yankees to be on the team. I have no defense in this regard, except that I do feel more of a sense of proprietorship with the stories I specifically chose to run in the *Gazette*. That especially applies to the one story I've included that was penned by, well, me. The way I look at it is, when you get the chance to have your work included in a book alongside names like Abbey, Roberts and Nichols, you jump at the chance and deal with the accusatory fallout later.

    Bill Cosby once said that he didn't know what the secret of success was, but he knew for certain the secret of failure was to try to please everyone. I put every issue of the *Gazette* together with the understanding that not every reader will like every story. I strive for diversity the same way I believe Mike Moore and Gaylord Guenin did, and that effort will surely manifest itself in this book: Not all of you will like every chapter. Hell, I don't like everything we publish in the *Gazette*. But it's my guess you will find plenty of material in here that you will not only like, but that you will remember. If not, please send your copy of the book to someone you hate. You may even convince me to pick up the postage, but it'll cost you some beer.

    Second to last: The *Gazette* has always used art—photos, illustrations, overtly plagiarized clip art—rather loosely. A significant percentage of the art accompanying *Gazette* stories (or is it the other way around?) has little if anything to do with the words. There are many reasons for this "design philosophy," not the least of which being disorganization on the part of many of our writers and (at least in my case) editors. It was tempting to engage in some revisionist history in this book by way of making certain that the art included herein actually matched up in some conceptual way or another with the story. Books are, after all, supposed to be loftier in their self-perception than mere magazines. But we nixed the notion because it seemed like cheating, not that we have anything against cheating per se. So: Par for *Mountain Gazette's* course, much of the art in this book has little if anything to do with the story it's attached to (although I should point out that many of the photos and illustrations did actually accompany the story when it first appeared in the *Gazette*, which when you get right down to it, is apropos of nothing). And much of the art was done by long-time *Gazette* contributor Rob Pudim, specifically for this book. If any of the art goes with the story it appears with too well, I apologize and promise to correct the problem in future editions.

Last: I made every reasonable effort to contact all of the writers and photographers whose work is included in this book. I succeeded in the majority of cases. A couple, sad to say, had passed away. A couple I simply could not find, despite my best efforts. To any of these writers or photographers who may stroll into their local prison library and see their work used herein without permission, I offer my most profuse apologies. All the people I was able to contact were happy to let us use their stories and photos (for which I thank them from the bottom of my heart), and I would like to think that, had I been able to find you, you would have also.

Last (seriously): In addition to all the contributors, I'd like to thank Laurel Wehrman, art director for *Mountain Gazette*, for helping us get words and pictures to the publishers, and Lynn Amstutz for all her transcription help.

I hope you enjoy this book, if for no other selfish reason than that I consider it one of the best ideas I ever had.

—M. John Fayhee, Editor & Publisher, *The Mountain Gazette*
The Dillon Dam Brewery, 11 February 2002

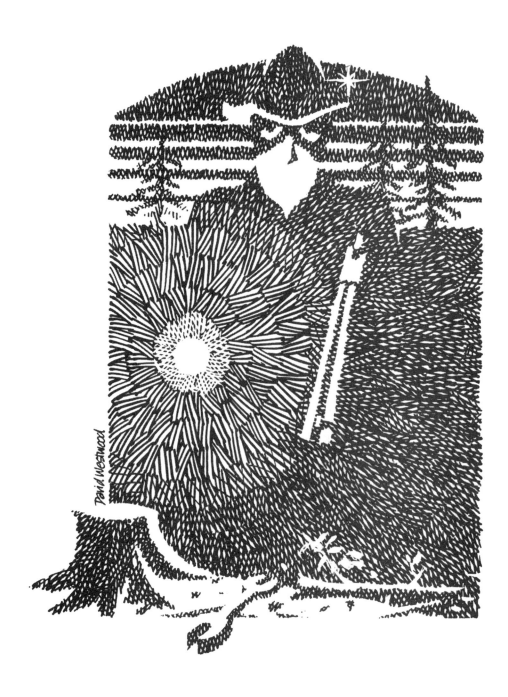

Where's Tonto?

BY
EDWARD ABBEY

Hayduke parked his jeep out of sight among the pines, near the entrance to the Georgia-Pacific logging area. He stationed Bonnie on the hood of the jeep with instructions to keep her eyes open and ears clean. She nodded impatiently. Yes, yes, I know what to do; I'm the best damned lookout you'll ever have, Hayduke.

He put on hard hat, coveralls, gunbelt, gun and leather work gloves, took a small flashlight and his other tools and disappeared from Bonnie's ken into the twilight of the cut-over site, fading like a shadow among the giant machines. She wanted to read but it was already too dark. She sang songs for a while, softly, and listened to the cries of little birds, off in the forest, retreating to their nests for the night, heads nestled under fold of wing, retiring into the simple harmless dreams of avian sleep. (A bird has no cerebrum.)

She was aware of tall presences around her, the brooding and transpiring yellow pines, the dark shaggy personalities of the Engelmann spruce and white fir—their high crowns pointed like cathedral spires toward the fireball array of the first-magnitude stars—and off by themselves, an exclusive group, a grove of aspens, slim and white, delicate, gay and ladylike. So I see them, she thought. What is their consciousness of me? Do they give a shit? Bonnie Abbzug, the metaphysical arborologist, philosopher of the psycho-morphology of vascular plants. And other properties. She rolled and lit a joint. Weary and warm in her down-stuffed parka, she nodded for a moment, dozed, woke with a start to find that nothing had happened, nothing had changed except the map of the stars becoming slightly more elaborate.

A small wind rustled through the trees. It sounded so much like human voices that Bonnie looked around for a moment in surprise, expecting to see someone near. Nobody there; only Hayduke's little light appearing now and then from beneath or from the hulk around the silhouette of some machine.

She sucked on her weed. The wind continued talking with the trees; like the dialogue of dreams—voices far away, not in space or time but out of reach, on the other side of an invisible barrier. Bonnie smiled, subsiding into a warm reverie of oceanic sympathy. We are all ONE, she thought. One what? Who cares? Hayduke however, under the belly of the bulldozer, was tugging at an oversize spanner, trying to open the drainplug in the crankcase of an Allis-Chalmers HD-41, which is merely the biggest tractor Allis-Chalmers makes. His wrench was three feet long—he'd taken it from the tractor's toolbox—but he couldn't turn that square nut. He reached for his cheater, a three-foot length of steel pipe, fit it like a sleeve over the end of the wrench handle and tugged. This time the nut gave, a fraction of a millimeter. All he needed: Hayduke yanked again and the nut began to turn.

So far he'd done nothing dramatic, merely followed routine procedures: cut up wiring, break off cylinder heads, cut hoses, smash instrument gauges, pour shellac into crankcases, destroy air cleaner and oil filter elements, slice fan belts, crack batteries, smash lights, sprinkle emery powder into gearboxes, puncture tires (where applicable) and chisel-punch a few holes into radiator cores. Nothing special. Where possible, as in the case of the HD-41, he drained the crankcase oil as well, planning to start up the engine just before leaving. (Noise factor.) He had no keys but assumed he would find what he needed by breaking into the G-P office, a small house trailer close by.

Another possibility, of course, was fire. Why not as a farewell salute set fire to the tractors, loaders, skidders, et al., et cetera, once and all? Hayduke was a pyromaniac, fond of fire. He liked the warmth and he liked the purity of it, he appreciated fire's quick cleansing action. But he couldn't do it tonight. Not here. Why not? Because George Hayduke, like Smoky the Bear, had a horror of forest fires. Because he, Hayduke, had worked too many summers as a firefighter in too many national parks and forests. The idea of deliberately setting fire to the number of large oily paint-coated objects upwind from a forest of living trees—even though these objects were set in a clearing, even though he knew the loggers planned to cut most of the trees down anyhow, even though he knew that fires are really good for forests (hadn't Doc Sarvis himself said so and explained, at great and technical length, why it was so?)—despite these considerations, George Hayduke could not do it. Could not bring himself.

Another turn on the plug and the oil would begin to drain. Hayduke eased his body out of the way, regripped his pipe-handled wrench—and froze.

"How you doin, pard?" said a man's voice, deep and low, from more than 20 feet away.

Hayduke reached for his sidearm.

"Naw, don't do that." The man flicked a switch, training the beam of a powerful electric torch directly into Hayduke's eyes. "I got this," he explained, pushing the muzzle of what certainly looked to be a twelve-gauge double-barrelled shotgun into the light, where Hayduke could see it. "Yeah, it's loaded," he said, "and it's cocked and it's touchy as a rattlesnake."

He paused. Hayduke waited.

"Okay," the man said, "now you go ahead and finish what you're a-doin' under there."

"Finish?"

"Go ahead."

"I was looking for something," Hayduke said.

The man laughed, an easy, soft and pleasant laugh. "Is that right?" he said. "Now what the hell is anybody lookin' for under the crankcase guard of a bulldozer at midnight?"

Hayduke thought carefully. It was a good question. "Well…"

"This oughta be pretty good."

"Yeah. Well, I was looking for—well, I'm writing a book about bulldozers, you see, and thought I ought to see what they look like. Underneath."

"That ain't very good. How do they look?"

"Greasy."

"I coulda told you that, pard, saved you all the trouble. What's that three-foot end wrench for you got in your hands? That what you write your book with?"

Hayduke said nothing.

The man said, "You go ahead and finish your job." Hayduke hesitated. "I mean turn the plug. Let the oil out."

Hayduke did as he was told. The shotgun, after all, like the flashlight, was aimed straight at his face. A shotgun at close range is a perfectly logical argument. He loosened the plug; the oil streamed out, sleek, rich and liberated, onto and into the churned-up soil.

"Now," the man said, "drop the wrench, put your hands behind your head and kinda sidewind outa there on your back."

Hayduke obeyed. Wasn't easy, wriggling out from under a tractor without using the hands. But he did it.

"Now roll over on your face." Hayduke obeyed. The man rose from his squatting position, came close, unholstered Hayduke's gun, stepped back and hunkered down again. "Okay," he said, "you can turn over now and sit up." He examined Hayduke's piece. ".357 magnum, Ruger—that's power, boy."

Hayduke faced him. "You don't have to shine that light right in my eyes."

"You're right, pard." The stranger switched off the light. "Sorry about that."

They faced each other in the sudden deep darkness, silent, bashful, each wondering, perhaps, who had the quicker and better night vision. But the stranger had his left thumb on the switch of the flashlight and his right forefinger on the forward trigger of the shotgun.

The stranger cleared his throat. "You sure work slow," he complained. "I been watchin' you for seems like an hour."

Hayduke didn't know what to say.

"But I can see you do a good job. Thorough. I like that." The man spat on the ground. "Not like some of them half-assed dudes I seen up on the Powder River.

Or them kids down in Tucson. Or them nuts that derailed—what's your name?"

Hayduke opened his mouth. Henry Lightcap? he thought. Joe Smith? How about—

"Forget it," the man said. "I don't want to know."

Hayduke stared harder at the face before him, ten feet away in the starlight, gradually becoming clear. He saw at last that the stranger was wearing a mask. Not a black mask over the eyes but simply a big bandana draped outlaw-style over nose, mouth and chin. Above the mask the dark eyes, vaguely shining, peered at him from under the droopy brim of a black slouch hat.

"Who are you?" Hayduke said.

The masked man grinned. "You don't really want to know that," he said. "But I'll tell you this much: they used to call me Kemosabi."

"Who did?"

"Oh that fool of an Indian used to run around with me."

"Tonto?"

"Yeah, that asshole."

"Tonto means fool in Spanish."

"Yeah, he finally caught on. About the same time I learned what Kemosabi really means in Paiute. So we split up. Last I heard old Tonto was hangin' around the United Brethren mission at Elko, hitting the Ripple and the Thunderbird pretty steady. He never was any damn good." The stranger paused, reminiscing, then chuckled. "Bet you thought I was the watchman, didn't you? Made you sweat a little, huh?"

Hayduke was beginning to wish it was the watchman. "Where is the watchman?" he said. (Help!)

"In there." The stranger jerked a thumb toward the office trailer, where a Georgia-Pacific pickup stood parked.

"What's he doing?"

"Nothin'. I got him hogtied and gagged. He's all right. He'll keep till Monday morning. The loggers'll be back and turn him loose."

"Monday morning is tomorrow morning."

"Yeah, I reckon I oughta mosey on outa here."

"Still got your white horse?"

"No, I got rid of him long time ago. That old Silver, he was just too goddamned conspicuous for this line of work. Got me a big old black gelding now. You wanta see him?"

"What do you mean," Hayduke said, "by this line of work?"

"Same thing you're doin'. You wanta see my horse?"

"No. I want my gun back."

"Okay." The stranger handed it back. "Next time you better keep your lookout a little closer. That girlfriend of yours never seen me nor heard me a-tall."

"Where is she?" Hayduke reholstered his weapon, reluctantly.

"Right on the jeep where you left her, puffing on one of them little mary-jane cigarettes. Or she was. Probably out there in the dark somewhere now, wonderin' what the hell's a-goin' on here." The stranger waved one hand at the surrounding night. "Here's something else you want too," he said handing Hayduke a bunch of keys. "Now you can start them starters and burn up them engines real good."

Hayduke looked toward the trailer. "You certain that watchman is secure?"

"I got him handcuffed, hogtied, gagged, dead drunk and locked up."

"Dead drunk?"

"He was half drunk when I got here. After I got the drop on him I made him finish up a pint of bourbon he was suckin' on. He passed out scared and happy."

So that's why nobody squeaked when I knocked on the door. Hayduke looked at the mysterious masked stranger, who was shuffling his feet, apparently eager to leave.

"Where you heading for now?" he asked.

"You don't wanta know, pardner."

A high voice, strained and frightened, came out of the dark. "George—are you all right?"

"I'm all right," he shouted back. "You stay out there, Natalie. Keep watch. Anyway my name is Leopold."

Hayduke jingled the keys, looking at the dark hulk of tractor at his side. "Not sure I know how to start this thing."

The masked man said, "I'll give you a hand. I ain't in that big of a hurry." Off in the woods somewhere a horse stomped, shuffled, nickered. The man listened, turning his head that way. "You be quiet, Sam. I'll come and git you in a minute." He set the (watchman's) shotgun down and turned back to Hayduke. "Come on."

They climbed to the driver's seat of the tractor. Hayduke found himself faced by an impressive battery of switches, dials and levers.

"Okay, hokay," the stranger began, "What do we got here, a HD-41, right? Okay. First put this here lock lever in neutral position. That ties-in the starter switch circuit."

"I know how to operate a Cat," says Hayduke, "but this one's different."

"It's different, all right. This ain't a Cat, this is a Allis-Chalmers. But these new units is simple to start. Now we pull up this here shutoff knob to RUN position. Got a little release button in the middle, see. You gotta press that first. Yeah. That's right. Now…here's a little light switch under the cowl. Might as well see what we're doin. Okay. Now the starter switch is this little button right here by the speed shift."

"That's what I thought," Hayduke said. "But when I tried to start it nothing happened."

"I'll show you why," the masked man said. "You didn't have the master switch turned on."

"The master switch?"

"The master switch. Real tricky, them Allis-Chalmers folks."

"Well, where's the master switch?"

The stranger grinned. "Patience, old buddy. I'll show you. Give me the keys again."

Hayduke gave him the keys. The stranger examined them under the cowling light, chose one, bent down and unlocked a padlocked access plate on the steel floor of the operator's compartment, behind the braking pedals. Lifting the hinged plate, he showed Hayduke the master switch and turned it to the ON position.

"Now," he said, "the batteries are connected to the electrical system. Now we can start the engine. Push that starter button again."

Hayduke pushed the button. The starter solenoid engaged the starter pinion with the flywheel ring gear: the 12-cylinder 4-cycle turbocharged Cummins diesel coughed into life, 1,710 cubic inches of packed piston power. Hayduke was delighted. He pulled back on the throttle lever and the engine revved up smoothly, ready to work. (But heating rapidly.)

"I'm gonna do something with this machine," he announced to the stranger.

"Yes, you are."

"I'm gonna move things around."

"You better move quick then. It ain't goin to last but a few minutes." The stranger was eyeing the instrument panel: oil pressure zero, engine temperature rising. An odd unhealthy noise, like the whine of a sick dog, could be heard already.

Hayduke unlocked the lock lever and pulled the speed shift lever into gear. The tractor bucked forward against the lowered dozer blade, shoving a ton of mud and two yellowpine stumps into the Georgia-Pacific office.

"Not that way," the stranger shouted. "There's a man in there!"

"Right." Hayduke stopped the machine, leaving his load piled high against the buckled trailer wall. He shifted into reverse and the tractor backed over the Georgia-Pacific pickup truck; the truck collapsed like a Coors beer can.

So who's next? Hayduke looked around through the starlight for another target.

"See what you can do with that brand-new Clark Skidder over there," the masked man suggested.

"Check." Hayduke raised the dozer blade, turned the tractor and charged at full throttle—five miles per hour—into the skidder. It crumpled with a rich and satisfying crunch of steel flesh, iron bones. Hayduke buried the wreck under the tracks, squashing it deep into the mud. Now what? He pivoted the tractor 200 degrees, and started for a tanker truck full of diesel fuel.

Somebody was screaming at him. Something was screaming at him. Full throttle forward. The tractor lurched ahead one turn of the sprocket wheels and stopped. The engine block cracked; a jet of steam shot forth whistling from the fissure. The engine fought for life. Something exploded inside the manifold and a gush of blue flame belched from the exhaust stack, launching hot sparks at the stars. Seized-up tight within their chambers, the 12 pistons became one—wedded and welded—with cylinders and block. All is One. One what? Why, one unified immovable white-hot entropic molecular mass, what else? The screaming in the night.

"She's foundered," the masked man said. "There ain't nothin' we can do." He clambered off over the rear, under the eight-ton rippers. "Come on," he shouted, "there's somebody comin'!" He melted into the darkness.

Hayduke pulled himself together and got off the tractor. He still heard somebody screaming at him. Bonnie.

She yanked at his sleeve, pointing away into the woods. "Can't you see?" she screamed. "Lights, lights! What's the matter with you?"

Hayduke stared, then grabbed her arm. "This way."

They ran across the clearing, among the stumps, toward the sweet shelter of the forest as a truck came rumbling into the open area. Headlights flared; a spotlight swept across the open and almost caught them.

Not quite. They were in the woods, among the friendly trees. Feeling their way through the dark in what he thought was the direction of his jeep, Hayduke heard a thunder of mighty hooves and a cry that rang through the night:

"Hi-yo Samuel, awaaaaaaaaaaaaaay...!"

The mysterious masked stranger galloped past, his big black horse in a full run. The truck, which had come to a stop beside the whistling bulldozer, discharged some men—one, two, three, impossible to count them in the dark. Hayduke and Abbzug watched the spotlight probing the clearing, the trees, seeking the horse. Again too late: one glimpse of the horseman and he was gone, into the forest and down the road off to the end of the night. A gun barked once, twice, importantly but futile, and relaxed. The hoofbeats faded away. The men at the truck moved to the assistance of somebody inside the office-trailer, who was kicking at the walls. They'd have a tough time getting him out with the load of rubble banked against the door.

Bonnie and George got into their jeep.

"Who in God's name was that?" she demanded.

"The watchman I guess."

"No, I mean the man on the horse."

"Men call him...Kemosabi."

"I'm in no mood for bullshit."

"That's what he said."

"Who was he, goddammit?"

"A man from the past. Shut your door."

Hayduke started the motor.

"They'll hear us," she said.

"Not with that bulldozer howling they won't." He drove without any light but starlight out of the trees, slowly, and onto the main forest road, heading back toward the highway. When he felt he had gone a safe distance he turned on the headlights and stepped on the gas. The well-tuned jeep purred forward, guided by its burning eye-beams, at maximum cruising speed, 55 mph. Sure could use overdrive on this vehicle, Hayduke thought, as he had thought a hundred times before.

"I still want to know who that man was."

"I don't know, sweetheart. All I know is what I told you. He said his name is Kemosabi."

"What kind of name is that?"

"It's a Paiute word."

"Meaning what?"

"Shithead."

"That figures." She huddled closer. "Who was in the truck?"

"I don't know and I didn't want to find out, did you?" He decided to stick it to her. "Did you, my hotshot lookout?"

"Listen," she said, "don't give me any hard time about that. You wanted me to stay at the jeep and that's what I did. How was I to hear some lunatic on foot creeping around in the dark? I was watching the road like you wanted me to. So shut up."

"Okay."

"And amuse me, I'm bored."

"Okay, okay."

But why, he wondered, did they call him the Lone Ranger, when he had that faithful dumb Indian always at his side? Why? And then the answer came, obvious as the sneer on a racist's face: because a white man with only an Indian for companionship…Yeah…Of course…No wonder they don't come to our Thanksgiving picnics anymore.

"Amuse me," she said again.

"All right," he said, "consider this conundrum."

"Where?"

"Where? What the fuck are you talking about? Now listen carefully: What is the difference between God and the Lone Ranger?"

The jeep bumbled through the dark of the woods. Bonnie Abbzug thought and thought. At last she said, "What a stupid conundrum. I give up."

Hayduke grinned. "There really is a Lone Ranger."

"I don't understand."

"Forget it." He drove slowly now, looking for a place to camp. He pulled her a little harder against his side. "Those new sleeping bags, you know…"

"Yeah?"

"The fuckers zip together."

"No kidding!"

—MOUNTAIN GAZETTE, #28

REVIEWS:
The Monkey Wrench Gang

BY
EDWARD ABBEY

REVIEW BY
GEORGE SIBLEY

"Revealing my desert thoughts to a visitor one evening, I was accused of being against civilization, against science, against humanity. Naturally I was flattered and at the same time surprised, hurt, a little shocked..."
—from *Desert Solitaire*

I suppose a safe way to begin a review of this new book of Ed Abbey's would be to call it a good rollicking action story, a fast-paced, edge-of-the-chair adventure novel. In the pure categories of literature, what this book reminds me of a lot is the *Great Escape, Guns of Navarone* type of story, or the novels of Alistaire MacLean, Helen MacInnes, Graham Greene (in his lighter moments) in which a—well, not exactly an anti-hero, but what you might call a non-hero—a person probably a lot like the average reader is somehow suddenly precipitated into a slightly contrived situation of big-league intrigue and harmless violence (harmless, because you sort of know beforehand that the principals, the good guys, your vicarious self, will survive with at worst a bad flesh wound and a handsome and wholesome heterosexual nurse to take care of it).

If I were to review this book as being of that genre, then I would tell you that this is the story of four unlikely and, in varying degrees, normal Americans who find themselves whirled into the vortex of a plot against their country and who, against the magnificently-evoked backdrop of the Southwestern desert, unite in an effort to confound the aggressor and preserve American values in the face of this threat from the forces of moral and physical chaos and destruction.

But then on the other hand, I could also locate the book in the category that produced, say, *Day of the Jackal* or some of the Eric Ambler–type stuff, where the reader follows, with growing tension and even horror, the cold, deliberate machinations of some avatar of neo-fascism who manages through the clever exploitation of chance, bureaucratic inefficiency and corruption to stay just a step or two ahead of the righteous pursuers in working toward some evil intent. If I were to carry through on *that*, I would go on to tell you that this is the story of four anarchists posing as normal Americans who, against the magnificently evoked backdrop of the Southwestern desert, work in an infernal coalition to subvert American values and put the land under the sway of the forces of moral and physical chaos and destruction.

If I wanted to just play it safe, maybe what I should do is just say the hell with the story and dwell on the magnificently evoked backdrop of the Southwestern desert. Yes folks, Ed Abbey is writing about the Southwest again, that tangled and chartless labyrinth of canyons, arroyos, slickrock, dead ends, no ends and confusions that he loves so well and articulates so splendidly. Fully aware of the fact that it doesn't flatter me and might not flatter Abbey, I'd as soon read him writing about the desert as go there myself. He makes it such an interior landscape—as if he himself were no more than a perceiving interface between that harsh and tortuous terrain and a mind that was a continuation of those surrealistic forms and features…

Water. All they can seriously think about is water. Yet it's all around them. Carloads of water. High over the plateau rims, three thousand feet above Land's End and all across the canyonlands float huge, massive clouds trailing streamers of rain, all of which evaporates, it is true, before reaching earth. Two thousand feet below and only a few miles away, as a bird flies, deep in the trench of Cataract Canyon, the Green and the Colorado pour their united waters through the rapids in roaring tons per second, enough to assuage any thirst, drown any sorrow. If you could reach it.

Abbey's novels are always worth reading for that kind of work alone.

But it would be very difficult to read this book for that alone; the story is a very intrusive presence and one can't just ignore it. Infiltrating this landscape from every direction, in every kind of contraption, with every kind of purpose and mischief in mind, are all these people: On this side we have these four people very much like ourselves or someone we know trying to defend and preserve American values; on the other side we have a whole army of people very much like ourselves or someone we know trying to defend and preserve American values; and they are basically both trying to destroy each other's American values, not to mention each other.

The four against the many are: a former Green Beret with lots of good government-sponsored technical knowledge and an abiding love of the desert named George Washington Hayduke: "Though still a lover of chipmunks, robins and girls, he had also learned like others to acquire a taste for methodical, comprehensive and precisely gauged destruction." There is a jack-Mormon river rat with three wives and a Colorado River expeditions business named Seldom Seen Smith: "The tangible assets were incidental. His basic capital was stored in head and nerves, a substantial body of special knowledge, special skills and special attitudes." There is an aging but still very capable doctor named A.J. Sarvis: "Dr. Sarvis with his bald mottled dome and savage visage, grim and noble as Sibelius, was out night-riding on a routine neighborhood beautification project, burning billboards along the highway." And a mouthy, bosomy, intelligent, beautiful Bronxite named Bonnie Abbzug (no relation): "Something inside, deep within her, longed for a sense of what lay ahead."

This unlikely mixture meets on a raft trip down the Grand Canyon, hosted by Seldom Seen, and discover that they share a consuming hatred of "blind technology (not uncommon today) and (more uncommon) a willingness to (a skeptic might say) use this blind technology in a creative orgy of blind destruction."

Arrayed on the other side are the local representatives of the Bureau of Reclamation; the Park Service; those village collaborators we also know under their guise as progressive businessmen, multiple-users and real estate dealers; and the free and wheeling American equivalent of the brownshirts, the "San Juan County Search and Rescue Team, Blanding, Utah."

And then there is a third, you might say neutral, presence in this book, a product of the people and every person's symbol of something…"792,000 tons of concrete aggregate, cost $750 million and the lives of sixteen workmen, four years in the making, sponsored by the U.S. Bureau of Reclamation, courtesy U.S. taxpayers…a glissade of featureless concrete sweeping seven hundred feet down in a concave façade": the dam, Glen Canyon Dam, symbol of progress, nature-in-harness, man as master; symbol of destruction, nature raped, man rapacious. And for the Monkey Wrench Gang, our feckless foursome, perhaps a new emerging symbol: symbol for everything that stands between man and—well, between man and what? Mostly, I guess, a personal shot at a fresh start, a chance to be back at "the bare skeleton of Being" looking for a new direction to strike out in…even a fresh batch of innovative fuckups would be better than the same old threadbare set we play over and over.

But of course *The Monkey Wrench Gang* doesn't start right out with an attack on the dam itself—what do you do about 792,000 tons of concrete? They, uh, well, practice a lot. They start small, on bulldozers, and work up to bridges. There is a lot of enjoyable writing here (if you can maintain a certain aesthetic distance); Abbey stretches the rubber band of credulity right to the elastic limits and keeps it there, never quite going too far (almost never, anyway), and never letting the pace falter either (although I did get a little tired of the fact that Hayduke just couldn't pass up one bulldozer). He shows great diversity in coming up with ingenious ways to ruin things—and, I might add, what appears to be a respectable amount of specific technical knowledge (such as detailed directions for the starting up of an Allis-Chalmers HD-41), not to mention a certain amount of what the Army calls field-expedient know-how (like remembering to always carry some Bosch and Eisman rotor-arms in case you find a bulldozer at night that wants driving into the lake).

Of course the folks who brought all these improvements to your basic desert don't take all this lying down; a sizeable manhunt gets organized, featuring Bishop Dudley Love and the San Juan County Search and Rescue Team mounted in Broncos and Blazers, determined to ride down, rope up and drag to dust these devastators who leave

"Rudolph the Red" as their calling card. The chase scenes are better than in the movies.

The whole book, in fact, is pretty damn good. Abbey leaves no real doubt as to where his own heart lies in this guerrilla war of those we know against those we know; but he isn't so in love with his non-heroes or so spiteful toward his enemies that they dance on obvious strings. They all seem imbued with the empathetic qualities of life the lack of which (relatively) has made Abbey, up to now, a better writer about "non-human nature" than a novelist. To say that his characters begin to show expansive life on a level with their setting is about the finest thing I could say about Abbey's novelistic work, and I feel that way about this book.

The thing about this book that is, of course, going to bother a lot of people is the general scope and descriptive particulars of what really does amount to a blind attack on blind technology. I'll admit it bothers me. Disgusted as I might personally be by the excesses in which bulldozers participate, I will more likely than not stop and watch one at work whenever I can. And while my conscious mind is evaluating the project in probably derogatory terms, another part of me is quite frankly fascinated and not just a little impressed at this great belching and clanking vector of rationality steadily rearranging the careless sprawl of nature…I can't read about some dude "pouring sand into each crankcase and down every opening which leads to moving parts" without a part of me twitching like a frog under low voltage. Sorry, but I just can't.

What I start to say is, goddammit, it isn't the bulldozer's fault! But (partially because this desert rat, this canyon crawler hit me when I was vulnerable with *Desert Solitaire* and has had me thinking ever since) I almost immediately anticipate the counter argument, or at least a good counter argument: It doesn't really matter that the bulldozer isn't responsible for what it does. You can only evaluate a tool in terms of the job it does, a means in terms of the end it works toward. In a certain isolated way, a bulldozer is indeed a marvelous and beautiful thing. But if it takes part in shitty work that cannot be justified, then it is not beautiful in its context. It's like being an artist, or a writer: The hardest thing of all to learn is to throw away your mistakes when the common mud therein is spotted with "a few pretty nice things."

In *Desert Solitaire*, Abbey makes a distinction between what he calls "civilization" and "culture": "Civilization is the vital force in human history; culture is that inert mass of institutions and organizations which accumulate around and tend to drag down the advance of life…Civilization is mutual aid and self-defense, culture is the judge, the lawbook and the forces of Law & Order…Civilization is tolerance, detachment, humor, or passion, anger, revenge; culture is the entrance examination, the gas chamber, the doctoral dissertation and the electric chair…Civilization is the Ukrainian peasant Nestor Makhno fighting the Germans, then the Reds, then the Whites, then the Reds again; culture is Stalin and the Fatherland…" Never mind that he is using two grab-bag words that everybody defines in a thousand contradictory ways; the distinction he is trying to make should be clear, and I think it is valuable. To the extent to which we are reluctant to—or refuse to—separate the "civilization" from our "cultural" chaff, our society is very aptly represented, I think, by the freak who doesn't yet know about the invention of socks, yet displays his pathetic little array of clumsy coffee mugs and Budweiser belt buckles. Or the Search and Rescue Team with its "Jeeps, Scouts, Blazers, Broncos, elaborately equipped with spotlights, hard tops, gun racks (loaded), winches, wide-rim wheels, shortwave radios, chrome-plated hubcaps, the works..." No taste, no style, no civilization, no art to their (visible) existence...Jesus (not the plaster patron but the one who was deeply distressed at the lack of civilization in the cultures about him) was recorded as muttering something about "fools and hypocrites, who strain off the gnats, then swallow the camel whole."

The question of some moment at present is whether or not it is possible to establish an island of civilization in the middle of a sundering sea of mindless culture. Can a person grow up in the middle of twelve thousand million automobiles, hemmed and directed by their twelve thousand million miles of roadway, and still be independent of the automobile culture? You might as well ask if Edward Abbey could sit on top of a mountain and develop the same basic thought patterns and philosophical directions as the desert has bred in him.

The fact that I am a part of the internal-combustion culture was no conscious choice of mine; it is practically impossible to lead an active life in this society without noting the pervasive influence of the automobile, the bulldozer, the diesel in just about everything you do. To a certain extent, art is possible within this culture—although to a certain degree we are blind to our own art, continuing to bore ourselves with the spectacle of painters flying up their own asshole, as Abbey might say (me too), and ignoring such things as the BMW motorcycle, the early Volkswagen Beetle, the smooth confluence of U.S. 6 and Interstate 70 near Denver. (Abbey might not say that.)

But the elements of our culture are so numerous, overwhelming and pervasive that any kind of an alternative culture, let alone any new and unique aspects of civilization, are all but impossible. (And don't confuse movements like the "back-to-the-land" movement today with genuine alternatives. In the canyonlands of abstraction they only go back to an earlier point in the same old drainage; they don't change drainages.) There is probably an infinitude of gadgetry and chrome fluff that can be added on to our culture, but it is not to be confused with civilization, or art, or necessity. So we end up with a mind-boggling accumulation of culture, some of which has been carried through as part and parcel and contributor in "the vital force in history," but most of which is entirely unredeemable and nothing but an incredible garbage problem. What do you do with a world like that, if you aren't really excited about next year's modifications on what you were already tired of a decade ago? How are you going to build the new Jerusalem with the old one still in the way? Ask George Washington Hayduke; he has some ideas as to what to do with the old one. Not very creative, you might say. But Hayduke might answer that you can't paint a new picture on old canvas until they paint out the old picture first. Or as Abbey said in *Desert Solitaire:*

> I am here not only to evade for a while the clamor and filth and confusion of the cultural apparatus, but also to confront, immediately and directly if possible, the bare bones of existence, the elemental and fundamental, the bedrock which sustains us. I want to be able to look at and into a juniper tree, a piece of quartz, a vulture, a spider, and see it as it is in itself, devoid of all humanly ascribed qualities, anti-Kantian, even the categories of scientific description. To meet God or Medusa face to face, even if it means risking everything human in myself. I dream of a hard and brutal mysticism in which the naked self merges with a non-human world and yet somehow survives still intact, individual, separate. Paradox and bedrock.

The artist—artist at just living if nothing else—has to first of all be a nihilist, atheist, anarchist. Then has to try to remember to be an artist, and not get lost in nihilism, anarchy, atheism. But somehow all that is easier to say like this than it is to do...and just as a person aware of the dangers of cancer will put off seeing a doctor about the tiny wart that might be melanotic, so there is a long, long psychic trip between the abstract and passive dislike or even hatred of something and the active process of beginning to do something about it. Because when a person tries to excise something like "blind technology" from his or her life, it becomes very much like cutting away a pound of personal flesh: Bad as it all is, it is what we have lived with all our lives, it is the only world we know for sure, it is us.

This would seem to be at the heart of the paradox that runs tangled through our thinking, wilts our resolve, takes the heat out of our fire. We are not even any longer in the position of pretending like Thoreau that technology is a rampant social disease against which we have to protect ourselves; we are already infected, invaded, pervaded, overrun; at this point, what we are is what is killing us. And Abbey seems increasingly unequivocal about "the old road that leads eventually out of the valley of paradox." Like the man in the wilderness with the gangrenous foot, either the patient becomes his own surgeon (with attendant risks and no anesthetic) or the patient dies.

It isn't that easy to pick up the knife and start cutting. Take the case of Bonnie Abbzug, during the plot to go "to work on the railroad" that runs the coal to the Four Corners Generating Station... "One other thing, and this is serious, men. What in the fucking name of sweet motherfucking Christ is the use of blowing up a railroad bridge and a coal train if we're not going to be there to watch it happen?"

A rough tough creampuff, that Bonnie. But here we are, in full panoramic view of the bridge with train approaching; for her good attitude, Bonnie has been given the honor of pushing down the detonator handle...

Smith, twenty yards away and helpless, stares at them, at Bonnie stooped over the infernal machine, at Doc Sarvis stooped over her. She clutches the uplifted handle, her knuckles blanched with strain. Her eyes are shut tightly, squeezing forth at the corner of each eyelid one jewel of a tear.

"Bonnie: push it down."

"I can't do it."

"Why not?"

"I don't know. I just can't."

Yes. A very interesting book. Not only what to do, but why it's so hard to do it.

—MOUNTAIN GAZETTE, #37

Ascentuality

BY
MICHAEL CHARLES TOBIAS

As with all expedition training, I'd begun some six months prior in San Francisco making girdle traverse reconnaissances around the Trans-America Building, climbing up the drain pipes of the Esalen Institute on Union Street and doing pullups surreptitiously from a mid-point hanging under the Golden Gate Bridge. Then to Colorado and the sandstone buildings in Boulder and after that, with time fast dissipating, to the Himalayas themselves for final preparations. In the company of Tenzing Norgay, a veritable Don Juan who managed (despite his good looks) to climb Mt. Everest in 1953, I was able to acclimatize both to altitude and vertigo, the latter an ailment I'm particularly vulnerable to and one which has never fostered much confidence on climbs. I was able to scale a mosque in Northern India and several religious shrines in Bhutan. Mind you, I have a long, enduring history of such endeavors, including a partial ascent of the Wailing Wall, of erotic ruins in Kashmir, on Stone-Age shelters in Northern Israel, minarets in Bangladesh, train stations in Switzerland and up the Lion's Gate of Mycenae, a city bank in St. Louis and several tall turrets in Amsterdam.

After a month in the area of Kanchenjunga, the third highest mountain in the world within Sikkim, and some additional days lingering up in Wales, I believed myself ready.

"There they are," I hailed to a woman beside me on the plane as we circled Kennedy for landing, looking down at the Battery and Midtown: Those glittering ranges, a wilderness of sheened sides, a metropolis of arcs, angles, incomprehensible projection, citadels, darkened with soot or shimmering with metallic perfection. Domes and underground passages, stones and spires shooting up like ruthless flares from gorges laden with Manhattan's strolling Montagnards.

A whole network of surfaces given deliberately to the waiting and the muteness inherent in its hidden volumes. All the glaciers of Europe, South America, Baffin Island, the Arctic and Baltoro were there beneath me with the whirring of populations and the skittering bundles of haze sitting in and out between the monolithic cirques. New York City is the finest mountaineering playground in the world as far as I'm concerned. Where else can climbing at such heights be a performing art? In Yosemite Valley one can occasionally expect to have onlooking tourists with telescopes gawking from the roads deep beneath the walls, and if one is clever and high enough he'll try and let them have it with excrement that sails sometimes several hundred yards out into space before collapsing, with luck, on one of them.

In the Alps or Cascades or Taurus Mountains, the only companionship comes from mountain goats, rock bugs or hawks, and if one falls his or her body turns blue quite alone on the talus. But in New York, there is a human component utterly staggering. Each human being has his or her own private access to revelation. Here in the heart of the most intricate civilization a person can wake up one day, stop aspiring through life horizontally and embark on a mythical journey. The buildings are beckoning to be climbed. It only takes one a few quick moves over a door, across a lintel, up the side of a sidewalk window to appreciate the possibilities. These walls are mountains and adorned with equally interesting surprises. The German philosopher Ludwig Feuerbach believed in the imagination as one more potent force of nature, and these great towers of steel rose in my mind, erupted with the skill of thousands mingling over their frames and designs, the way a deity might hover over Aetna. Each window holds back a compelling enigma, a business executive in the closet with his mistress, a hysterical conference in progress to promote a birth-control deodorant, rapacious secretaries, computers, editorial offices, advertising offices, dark rooms, bathrooms, steam rooms and people in all of them doing their business wholly unaware that a mountaineer is outside engaged in stalking an epic first ascent.

The Upanishads and various Zen treatises speak of "mountains that meditate" and I know from experience that buildings also meditate. Certainly some of them move back and forth in the wind, and I've always taken this to symbolize the divining of matter. As for the steam which rises elliptic from the summits and the rumblings from down within, these signs too I look at religiously. And the glaciers, those mighty bustling streets which flow like lava between the towers, they too are significantly mystical, particularly around the Avenue of the Americas, my destination. Mark Twain once attempted to take passage on a glacier from one village to another in Switzerland, having heard that glaciers move. When he failed to get anywhere he evidently threatened to sue the Swiss government. I had an equally difficult experience getting to my "Base Camp" near Lexington and 61st Street on the fourth story. With an 85-pound duffle bag containing ropes, pitons, double boots, down gear and chock stones, P.A.'s (special smooth bottomed, galvanized rubber, balletical climbing boots) and gymnastic chalk, a bulging rucksack on my back with medicines, guide books, dried potato latkes, film and Palinurus' *The Unquiet Grave*, I scrambled down Fifth Avenue asking fellow climbers, in no light urgency, directions to a public lavatory. Fortunately I chose an ideal time to land in America for it put me in Midtown precisely at the moment when streets were impassable with all the people. So I had no trouble making contacts and eventually was able, in private, to take a leak. Some kid accosted me outside the john with a brush and polish in his hand but, to my enjoyment, was unable to do much for my climbing shoes. I gave him a dime for his efforts nevertheless, which he elegantly refused, a meekness unprecedented in India.

Now the French and Dutch, even the Japanese and Polish to some extent, believe in attempting the first ascents of lesser known but more difficult peaks, and this was my desire. Something high, but not the highest; perhaps 50 stories, smooth, vulnerable, cold and centrally located in the range. A wall with wide boulevards beneath it, with an escape route, and other options. A wall, in other words, with class. I was not interested in the Empire State Building, The Pan Am, the Chrysler, Time-Life, U.N., Newsweek, or Rockefeller Center faces. I was after a wall that embodied the "Ideal of the State," the "Myth of the State" and, of course, the "American Dream." A precipice which offered the greatest diversity of holds and the finest examples of workmanship. I wanted a total piece of architecture in which nature had assembled her greatest forces and talents to resurrect this one colossus from the moraines of delicatessens, bookfairs, barbershops and whorehouses underneath. Something of the quality of the Porcelain Pagoda near Peking, of Shiva's Temple at Perur, of the columns of Persepolis, the Mayan ruins at Tikal, the Great Wall of China and the Mosque of Smarra. New York is the only wilderness area in the world that boasts of such a monster. Tales had filtered down to me of a building which, never ascended on the outside, met all the conditions for an enlightening encounter, for a climb in which man, microcosm, metal, mist and 10,000 eyes beneath could nobly, lasciviously cavort aerially. A building of great repute as well, housing one of the nation's largest publishing corporations within its labyrinth of marble, brass, steel and tinted glass.

As with all other major expeditions, such ascents require a certain amount of initial acculturation. To solo Everest, for example, one would want to spend, say, six months among the Sherpa, eating their food, making love to their women, and so on. Acclimatized thoroughly then, one could slowly begin ascending from camp to camp. The situation in New York City is no different, I supposed.

So I invited a dancer over for salad and left the roll-away bed exposed and neatly arranged beneath my five-pound Antarctic sleeping bag. That afternoon I visited some ten assorted food rip-off stores searching for ripe avocados. I finally settled for frozen scampies instead, rushed home and constructed a special Sudanese mushroom-bread, brick-cinnamon salad tossed with ice water. Naomi arrived at seven with her carbon paper and a diaphragm. She was just the female I required, for she's lived in Brooklyn all her life, taught dance therapy in Manhattan, knew the buildings well around town, was lewd, lean and neurotic. In fact, we spent the entire evening discussing her warped childhood affairs with 53-year-old poets whose pictures would appear once or twice in a *Times Book Review*, after which they'd become grossly impotent. I thought about asking

Naomi to come to Mongolia with me this summer to attempt climbing nomad tents, but feared she'd take it the wrong way. Then she started vomiting all over the paper, no doubt from the salad, and left. It just so happened that she had disgorged on the movie pages and I noticed, under it all, a listing for Jodorowsky's *The Holy Mountain*. It was being shown at midnight near NYU, and I didn't want to miss it, not that I had the slightest idea how to get to NYU.

So at approximately 11 p.m., to the tempo of a siren traversing some nearby block, I left the comfort of my sleeping bag, opened the window of the apartment and crawled out into the rain. Hanging there, consumed with that old horror-ecstasy of space beneath my feet, I began down-traversing the brant and the slape above the street, this way both honing my fingers for the big climb ahead and avoiding a very suspicious if not obnoxious doorman at the entrance to the apartment building. Stealthily reaching the street, I was instantly greeted by a nice old man with a *Wall Street Journal* rolled up in one hand and toy poodle rolled up in the other. "Good work, kid," he mumbled. "Thank you, sir," I said, gratified. "Care for some beef jerky for your dog?" reaching into my blue cagoule. "Oh no, he's a vegetarian, but thanks." I ran down the street to catch a bus.

I was the only person on this bus, save for the driver who kept eyeing me funny in the mirror. Perhaps it was because I didn't have a shirt on and was busy shaving hairs under my arms, or perhaps because I was just in my socks. It was cold out, granted, but the great, mad German climber Herman Buhl used to urinate on his hands and carry snow balls during winter nights to toughen them up, for training of course. I told this to the driver. "What you sayin, man," he answered reprovingly. "You mean," I exclaimed, "you've not heard of Buhl the Bavarian who fell 300 feet onto his head and giggled ever after though the world thought he was dead? Well, what about the lunatic Italian Caesar Maestri. Before repeating his tremendous route in Patagonia, he'd refused to hump naturally during fornication but rather via pushups so as to stay in shape?"

"That man was crazy, I mean it, he was crazy on his ass. Now where you all goin, cold boy?"

"*The Holy Mountain*," I replied.

I got to the theatre where I had to wait some for the first film to end (a porno film but done in "bad taste," I later heard). There was this lovely, emaciated looking podiatrist's wife-type standing across from me whistling and peering somberly across the street to a McDonald's. I'd seen that same, desperate look in Dacca before and I ran across, purchased several Big Macs, some flowers, ran back breathless to the woman and handed them to her.

"Here, I knew you'd want this."

"No, oh god no. I stick strictly to rhubarb and poison pen letters, sour cream and biscuit. Now please...I'm rehearsing. I play a 95-year-old Costa Rican mistress to a 100-year-old revolutionary who is in jail, but is released once each evening for his elimination.

Instead he comes to me. They are our only moments together."

"So where is he?"

"Oh, well he's down there, in the subway searching for animals."

"Animals?" I exclaimed. "What kind of animals?"

"I don't know, just animals. He does this every evening before we meet. Something having to do with the lunar penumbra and the grease that drips steadily from the trains. He tempts the critters with animal crackers of course and sings some Costa Rican oratorio. Last night he brought up a swordfish that he believed to be from the Gulf of Mexico. Once he found a kangaroo that had had a hysterectomy. He's strange that way."

"Yes, aha..." I choked.

The theatre emptied out, I left the woman waiting for her friend and took a seat next to a rabbinical student right under the screen. *The Holy Mountain* is not a film during which one should eat Big Macs. That's about all I remember that night except that coming home, the bus driver fell asleep or something and drove some 35 blocks further than I wanted to go. But it gave me the opportunity to climb several stories worth of Park Avenue concrete at 3:30 in the morning. By sunrise, I was back in my apartment, listening to the dawn.

After lunching in a well-known Armenian restaurant with a well-known editor, I trekked in an unknown part of Manhattan, this time in my favorite seedy lederhosen, to the decided upon precipice. I'd conferred with all the hot window washers in town and their wives and had obtained tips regarding which window sills to avoid, the best time of day to attract the finest notice of people on the street, and which particular windows to look into. They told me of the 36th floor.

It rises out of a colorful book store. Once the first series of moves has been worked out, or "wired" as we say, the entire building has virtually been mastered. Rene Daumal in *Mt. Analogue* had said, "The first step depends on the last," and the Greek climber, Nikos Kazantzakis once wrote, "There is no summit, only height...Happiness is the same height as man..." The only danger involved is the risk of slipping off the top. This has happened to me on several occasions though only off of yurts in the Sierras. But the general tendency to slip during climax is a formidable one, and from 750 feet it can cause great bitterness. My father particularly would feel guilty for not having said, "We gotta keep that kid off the street" enough times. My mother would run out of hard-boiled eggs for the mourners sitting Shivah and my senile great-uncle would no doubt send my dad *The Principles of Architecture* as condolences.

Entering the front mezzanine, I moved over to an empty elevator and rode it to the 36th floor. My purpose was to scout the building and the 36th floor was my escape option. As disguise I brought along old ropes and rusty buckets, an Italian squeegee, German galoshes and two bottles of squirt-a-pleasure ammonia.

The elevator stopped and I moved into the hall and then into a vast empty room the size of the entire building. The ceiling was ripped up with wires and circuits hanging loose, the carpets and room partitions uplifted. Imagine, right in the middle of this lavish kingdom of a building, an emptied, unknown space. Taking off my paraphernalia, I strode around the room along the big bay windows. All of New York's walls were rising immaculate on all sides. There were dreidel factories, hashish dens, Park Bernet, a Russian olive tree, Madagascar typewriter outlet stores, Iranian garbage collectors, fellatio, the Brooklyn Bridge and *Commentary*. And a pigeon pooping on the window sill right before me. I opened the window and the wind and now augmented buzzing from the streets below blew in.

The pigeon hops inside. Then, carefully, very, very slowly, I crawl outside onto the ledge. This breaks the pigeon up, my last friend in the world: It looks across at me as if to say, "Idiot!" His perspiring, grey, fleshy chest is resonating with excitement. Suddenly I hear the humming of an elevator rising through the building. It's slowing down, oh Jesus Christ, it's stopping on my floor. I squirm frantically to hide under the ledge. It's a desperate hold and there on the inside are all my discarded clothes, my squeegee and a pigeon sitting there with diarrhea, dumbly staring at ten frantic fingers outside, hushed above the emptiness. A group of people saunter in. A woman with a voice like Peter Falk is explaining the new floor plans to a group of businessmen. They are walking around the periphery of the room. They are coming towards me, vei, I steady myself. They spot the pigeon and my stuff, then my fingers. "What the hell," one of them shouts splenetically. Now the woman is screaming like something from a Fellini bedroom scene.

"Shiiiiiit..." I screech. "Get that pigeon away from me. Get it away from my clothes. I'm allergic to pigeons. He's already dew-dewed twice on my squeegee." The bird doesn't move. It's a pregnant pigeon. The people pull me in.

"You alright, fella?"

"Goddamn pigeon. They're all alike. My friend was killed by one. The pregnant ones are the worst hazards. You better do something with this bird or you'll have a maternity ward worse than Lake Titicaca. I'll just get my things and be going. There's other buildings to wash, you know."

I quickly enter the elevator and descend to the lobby. Down the burnished brass escalators, past a lovely garden out to the street. People are swarming by me, my flabbergasted fingers, my dripping squeegee. Above, looming onerously glorious, the McGraw-Hill building soars, still unclimbed.

I march back to Base Camp, exhausted, humiliated, my bones aching, my head throbbing. I am dreaming of a shower, of sleep, of obelisks. But that sonofabitch doorman refuses to let me in.

—*MOUNTAIN GAZETTE, #50*

Royal Robbins contemplates his pint. JOHN CLEARE

A Dream of White Horses

**BY
ROYAL ROBBINS**

A Dream of White Horses—one of the great names, fulfilling Geoffrey Dutton's dictum that a name should tell you something about the climb or the way in which it was done. One glance at Leo Dickinson's masterful photograph explains it: that great sheet of spray leaping from the sea, rearing from excited waters like a splendid white stallion, and the two figures fastened to the rock just out of reach of the tormented foam. With its whimsical and romantic overtones, the name appeals perhaps more to the American climber than to his British counterpart. A Dream of White Horses. Drummond, who made the first ascent with Dave Pearce, has a talent for verbal imagery, for metaphorical and poetic phrasing. It's not surprising he should produce such a name. But still—A Dream of White Horses— five words! Turgid with subtle meanings. One of the few long names that succeeds. Although I knew nothing of the climb, because of that name and that sensational portrait by Dickinson, it was a route I had to do.

A chill, heavy wind flowed from the North Atlantic beneath dark masses of cloud, not rain laden, but casting showers on the soul, dampening the urge, as we stubbed in tight shoes across the barren, heather-covered crown of Holyhead Mountain. The green sea was agitated.

"Lots of white horses today," said Whiz.

"Yea, verily," I replied, suspended in heavy-handed mockery between Shakespeare, St. Paul and Zane Grey, " 'tis a right stam...*peed*."

"You know, you shouldn't try to be clever," my friend advised. He has my interests at heart. "You lose points that way. You don't have any judgment about what's clever and what is merely cute. You'll probably put something silly like that in your next article. You're one who should never digress in your writing. The way you ought to write is this: Open your story in the middle of a pitch, write about *that* for six pages, and close with the hero (*yourself*, presumably) ten feet higher. Then you might have an article that people would read."

"Thanks, pal. I'll let you know what's wrong with you, too, as soon as I figure it out."

We were soon at the edge, and I was startled by how sharply the green and brown slope dipped towards the sea. There was a break in the clouds, and the sun flashed on olive sea and rich, well-watered grass. From the west, a host of albino chargers rushed landward. It was a wind-and-light show and I looked forward to more. The approach was down steep, muddy slopes, the precipice just below. The signals of my mind flashed caution. This wasn't like grit, where one can relax when not actually climbing. On these sea cliffs, you can't afford to stumble.

We came across ropes and packs left by others. Whiz was concerned. "They'll be up to the same thing we are," he assured me. "They've just gone down to check out the wind and wet. We can beat them to the ab and get on the route first."

He talked on in his unique and entertaining fashion, about the likely identity of the others, and about how they didn't know the code, for they had parked in the wrong place. Then he digressed to other subjects: political scandals, rock music, architecture, climbing techniques, photography and the organization and conduct of international mountaineering expeditions. Whiz is an authority on many things. Delivering quick, sometimes harsh judgments, dogmatic, impulsive, abrasive, but often accurate, he is very much alive. So alive that he seems, at times, about to burst with suppressed energy. A weaker man might have drowned that much painful life force in a sea of bitter. But Whiz has it pretty well harnessed. He is irritating at times but inspires respect and somehow even a certain affection. But he is a steamroller on the bumpy road of sentimentality, a wine crusher of maudlin grapes, impatient of weakness, scornful of incompetence, not over-worried about people's feelings, a skillful and agile polemicist who usually wins. He comes on strong, and sometimes puts people off with his aggressiveness. But he is not impervious to a slight, or to a well-chosen phrase focused upon a weakness. And he is keenly sensitive to the nuances of status, insisting upon his turn in the front seat.

"Let's just whip over and set up the ab," said Whiz, tripping off. But first I wanted to look at the route, so I scrambled down the steep slope. Four climbers were coming up.

"What are you going to climb?" one shouted through the wind.

"White Horses," I threw back.

"That's what they're doing," he replied, indicating two figures on the wall, halfway between water and grass. I continued down to a promontory which curved and faced the climb across an impatient, foaming gap. Sea spray was dashing against the wall. What a picture! What a time to be out of film!

Whiz was waving wildly over the shoulder of the slab. I surmised that he desired my presence. I knew he would be thinking: "We mustn't let those 'nods' get ahead of us." (Except for occasional uses of "twit" or "freddie," Whiz, to express disdain, favors words starting with "n" sounds, words like "nod," "gnome," "nurd," "nob," "nibbler," etc.)

I arrived to find him taking great care setting up the "ab ancs." A close friend had recently been killed abseiling, and Whiz didn't want to go the same way. Nor did I.

"This is a dangerous place," I said, clipping to the 11-mm runner he had looped over a sturdy block and backed up with two more anchors.

We were now directly above the copulating waters. Whiz threaded the ropes through his brass figure-eight, and was off. The place provoked in me the same feeling I get descending to the notch of the Lost Arrow Spire in Yosemite, with the prospect of falling into the Arrow Chimney. In such places, the recurrent theme of abseiling, that one's eggs are all in a single basket, comes home with special force prompted by the hideous aspect of what one would fall *into*.

I dulfered to Whiz who was waiting on a narrow, irregular ledge, just the sort that someone, someday, will fall off. "Just look," he said in a jocular tone, "at that tangled and barking ocean." I missed the allusion to Yeats, but his metaphors seemed *nearly* right. The route properly started 70 feet lower, but as the water was only 60 feet down we had to forego the pleasures of the first pitch. We were on Wen Slab. "Slab" in this case doesn't mean low angle, for the rock averages about 80 degrees, and is vertical or overhanging in many places. It is a slab in that it is a single slice of compact rock, forming one wall of a zawn. A zawn is a yawn in a sea cliff, the bottom filled with water.

The outside wall of this zawn is a promontory that juts into the sea and curves toward the south, a sort of pillar the top of which is directly across from, and halfway up, Wen Slab. This pillar is a natural bridge. The waters in their ebb and flow have surgically cut a tunnel through the pillar in an attempt to isolate it and transform it into a sea stack.

Through this tunnel the waters burbled and boiled, forming a counterpoint to the larger boiling flow from the natural mouth into the ocean. I was struck by the savagery of the place, the frothing and sucking, the wind blowing, spindrift flying, the water churning and surging onto the cliff, and the surges breaking and lapping up the wall to hang like lace curtains while slipping back into the sea. Aphrodite, they say, was born of sea foam. But this was a rough sort of love, this mating of water with rock. It was elemental, and affected me as do thunderstorms or raging blizzards. I loved the fearful violence of it. In the clash and dash of water, the thundering and pounding, the crashing, breaking din, there was great power, but no evil. It gave pleasure similar to boulder trundling, but touched one's emotions at a deeper level.

I turned my back to the sea and started climbing, moving vertically, following a crack on the right side of the slab. It was steep, but not difficult. All the same, I slotted a wedge and continued up on toes and fingers. The rock there is good for climbing: It doesn't have holds everywhere, but might have holds anywhere. In this respect it is like the rock in the Lakes or the Welsh mountains, but it isn't granite. Such rock offers very good face climbing, but is comparatively rare in the United States. Most of the climbing in the U.S.A. is on "young" granite or sandstone, and generally follows crack systems. But even in the Shawangunks, where there are few vertical cracks, the complexity of this sort of British rock is lacking. In the Shawangunks there is a certain predictability about what lies ahead. This is not true of places like Craig Gogarth, a fissirostral sea cliff, where it is very difficult to tell from 50 feet away whether a passage is easy or impossible. Or from even closer than that.

I was 40 feet above Whiz. At that point the route traverses left two-and-a-half pitches straight off the cliff. I started, aiming for a crack 50 feet away. There was nothing obvious. A solution had to be found for every foot of progress. The route wasn't going to yield to an aggressive approach. I advanced, tentatively, and the rock revealed some of its secrets. I made further advances, careful but firm. And the rock again yielded, unfurling more, exciting me, raising my hope. There was a bit of a struggle, a setback, a renewed onslaught, and I broke through, confident now and eager for success. But I was stopped cold. False hopes; the damning disappointment of hoping too soon. The next bit was difficult. Should I take a chance and push forward? No, I would have to find the key. I was almost ready to give up. I considered it. But what would Whiz think when I admitted that I couldn't handle this bitch?

Where was the fire? It was cold on the shady slab, with the constant cold wind and the water running down over the holds. The climb was proving cold and contemptible, leading me on, acting easy, then closing the door in my face. Rebuffed, I knocked again. My eyes scanned the rock, noting every detail. I could take a high line or a low one in an effort to traverse to good jugs and a crack. I chose the low way and descended a bit, fingers feeling the rock, sensitive to every minute variation of texture and contour. I had to be careful, for I was now committed. There was an awkward piece before reaching the crack. I had to restrain myself, to hold back from lunging to escape the tension. But I was getting sick and tired of fiddling around. Finally I found the key and moved into the crack, and up a bit, to a belay where I was to spend the next hour suspended from two chocks and a spike.

I took in the rope. "OK, Come on up." He came up neatly, not fast, but controlled and competent, and soon reached me.

"Found that troublesome, did you?" he asked.

"Rather," I admitted. "But good. Real rock-climbing."

"I led that last time," he continued, "and I daresay I found it a sight easier than you did."

"That's nice," I responded. "That's why I come to Britain, you know, to ease the inferiority feelings of you D-team men." Whiz always insists he's strictly D-team. He's kind enough to classify me as B, so it greatly pleases him when he climbs better than I do.

"Well, there's a lot to be said for age and experience," said Whiz, encouragingly. "But I've yet to learn what it is."

Leaving these words of "whizdom" to rattle around my head, the mountain *herr* grabbed the next lead. It was a classic. I longed for my camera. Whiz followed a flake that ran up left, offering good grips, but little for the feet except friction. It was vigorous climbing, different from my tippy-toe balance lead. The protection wasn't brilliant, but he made quick work of it, and then took a long while arranging his belay anchors. I like that sort of care.

I followed the flake up and then descended a bit to Whiz's belay and took a gander at the last pitch. I would not have guessed it went *that* way. It looked like rubble. But not so much rubble, because rubble implies angularity; it was more like Dalinian rubble, rubble gone soft like Dali's watches, rubble that had melted and then refrozen, a great dripping morass of melting vanilla ice-cream refrozen in its slopping and plopping descent into the sea. Frozen magically hard. It was uncanny the way some force had cemented those sand grains. First it was sand, then sandstone, and then it was heated and squeezed into super-sandstone: quartzite. Picture the outer walls of a sandcastle that a wave has overrun. The soaked edifice crumbles and melts, but in this case it was caught in mid-melt and rehardened into an extremely durable battlement.

The traverse proved a rare treat. I climbed slowly, savoring it. Whiz gave me pointers on double-rope technique. Above, the rock overhung, forcing a leftward passage. It overhung below as well, so much so that an unroped climber, slipping from the traverse, would fall 200 feet directly into the water and perhaps escape serious injury(!?). I finished in dazzling sunshine, hot in a heavy sweater. Before disappearing up the grass to a belay niche, I looked back along the double ropes across the crumbling wall to Whiz, who had been enjoying my enjoyment of the rock. It was spectacular, and the rays bouncing off the white rock made me squint. White rock. White horses. This route will be a classic.

Whiz walked up the green hill and took off the rope. We coiled it and packed our gear.

"Let's go," said Whiz, his eyes suddenly lighting up with boyish enthusiasm. "I know a great place near here for trundling."

We rushed off, eager for more tumult, forgetting in our haste to retrieve the runner and carabiner we had left for the dreadful abseil.

—MOUNTAIN GAZETTE, #28

ROBERT CHAMBERLAIN

Mountain Towns

BY
TED KERASOTE

I've lived in a lot of places during the last seven years, probably a dozen towns, many of them in Colorado. Aesthetically, they've been as different as Leadville is different from Aspen, but they've all had one thing in common: somewhere within them I found an apartment, a restaurant, a bar that had a touch of romance and made my life special for a while.

In Aspen it was one room above the Epicure Restaurant at the corner of Mill and Main. From my window I could see the slopes of Aspen Mountain and a good way down the valley of the Roaring Fork. I wrote in the morning and worked in the kitchen of the Hotel Jerome from three to midnight. It was a fine life, though the room was often so cold that I could see my breath, and I would have to wear a down vest while writing. I had a single electric hot plate on which I cooked my meals, and one day I got the bright idea to put it under my chair as I wrote. Fine idea. The rest of the winter my bottom and the backs of my legs were toasty as I scribed noble stories.

Aspen was a very special time for me, and the hot-plate heater in a freezing writer's garret was only part of the romance. The other part wasn't the skiing, the magical Hansel and Gretel houses, nor the high-stepping ladies. It was the Mesa Store Bakery. On my one-day-a-week off, I would sometimes write 15 to 20 hours a day, trying to make up for lost time, trying to get out my masterpiece before spring and the melting snow stole the aura of privacy that surrounded me. It was a happy delusion.

At 11:30 on my night off, I would appraise what I had written, and if I deemed it good enough, I would take some change from the ash tray on my dresser and walk slowly up a deserted, snowy Main Street to the bakery. I would buy a large cinnamon cookie and a cup of hot apple cider—my reward for the day—and a small carrot to keep the dream going. Sitting there with my cookie and cider, the backs of my legs still faintly warm from the hot plate, I was the happiest creature on earth—impoverished, proud, my life filled with youthful promise.

Then it was Crested Butte. C.B. and the Wooden Nickel Bar with its worn plank floor, its potbelly stoves, the Victorian sofas and spinsterish lamps out of the age of F.D.R. It also had a greasy kitchen, a cook named Mo, a barmaid with a diamond pin stuck through the left side of her nose, and the best draft beer in Colorado.

That was before skiing began in earnest, before they paved the streets, when the dust still settled behind the jeeps that came down from Kebler Pass. A band would play in the Wooden Nickel during the hot afternoons of September, a band whose name I don't remember, though I recall that there was a drummer and two guitarists. The lead guitarist was black, and his voice was made of silk and sequins and rolling waves.

I used to sit for hours, eating a hamburger and drinking Coors, and listen to him sing of oceans and islands. Then I would go up the mountain to help build the success that ruined the town.

In Silverton it was the train, and it always will be the train, no matter how camp, hokey and obnoxious a tourist magnet the train becomes. Once in your life you ought to be on the summit of Arrow Peak in the Grenadiers and hear the train whistle from the valley of the Animas, 6,000 feet below you and ten miles away. You'll be looking out to the deserts of Utah, the Henry Mountains floating on the horizon like the castle of Moroni, and from deep in that valley with the loveliest name in the state, Animas, will come this black eight-wheeler peal, strung out in Doppler, haunting like an elk's bugle, like wind in the crags. That train is how you get into the Grenadiers—it leaves you off at the Needleton Water Tower—and its engineer in coveralls, giving the high sign, its fat conductor calling, "Board," are the lost memories of a time before the automobile.

In October, when the hillsides along the Million Dollar Highway turn to riot, and the train blows its last whistle of the season, Silverton becomes just another mining town—a bit rundown, slightly picturesque, a touch of rustic poetry caught on a weathered sign—in the backwoods of the state.

Leadville, on the other hand, is a most unpoetic, unromantic town. A train couldn't save it. The impressive view of the Sawatch has never saved it. A new strike of the richest, purest gold ever found wouldn't save it. The only thing that saves Leadville is the El Perdido Bar and the El Metate Restaurant.

They're both under the same dilapidated roof: the lost one and the grinding stone of eternal labor. The Mexican cuisine is not exceptional, no longer even filling. The green felt on the pool tables is getting thin, and there's no more draft beer. In the winter it's cold, in the summer hot. But the atmosphere is so plebeian, so base and no-nonsense, so absolutely and eternally memorable to the passing traveler used to the Ramada Inns of the land, that it couldn't be replicated by ten Hollywood directors with ten years and ten million dollars to spare.

It took Leadville to build the El Perdido Bar and El Metate. It took lots of muddy boots, broken bottles, two-time losers, drunken miners and fighting Chicanos and Mexicans to put the smell of sweat and scams in the backs of the leather chairs. It took a lot of spilled beer and vanished hopes to build the lost one and the grinding stone: the only place in Leadville from which you can look at the shantytowns, the molybdenum mine, the thermometer that says 36 below, and smile like a demon.

One more place: Boulder. More fine running, cycling and climbing, more eateries, more good weather and handsome women than any other town in the West—maybe the country...maybe the world. Boulder and Leadville: the Beautiful and the Damned.

It's so chic it's obscene. I think of browsing through plant-filled bookstores, of skiing by deer in Chautauqua Park, or Christmas carols on the mall, of the Green Slab Direct on a spring day, of Eggs Benedict and Fettucine Alfredo, of the best $1.80 Bordeaux purchasable in the United States, of Masi bicycles gleaming with Campi gear.

But, in the end, when I've been on an expedition for a long time and am dreaming of civilization, it's not a single one of those luxurious things I remember. I remember Saturday night and my room. Snuggled under an elskin bedspread, I read the *Atlantic* and hear over the stereo the delicate bars of KVOD's "Down to Earth" theme music: the pure notes of a flute and guitar drifting me to moody, cloud-hung places—the sea coast, bracken-covered moors, ice mountains—all far away.

And from my warm bed, close to 90,000 people, I smile nostalgically: the wilderness, like a loved town, will always be sweeter in counterpoint.

—MOUNTAIN GAZETTE, #76

The Ride

BY
KATIE LEE

It was the hottest day of the year when I decided to do it. Sure as hell it wasn't anything planned in advance. Intuition works best in a case like this. I remember dreaming up something similar a year or so back and threatening that I would do it someday, but the notion didn't even get out of bed with me that morning. I was still deep into the sorrow of loss.

He was such a wonderful friend, such a joyful man, full of life and living it, of love and giving it. He could torch you with his sense of humor—fire your laughter until you peed your pants to put it out. An artist, a sensualist and, I suspect, a creative lover as well. Gentle. A listener. A man who treated women the way he handled his sculptures—molding, caressing, teasing them into creatures of beauty and supple grace until they glowed with a life they didn't know they possessed. I watched that happen with more than one of his many women friends—women he'd never made love to, women he saw every day in his shop, in the shop next door, all around town, really. Everyone knew him. It's a small town, less than 500 of us.

Then, about three years after his move from the Big Rotten Apple to this little burg that he so loved, tiptoeing quietly toward another peak in his creative talent—not with money in mind this time, only love—and with all the freedom to do whatever he chose whenever he wanted and with whomever, his generous heart betrayed him. Harvey up and died!

And the heat moved in.

We held a memorial for him in the park. A big one. Everybody came, brought food, drink and things to say and remember about him. Our town is still one that lets its folks express their grief in their own way, lay it out and mix it up so it doesn't hurt so much. No preachers. Just all of us talking about him, telling stories of our time with him, things he said, things he did, what he meant to us. Harvey's spirit was right there that day, moving among us, telling us to get on with it, laughing at us, caressing us. And we all knew it. For me it was really tough because I had to sing his favorite song. I ain't no Judy Garland, I can't sob and sing at the same time and I was holding tight to a big aching bubble as I tried to get the words out. Then, I felt Harvey pat me on the bum, right in the middle, right in the hardest part—hardest part of the song, not my bum—and through my tears I almost ended up laughing, which is no better than crying when you're trying to sing.

Then the heat bore down.

The first year he was here he was my next door neighbor. I won't forget the day he first walked down the street in front of my gate with a couple of his friends. I was out watering the nematodes that like to make my carrots into funny little men with penises and hair all over them, when he stopped laughing at whatever his friends had said, turned his flax-blue, slightly bulging eyes on me and supposed: "Ooooh! You must be the lovin' lady (lovin' as opposed to lovely I noted right away and he knew that I noted, which is what blew me away), the lady I'm going to live next door to?"

Harvey had class!

I am a lovin' lady, though I try hard to disguise it, which is why lotta people don't call me a "lady." I could care less. (I've always maintained—for those who believe in such messy-physics-moon-shine—that being a Scorpio, right there on the cusp of Libra, the balance of the Scales keeps me from being a total bitch.) Gimme a break. Harvey did.

He sure enough rented the house next door. Every morning from his back deck when I came out with my cup of coffee, he'd wing his sweet morning greeting across my yard, "How-dee-doo, Miss Kitty Lu—you feeling fine today?" He never started working until after nine o'clock because his electric sanders and shapers and drills would make too much noise. (A caring man among his other talents—like he loved music, jazz, folk, Cuban, but he played it benignly, not at 2,000 decibels like some twits in our built-like-a-Greek-auditorium-town do.)

By the time I returned from my morning ride, I'd find him nearly smothered in clouds of alabaster and marble dust, tooling away at some beautiful sculpture he was creating. He'd ask me to come over and check it out—see if it looked all right. Wow! He sure didn't need my two-bits, his pieces were always elegant. Never mind that it was a media he'd never worked in before—his imagination was limitless, his gift, divine.

Then he bought a broken down little house up the mountain a ways with an outbuilding he could make into a studio, and began its renovation like a doting father building a dollhouse for his beloved daughter. I don't think a nail or a board went into that place without his kiss, or his blessing on it—that it be happy there—happy like he was, and "thank-you-very-much for being such a beautiful piece of wood and for coming from such a fine tree in such a lovely forest."

"But where, Harvey?"

"I don't know, but it's a fine tree, just look at that lovely, graceful and ooooh…sensuous grain!"

Key word. Harvey was indeed a sensualist.

He had an open Jeep, little runt of a thing it seemed for a guy as big-boned and tall as he was. Drove it with his girlfriend through rain or shine, snow or hail, through forest and desert. In summertime, all over the back roads, windshield down, canvas off, toodling up and down the mountains in his hirsute and shorts, his ponytail straight out in the slipstream—winter it was khakis, jacket and headband, maybe with the top up, maybe not. Rarely did I see the side-curtains except on the floor of his studio.

About a week before he flipped his coin for the "other side," I stopped by his shop with some friends to show them the lovely things he made—I often did that, especially after he'd grown a new wig-bubble and made something that nobody even dreamed about. It was ladders this time. Crazy ladders like some that might have come from a ceremonial kiva back a thousand years ago, except that they were so beautifully and imaginatively carved, they'd have to have been used only by a shaman for special initiation purposes—twining snakes and lizards slithering up the rails and whole Pueblo villages on the rungs that you stepped between as you went up from desert floor to mountains near the top. They were transcendental!

His blue bulbs seemed to be dancing to extra potent jazz that day. He pulled me into a corner and whispered, "Just got back from the Apple, baby, and I dropped a shitload of problems back there that I won't ever have to deal with anymore. I'm never going back. Wow—do I feel great!"

I was so happy for him, knowing how he hated going to New York for anything, except to show his rarefied Arizona girlfriend something she never need worry about missing.

After the memorial, the heat became grotesque.

Nothing short of our Main Street knee deep in rattlesnakes will keep the damn tourists out of here, but the weather that week was proving to be a deterrent of the same magnitude. Shopkeepers were kvetching and moaning "no business"—never mind that most of them came here as artists or Flower Children to enjoy life, grow a little pot on the back deck, and just incidentally make enough to pay the bills, before they opted for a Chamber of Commerce, after which violation, as Ed Abbey said, you can kiss your town goodbye. Friends were snapping at each other like looney birds in a tank of toxins and the humidity was a wet, down comforter under a 110-degree heating pad. Even at eight o'clock in the morning, pulling on my Lycra shorts and top was a sticky chore.

That's when I decided to do it.

I ride my bicycle up the mountain about three miles from the house five days a week, before the traffic gets repulsive, if possible. I've been doing it for almost 30 years, so people are used to it and pay me no mind. (Had the first mountain bike in town—1980, I think, when I was 60 years old—and I took all the outlying cowpaths with the same sort of joy and devil-may-care as Harvey did with his Jeep, 15 years later.) That morning was no different with regard to the joie-de-vivre. I always love the ride because of the canyon, once I get above town. The rocks are so beautifully cruel; deep maroon, red and orange and pale sheeny-green in ragged pinnacles and spires, spit from the pit of the earth into great ridges and crevasses that time can't seem to smooth over. The smell of juniper and mountain earth is heady under early morning sun before cars and motorcycles take over the highway with their noxious fumes, and there's a sometime-creek rippling below the winding road where canyon wrens sing their sweet song down the scale. All this adds up to perfection, or as close as you can get to it—on a paved road.

It's my meditation time, too, that ride. I sort out everything for the day, the week, sometimes the month. It's where I learn the lyrics to new songs—up, on up, to a rhythm of the pedal's turning—where I find inspiration for a story, a show, a letter or a melody. And the reward! All downhill at 25 to 30 mph—cooling the sweat, blowing through my helmet, down my back and neck, even through my shoes! Ya-hooooo!

So...8:00. I got on my Trek and pedaled up through town. Didn't stop at the P.O. for my mail, it wasn't in yet; besides which, I was looking very hard for special little nooks and hidey-holes that could assist me on the way back down. My heart was beating a bit faster than usual—the adrenaline of anticipation already started.

I must not have been paying close attention to everything like I usually do, because the "beep" from behind startled me. After 30 years I can hear cars coming both ways before they get anywhere near me—it's an acoustic mountain—I can identify locals from tourists and tune in the driving "mood of the day." There is one, you know—fast or slow, frantic or relaxed—it infects the whole road like a virus. That day it was relaxed, nobody on the road but me. The beeper wasn't a local—they don't beep—so I looked in my bike mirror and saw...Harvey's Jeep!

I'd been thinking about him so hard, my sadness turning to giggles as I rode, picturing what a hoot he would get out of this caper, that for a second I just accepted it. When I remembered him being gone, I nearly fell off my bike. But it wasn't his Jeep, just some slow-driving, far-more-than-ordinarily-polite dude trying not to run over me. I got to my rock under the tree, ducked in, sat down and poured half the water bottle over my head. Cooling down, I rested there in the shade for about fifteen minutes. Even so, my heartbeat was much faster than usual.

So...9:15. I had chosen my spot on the way up.

As I zoomed down the mountain and through the upper residential section to the turnoff that goes to the open pit, I was so exhilarated, so hyperventilated, that the wind made me shiver. No one in front of me—good. No one behind—even better. Very few cars on Main Street (the only level spot in town) and I didn't see anyone walking the street.

I darted in behind Robby's antique ore truck—yanked off my Lycra, everything but helmet-socks-shoes, mounted the Trek, buckass, and pedaled furiously through the center of town, past the P.O., past the shops, the bars, Town Hall, Police office...Oh-oh!...never had I seen our Town Marshal (as he liked to be called) out in front of the cop shop, but there he was, in a blur, talking with someone beside him. He looked up, automatically began a wave...Hull...ohh, (double take) Kay—t-e-e-eeee?!

I was gone!

Faster then, working the brakes, no more level ground. As I passed Harvey's empty shop, I looked up and cried out—"Harvey...this is for you, Harvey. Bye-bye!"

The last half-mile to my front gate, I was laughing so hard I could hardly steer and hoping to hell no one would pull out in front of me, when up came Wally, one of our town crew, in the frontloader. You got this? These are no-passing, two-lane roads, barely, and behind him was a whole string of tourists, ten or more cars long, chugging along at three miles an hour.

Hoo-eee! What an opportunity!

"Ole!" I yelled, as I sailed by, "Welcome to _____ !" (You cannot have the name, you won't like it here.)

In my mirror I saw arms flapping out windows and a couple got out and stared downhill, not at all sure of what they'd just seen.

My uproarious laughter had turned to streaming tears and coughing by the time I hooked a U-ie in front of my gate, hauled in, decked the bike and headed for the shower, where I sat down under the spray and howled.

Still wired and laughing, my tummy sore from it, I dressed, got in the car and drove up town. I knocked on the Police Chief's door, walked in and said, "You wanna arrest me, Ray?"

He just looked at me, shook his head and sucked in the corners of his mouth to keep from laughing. "I thought about that," he said, "but what exactly would I do?"

"Yeah, that could pose a problem," I answered—picturing him chasing me down with the cop car, getting out, yanking me off the bike, steering me onto the seat, nude-o, and driving me up to the office.

"Phones are ringing like crazy (as was his), you certainly gave the town something to talk about."

"That was the object of the exercise, Ray. Everybody's so damn glum they need something to distract them; besides, hardly anybody saw me, there weren't more than three or four people on the street."

"Enough. I took a consensus and asked them, 'Well, is anyone deeply offended?' Only the retarded son of one of our emporium owners raised his hand and said, 'I am,' so I told him to go chase you down the hill and tell you so."

His phone was still ringing. Ray ignored it. "You going to make this an annual event?" he queried.

"Absolutely not. I like to quit when I'm ahead...meanwhile I'm going to enjoy what all the hardnoses have to say—the old farts who need their blood stirred up. As for my friends, they'll just laugh their butts off."

There was, as expected, quite a reaction—notes, phone calls and letters—only one anonymous, which wasn't really bad. The next day, tacked up on the post office bulletin board, there appeared an Ode to me and my stunt, and folks were smiling again.

The town had lightened up, and I had purged the heavy loss of our dear Harvey.

Before the weekend, it rained and cooled the town down right smart. I ran into friend Mod-Bob at the P.O. and stopped to chat. Underplaying it, he eyed me sideways from beneath his brows, and sneaky-like, whispered, "I saw ya."

"Oh, yeah? I didn't see you—where were you?"

"Sittin' on the bench…in front…out there…by the Nellie Bly," he misered it out, one word at a time, "Hollered…but you weren't lookin'."

"Nope," I laughed, "I was in kind of a hurry."

"Uh-hh…ya think anybody got any photos?"

"Gawd, I sure hope not!"

"Yeah…uhh…right." Then he looks at me straight on, eyes dancing, his face nearly fractured by his smile, and says, "B'cause I thought you had…your backpack on backwards."

The nerve of that boy!!

—*MOUNTAIN GAZETTE, #78*

The Guardian of Sleep

BY
JEREMY BERNSTEIN

"The interpretation of dreams is the royal road to a knowledge of the unconscious activities of the mind."
—Freud

1. It is Paris. It is winter. We have just made love. It has been a failure. Our separation has been too long and our reunion too painful. It is early afternoon. She is next to me in bed, her face buried in a pillow and her naked back curled into a question mark. I know what she would say if I were to ask her what was the matter. She would say, "I have no feelings." She would not say, "I have no feelings for you." This would end our love affair and, as yet, neither of us has had enough courage to do that. How can one end so many months of experience shared? Each corner of Paris contains memories for both of us. I can hardly think of any street without thinking of her. I cannot imagine a future for myself which does not include her. She has gotten out of bed and I can hear the water running in the bath. I can imagine her settling herself in the bathtub and covering herself with warm water like a blanket. I find myself staring out the window at grey Parisian sky—the jigsaw roofs of Paris. I feel like a small child being punished for something he can vaguely perceive but cannot understand. What is it like to have "no feelings"? I can only think of people who die by freezing. First they are in pain and then they become numb. The danger for people exposed to extreme cold is that when they start to become numb they lose all will to resist. They do not want to return to life and they die peacefully. I have heard of such people in the mountains. They must be aroused—often brutally shaken and even slapped until they return to life, often against their will. Perhaps I should shake her and return her to life. But what kind of life would it be and would I be part of it? Freud once wrote that every sexual act was a "process in which four persons are involved." Who were the other two in bed with us this afternoon? Does she have a lover or were the others simply memories of what each of us had once been to each other?

2. I have been having a recurrent dream. In it I am to give a lecture or take part in a play. She will be there and I have promised to save her a place in the theatre. She has arrived and I am trying to find the place I have saved for her. My search takes me farther

and farther from the stage and becomes more and more futile. From time to time I remember that I am to deliver some sort of speech from the stage. The notes for my speech are in a briefcase that I am carrying. I have not had time to look at them and have no idea what it is that I am supposed to say from the stage. I will have no time to look at them because of the search for the missing place. As we get farther from the stage she becomes more and more scornful and eventually vanishes. Freud once described dreams as the "guardians of sleep." I have often wondered just what he meant. After having read Freud I picture dreams like watchdogs perched beside my bed as I sleep. But what are they supposed to frighten away? What is more frightening than the dreams themselves? What is it that the dreams are guarding us from?

3. It is winter. It is Chamonix. I have told my friends that I have come to Chamonix to go skiing. It is true that I have brought my skis to Chamonix. I had a friend once who came to Chamonix to "go skiing." He was an expert mountaineer and knew every glacier in Chamonix like a brother. He took the cable car to the top of the Aguille du Midi to ski down the Valée Blanche. He had skied down the Valée Blanche a hundred times and knew the glacier like a friend. The tracks of his skis disappeared at the edge of a large crevasse—perhaps 300 feet deep. They never found the body. They said it was an "accident." I knew better. A few days before he left he had come to see me. He had been having severe recurrent headaches for nearly a year. Much of the time he was unable to work. He told me he was going to Chamonix because even though the medical specialists had been able to find nothing wrong with him and no cure, he was sure that in the high altitudes and pure air he would discover a way to stop his headaches. As far as I know Freud never addressed himself to the physical aspects of skiing so that we are free to make of it what we will.

1. The water has long since stopped running in the bathtub. She has taken her bath, gotten dressed and is ready to go out. I am to walk her from the Left Bank where I live to the Right Bank where she works. It is a walk we have taken hundreds of times. We will pass by the Ile de la Cité and stop for a moment or two on the banks of the Seine. In earlier days we would walk arm in arm. She would, from time to time, nuzzle against my shoulder like a gentle grazing animal. We would say nothing because nothing needed to be said. Today we will say nothing because nothing can be said. We have taken the steps down to the banks of the Seine. From underneath a nearby bridge there emerges a small band of young people each carrying a flower. One of the girls is laughing but each time she laughs too hard she sits down in the middle of the walk and sobs uncontrollably. They are having difficulty walking straight and I am afraid that one of them will fall into the river. They have all passed by except for one boy who has stopped to stare at us his eyes wide open, like two great lamps.

He looks extremely cold, his thin body barely covered in jeans and a tattered jacket. First he studies one of us and then the other. Nothing has been said and we are all transfixed in each other's eyes. Finally he leans forward to whisper in my ear. He points at her with his finger and whispers, "Elle—*elle est trop bien pour toi*," and disappears after his friends. "What did he say?" she asks. "I couldn't understand his French," I lie. "He must have been taking drugs."

    2. In 1907 in his "Obsessive Actions and Religious Practices," Freud describes a woman of about 30 who had been separated from her husband for several years. (In these descriptions we are never told the color of the woman's hair or if she liked music or even what first attracted her to her husband, unless these are somehow specifically relevant to the analysis.) Several times a day this woman would run from her own room to another in which there was a large table with a tablecloth on it. She would arrange the tablecloth in such a way that a prominent stain on it would be most visible. She would then ring for the maid and send her out on some trivial errand, always making sure that the maid had seen the stain on the tablecloth. Freud discovered that on their wedding night the woman's husband had been impotent and had come running into her room several times to try to make love but to no avail. (Why, on their wedding night, they had separate rooms is not explained.) In the morning he was so ashamed of himself that he had stained the sheet with red ink so that the maid might think the marriage had been consummated. Freud concluded that the woman's obsessive act was still an attempt to justify to the maid the behavior of her husband on his wedding night. I find this analysis convincing but incomplete. I would like to know the feelings of the maid. Was she able to wash out the ink stain or did she simply throw away the sheet? What were her thoughts when she was summoned again and again to look at a stain on the tablecloth? If it were me, I would have taken the tablecloth and thrown it out the window. Although born in 1856 to a family in modest circumstances, by 1907 Freud had entered the professional middle class. He too, no doubt, had maidservants who came and went, clearing away the dinner dishes. We are all mired in the limitations of our environments.

    3. It is late afternoon in Chamonix. I have taken one of the last cable cars to the Aguille du Midi carrying my skis. There are no other skiers in the cable car. No one would run the Valée Blanche on skis at such a late hour. I explained, when asked, that I was going up for the view and taking my skis so that they would not be stolen during my absence from the valley. We have reached the top and I have walked through the small tunnel leading from the cable car to the snow outside. Lights have been turned on in the restaurant above the cable car station. The cold is intense, and my lips and cheeks burn like fire. The wind whips needles of snow and ice into my eyes. I look down over

the Valée Blanche. On the track there are long patches of blue ice with crevasses on either side. To ski on ice requires a special technique. One must traverse across the ice leaning down away from the mountain. This will make the edges of the ski grip and cut grooves in the ice. One must struggle against one's natural instinct to lean into the mountain—to hold it, to embrace it. This will only make the skis slip sideways along the ice—to become detached from the mountain and lead to a certain fall. Once a fall on ice begins on a mountain like this, one cannot stop himself.

1. Some days have passed since we were last together in Paris and some days are still to come before I go to Chamonix. We have not spoken since our afternoon in bed. From time to time I reach for the telephone to call her only to find that somehow one of the digits in her telephone number has slipped my mind. By the time I have looked up the number in my book I no longer have the courage to call. She has not called either. This morning a letter from her has arrived. As usual it contains no name or return address on the back on the envelope. She used to say that she did this so that I would be happily surprised when I opened the envelope. By now the handwriting is so familiar that it is its own return address. I feel, before I open the letter, that I know what it will say. For some time I do not open the envelope. I am tempted to write on it, "No longer at this address—present address unknown," and to put her address on it and send it back. Finally I open the envelope. Her handwriting is very neat and bold. She is given to short sentences and brief letters. "You must know that there is nothing more between us. During our last separation I met someone and have now decided to move to London with him. I do not regret the last two years. I hope you feel the same. Fondly..." A tidal wave of melancholy sweeps over me. I see an endless row of empty days and nights before me, aligned like unlit candles. My strongest urge is to sleep for days and to wake up somewhere else as someone else. I have told my friends that I am leaving Paris for a while to go skiing in Chamonix; to change air—"ça fait du bien."

2. In 1915 Freud wrote his celebrated paper "Mourning and Melancholia." The melancholic is often subjected to rages sometimes against others and frequently against himself. One of Freud's more famous patients—the Rat Man, so called because of a sordid fantasy he had involving rats—had been abandoned by a lady.

"On the day of her departure he knocked his foot against a stone lying in the road, and was obliged to put it out of the way by the side of the road, because the idea struck him that her carriage might be driving along the same road in a few hours' time and might come to grief against this stone. But a few minutes later it occurred to him that this was absurd, and he was obliged to go back and replace the stone in its original position in the middle of the road."

Freud points out that this behavior is more complex than might appear at first sight. To begin to comprehend it one must take into account the fact that the Rat Man replaced the stone at just the place where he thought the carriage of his lady "might come to grief." He was torn between the desire to injure his lady and to protect her. Frequently the melancholic becomes partially identified with the loved one now lost. His rage turns against himself in his identification with the lost object. In Freud's famous phrase, "The shadow of the object fell upon the ego."

    3. It is clear what I have come up here to do. All that is needed is to put on the skis, to point them down toward the Valée Blanche and to let myself go. Once on the ice there will be, there can be, no turning back. I cannot do it! It is not exactly fear of death. I envisage death in the snow as a long sleep; numb with no feeling. It is the fact that she, in London with her lover, might never know or hardly care. I have collapsed onto the snow and have begun to weep uncontrollably. Two men have come out of the restaurant, through the tunnel and toward me in the snow. Each takes an arm and partially carries me towards the warmth of the restaurant. As we go through the tunnel one says to the other, "*These Parisians—the thin air up here makes them a little crazy.*"

—*MOUNTAIN GAZETTE, #9*

EDGAR BOYLES

Hallucinations

BY
BARRY CORBET

I have a friend who saved for long years to buy a sailboat. The sailboat was to give him freedom. About the time he had accumulated enough money to buy the boat, he discovered that there was no place to go. That was a colossal disappointment, he said.

Last May, I had accumulated enough time to go somewhere for three weeks. I had no sailboat, so I pondered where to go. To my colossal disappointment, I could find no place to go.

In front of me is an Executive Planner. It has few entries because I normally am so overwhelmed by executing that I have no energy for planning. A notation for Saturday, May 25, says "Go to cabin?" I suppose that some crazy evening I must have actually decided to spend three weeks by myself in a one-room cabin in headlong retreat from phone, liquor, electricity, running water and relationships—all the scary things in my life.

I looked forward to this trip as I energetically met film deadlines, overreacted and, in general, indulged in a manic speed trip. No matter—I knew I was soon to be going no place with nothing to do.

The rules:

Take two serious books for edification and reflection, but not for entertainment.

No dope, pets or people.

No music-making, writing, painting, photography or other habit-forming activity.

Go somewhere and live with yourself.

I cheated. I took three liters of red wine, one for each week. I took my Executive Planner, in which I sometimes wrote a maximum of two column inches a day. The books I took were *The Lazy Man's Guide to Enlightenment* and *Cutting Through Spiritual Materialism*. I would prefer not to tell you that, but I have no choice. I cheated further by taking *A Field Guide to Rocky Mountain Wildflowers*.

I feel that I should tell you why I made this decision to disappear for a while: I felt pressed by professional and domestic conflicts, and I wanted to suspend them for long enough to see if either my problems or my sense of being victimized by them would go away. I also lusted after Sensible Alternatives. The following is inflated from my Planner.

Day 1: cabin. A stupefying hangover prevails, because friends had dropped in on the eve of my departure to celebrate my foolishness. Everything happened last night; nothing is happening today. I'm content, since I know how good things are going to get.

> *I worry about the overlay of Eastern mystique that hangs around things like this. I think there's both a mystery and a message in my solo, but it's all so damn trippy that it seems sort of precious. Mystery + message = mystique.*
>
> *I wrote the account of my 22 days at the cabin just recently, three months after the fact. To put the experience into a more "worldly" context, I've added another 22 days to the mix. There was no attempt at consecutive ordering made here. These are just some of my days, without the mystique:*

Day 1: world. A client finds out his contract doesn't include something big that he's already promised to his management. How could he be so naïve? Am I the keeper of my brother's illusions? I am.

I speed to the downtown office, attempt to reconcile my various prose styles, research all the ingredients carefully and write a letter that refuses to choose between alternatives. It scathes and consoles. It's an onslaught of good will, right thought, venom and threat. They'll love it because I'm in the right. So are they, but that's their problem.

Day 2: cabin. Evidently, nothing happens again. There is no entry in my Executive Planner.

Day 2: world. Yesterday's problem client calls to express his good feelings, and I wish I hadn't written. But he frees me to do what I do, which is edit film.

But this film appalls. All of our films are misleading in one way or another, but this one is 18k raunch. Funded by a Midwestern bank, its warped objective is to depict the good life as perceived by the bank's loan officers.

It's the good life for the teeming blacks and Polacks because of all the socially redeeming industry that exists to employ them into grateful consumers. It's good for industry because of all the toilet-trained labor that consumerism makes available. Highly-Semi-Skilled Hype. Good for the bank because of loans to labor and industry alike. How beautiful is the garden...

Day 3: cabin. Memorial Day. Visitors. A young couple on horses. A strident voice from a concealed man in a wheelchair (me):

"Would you mind going back the way you came?"

"Private property?"

"Yup."

They wheel their horses in acknowledgment of Private Property Rights. Aghast at myself and then at them, I return to the vigil. Repression.

More visitors. Two men and a daughter approach on horses. A voice from the woods once again requests withdrawal. This time, an adult in very full chaps detaches himself and lopes voiceward to parley. Intimidation: big horse, small wheelchair.

"You own this place?"

"Yup."

"I live in Paradise Hills."

"Great."

"You can ride from here all the way to Central City."

"Mmmm."

"You all alone here? Why do you want to be alone?"

"Religious reasons." (Holy, holy.)

"What's your religion?"

How the hell should I know? I decline to answer.

"There's no 'No Trespassing' sign up there."

He's right. "I know—I'm sorry to do this to you, but I really need to be alone."

"I'm sorry you feel that way."

"So'm I."

He leaves, without sensing that I admire his good balance in the face of hardship. He pauses to explain this unexpected turn of events to friend and daughter, then moves off toward the end of the trail.

It's still Day 3, a long day. More visitors. This time two four-wheel-drive pickups clatter importantly toward me. I am mindblown, outraged, undone—for the enemy is here. A redneck father and semi-longhaired teenage son occupy the first cab, and I'm out of words:

"I really don't like people driving over this land. I just don't like it."

"How'd you get here?"

"I drove."

"Well, I've been trying for three years to find a way from here down into Clear Creek, and I thought this might be the way," he says like some nouveau Vasco de Gama.

"It's not. U.S. 6 goes there."

"Listen, I'm willing to talk to anybody, mister, but I want him to meet me halfway. I mean. I want him to be friendly." He says this defensively, eyeing his young son as if he's looking in a mirror.

"Look, I'm really sorry that I'm not being friendly (true becoming false), but I just don't want trucks down here."

"Listen, we didn't hurt your goddam road at all. Not one goddam bit," he says in

53

triumph, and backs up to confer with his friends. They talk for a long, long time, produce no rifles and leave.

What should I learn from these encounters? Who was worse? Is it wrong to want to be alone? Is it wrong to hatefully and unjustly assume that all people in four-wheel-drive pickups are mobile insults?

I take a felt pen, a shingle and a 2x2; and I make a sign that says:

PLEASE
PRIVATE PROPERTY
NO HORSES, JEEPS, ETC.
NO THROUGH ROAD OR PATH
NO BULL!
NO OFFENSE?
THANKS

I drive to the cattleguard and plant the sign. I guess it's significant that I have no more gate-crashers during the remainder of the three weeks. Holy, holy.

I also have time to decide on this long day that there are only two kinds of films: Films to persuade. Films to open.

This consideration seems important. I also consider, with coaching from *Cutting Through*, that my thought patterns act as excellent ego-defenses.

Day 3: world. I ply my trade, converting footage into film. This film I struggle with is persuasive if you don't see the relationships involved, or naïve if you do. I derive perverse pleasure, an opening experience of sorts, by cutting the chrome and glass and steel of modern buildings to Bach's Third Suite. The cutting of visuals can reflect the tempo and mood (and even the intent?) of the composer so accurately that Bach seems to exalt what man hath wrought in downtown Milwaukee.

Alchemy.

Day 4: cabin. All-out war on overlapping thought patterns in an effort to be alone, to be nowhere. It's discouraging to find that the mind natters incessantly as if it had an audience.

I sit for a time in a swale filled with flowering dandelions and consider this thing. As I sit, I feel some love, although it may be only an affinity for dandelions. I call it a technical success, and take the money and run. Exploitation, beauty, peace and awareness have no real collateral.

Day 4: world. Trade-plying. Vaguely pleasant. I put off my partner, who is returning from New York tonight and who might want to stop for a drink. Why does that make me feel guilty? Wanting to be alone is tantamount to being unfriendly. And, for that matter, I don't even know why I want to be alone tonight. To write this? To be free not to cope? The fear of invasion of privacy is the fear of being discovered. I hide. The only thing worse than being discovered is being discovered hiding. The risks compound.

On Gaffer Tape

Gaffer tape is a wide, gray, sticky substance used to pack or secure film equipment. It resembles duct-tape, but is much stickier. Gaffer tape is exceedingly sticky. You can attach anything to anything with gaffer tape. It always stays stuck. Very sticky stuff. Insatiably glommy. It's a very comforting product, for it sticks your stuff together.

Now, as you become older and wiser, you watch your personal drama unfold with less and less cohesion. You lose what you had, you gain what you don't need, and you require more energy to remain in place. You acquire so much information and concomitant interpretive baggage and opinions and openness to new opinions that nothing can be guaranteed to be right. Absolutes fall away like flies, the reality depends upon the perceiver and nothing can be believed with confidence. This is a definite problem for beings.

The answer is gaffer tape. Gaffer tape enables you to enlist all the fine properties of stickiness in your own favor—just take all the fractured elements and stick them together in any old order. Doing that creates a cohesive entity that can be identified and therefore related to in safety. Maybe that's how we stick it to ourselves. We try too hard to keep it all together. There should be some way to neutralize the use of gaffer tape, or at least keep it judicious. Fly-paper is useful for catching flies, but you want to be able to step out of it if you fall in yourself.

The only clue I have to this dilemma, and I find the idea troublesome, is that Time is quite a lot like gaffer tape. They're both useful ways of arranging events.

Day 5: cabin. Excess baggage—more travail with the Dread Natterer.
   Natterer: Regretful Rehasher—"Look what I've done."
   Coach and Commentator—"Look what I'm doing."
   Future Fantasist: "Look what I'll do."
   Nasty Natterer.

Day 5: world. My enthusiasm for this piece has waned, and I've taken a week off from my book of days. Why publish this drivel? That's *Mountain Gazette's* problem. Given a publisher, one publishes. Who's at fault?

Day 6: cabin. Milarepa, a tenth-century Tibetan yogi, encountered demons that had "bodies big as thumbs and heads like plates." Demons, to Milarepa, were manifestations of his own negative energies.

My demons are also as big as a thumb, and they manifest as bumblebees. They cruise in and out in amiable stupidity. Battering from surface to surface like bumper cars, and I struggle with my noisy demons. I feel possessed.

With nothing to do, I try really and truly to do nothing. I can't. I'm trying to stop trying. I need a gap.

Day 6: world. Five separate films, and their sponsors, nag me with disruptive detail. This wall of trivia needs to be dispersed before craft can proceed.

If the mind is spared conflicting commands for a while, if it can relax and sense the overlapping qualities and similarities of the diverse material, then the elements of a film magically determine their own structure by being allowed to. Once the material makes its proper arrangement known to him, the editor is just a functionary who can splice. That's Phase One.

Phase Two is manipulative. After structured integrity is allowed to develop, then insidious intent runs wild, and the editor can challenge, threaten, stroke, alienate, seduce or coerce without either jeopardizing or enhancing the film's sound structure.

Structure is hardest, because it can depend upon tempo, mood, texture, story, shock, thesis, antithesis or nothing—or all of these things. It has to reflect what's there. It requires a discovery.

Interpretation is the most fun, because you can use technique—like putting English on your serve.

Workability is a mixed bag in which you jiggle the objectives, prejudices, persistence and clout of the various forces involved in the film. The biggest force is usually the guy paying the bill.

Then you stick all those little stop-frames together, twenty-four to the second, and you watch the photons of your ego bursting all over the silver screen until they bounce back, repulsed, seeping like spilled milk onto the shag rug of the screening room...such wretched excess.

Day 7: cabin. The Dread Natterer is perceived as an enzyme of insanity. Physical and mental defection. Eyes hurt and closely watched breathing scours the underbelly of my brain. I'm caught up with this nonsense and can't relax with it. This is the first time I've been worried. Claustrophobia. It's too deep to breathe in.

Wine day! Wine is nice. The flight of the self vindicator.

Wine is nice, and, being newly unaccustomed, it makes me want to try things I wouldn't always do, such as dive naked into ant hills. It makes me euphoric and communicative. Loving. It allows examination of concepts, or non-concepts, unavailable to me when I'm deep-throating the dharma. *It gives me time.*

Day 7: world. I took Willi Unsoeld back to the airport this morning and am left in contemplative shock. It's bizarre, after not seeing him for four years, to find how very much of our lives are common ground. Just one person acting out two fantasies.

It's been years since I've felt like discussing my magnificent and crummy past with anyone, but Willi's past is mine and mine is his, and talking about it is like one ego getting off on the antics of its various aspects.

We use each other as mirror, blackboard, stimulant and antidote.

It's scary to ask myself if I have this relationship, latent or developed, recognized or not, with untold numbers of people. Hideous. Our Karass is in a sling. Let us go then, you and I, for we have no choice.

Day 8: cabin. Still driving wedges into my brain. Painful, powerful and scary. I chastise another bumblebee. A CAT! I see an enormous Siamese, or a puny Puma, slinky and beautiful, gliding between trees.

Day 8: world. Nothing much happened today, which is a relief.

Day 9: cabin. "Try to be simple as you can," I tell my Planner. Minimum impact.

Day 9: world. Attending to busy-ness, and nothing gets done. I am irritated by the dedication required to maintain my speed and by the uselessness of speed maintained.

Day 10: cabin. At four p.m., I am stalked by CAT. While sitting, absorbed in other things, I have a peripheral flash that tells me CAT is crouched six feet away and working closer. I think of Castaneda and his friendly coyote. I think of madness and its messengers. For lack of communication, I think of conversation: "Hello, Pussy!"

Pussy bounds down the slope. I came to get empty, i.e., to become a suitable terminal for communication, and I can't even hold my tongue in front of the most quiet of animals. I have a sense of aggression and speed refusing to slow down. Or of hope and fear refusing to strike a bargain. The cat is out of the bag and the fix is on. We are separate.

Day 10: world. If alienation stopped being bad, would sunsets stop being good? Without expectation, can alienation exist? Without alienation, can dualism exist? Without dualism, can sunsets exist? Hello, Pussy.

Day 11: cabin. It's a quieter, better day. I'm doing nothing in a more businesslike fashion, accepting it as hard work (does nothing need to be done?) instead of harboring anticipations of the beneficial results of doing nothing. Nothing times nothing is nothing and yields nothing. Except that the Natterer editorializes. Simplicity is no longer a consideration. Business as usual.

Day 11: world. Solar extravaganza. Red skies, red earth; alpenglow lights the powerline and Mother Cabrini's shrine. Would sunsets stop being good? Enigma pie.

Day 12: cabin. Depressed. Technique and dedication have failed utterly. One cannot be dedicated to nothing. Each thought pattern has its friends and relations.

Wine. *Bad* wine. For a while, I sit outside with bottle and glass on a stump, trying to feel the dignity of the occasion, but it's just bad wine. I go inside in time to witness a tableau. A teenage couple is passing by the cabin. The boy prances and makes insistent animal sounds. The girl, unamused, repeatedly screams, "Don't!" The charade is real, though they seem physically apart. "Don't, don't!" she cries, pleasantly terrified. The boy sees me leering from the window, waves with a high good humor as if we share a conspiracy, and continues the ritual. "Don'ts" echo from the forest as the couple proceeds. I turn to serious matters.

Why, I thought, am I drinking lousy wine alone, trying to do nothing and indulging in voyeurism? These were my reasons:

1) I want to create or experience a gap in the constant rush of thought and emotions with which I fill my days. Sub-reasons:

a) To let intuition get an anti-thought in edgewise.

b) To develop a sense of space, which in turn is (I'm told) characterized by enhanced awareness and compassion for self and others. I especially like *Cutting Through*'s notion of making friends with oneself.

2) I also want to reconcile form, or phenomena, with undifferentiated space. This subject is appallingly vast, unless (I'm told) you understand it. I don't. Why do I call it a subject if the object is to transcend the subject-object? I know nothing. I can't do nothing. I can't even experience nothing. When I die, I'll be a beginner at that too. Where should I start?

Day 12: world. Counterpoint: empty times playing against the sting of being alive. I convince myself that this thought is creative, not morbid.

Day 13: cabin. I have a moral hangover from last night's excesses. Speculation is not my strong suit, and I really feel I'm getting too serious about all this.

I wheel to the spring, scoop a hole in the mud and wait for the little pool to clear. A robin bathes first, then it's my turn for a cleansing. It rains. Everything feels good, as if good things are going to happen.

Day 13: world. A nice piece of relationship takes place today, and I feel cheerfully uncritical. I'm alone tonight for the first time in a week, and I have a rush of expectation at having no expectations.

Day 14: cabin. OK.

Day 14: world. Lunched with a client. He was pleased with the commercials I'd made for him, and confirmed his euphoria with gin. Afterward, the color and whereabouts of his rental car had slipped his mind. I repeated directions to the airport and left him carless, for I am not my client's keeper.

Day 15: cabin. Snow all day long on June 8. A beautiful, nice day. Neurological Buzz. It's easier to do nothing when it's so busy outside. If phenomenal flux were more constantly apparent, perhaps emotions would subside. In a snowstorm, it's hard to know which snowflake to focus on.

Day 15: world. Last night's emotional encounter caused today's hangover. Didn't it? My goodness, I've been in the over-indulgence business for eighteen years, not knowing that I could do myself such a disservice.

Day 16: cabin. Wind and sun. Painful sunburn. I love it.

Day 16: world. Halloween. I masked off the screen on my editing table with gaffer tape to simulate a wide screen. I masked my feelings about Michigan spectator sports with devilishly clever cutting, and about a client who hasn't paid with hypocrisy. Children I know and love well arrived in monster masks clearly showing their true natures. Their chaperones arrived straight, unmasked, and seemed to be in relative drag. Who was that masked man?

Day 17: cabin. My eyes still hurt, as if I were breathing through them. If one seeks to prevent pain with each breath, one is busy as hell.
 The understanding improves, but the condition prevails.
 Last wine day. This wine is truly pigshit. As always, it triggers considerations, this time about my psychological set. This set is derived from the books I am reading, and from deep dissatisfactions that seem too close to the surface.
 As I am trying to learn to see it, here's the scam:
 ● We are alienated from ourselves (i.e., rendered neurotic) by dualism, a mode of perception and intellective thought that characterizes Ego. Neither dualism nor ego exist as things, but like a politician's media campaign, once they're postulated, their notions serve to influence and confuse.
 ● Ego perpetrates the dualist mode. Dualism serves ego. A circle jerk.
 ● At another level, dualism is the description of phenomena, space and whatever else exists, as either subject or object. Since this arbitrary separation helps ego see itself as discreet from whatever is not included in its own definition of self, ego and the dual-

istic mode become inseparable. Ego is the tendency; dualism is the process. Perhaps.

• Since ego constantly changes its estimate of what's inside self and what's left over—the penchant being to assimilate as much as possible and to increase the Empire—the dualistic effort of data-processing becomes confused and erratic. The Peter Principle, over-extended.

• This erratic confusion generated by tendency and process creates malaise, which preoccupies us so intensely and painfully that we fortify our Private Ground (holy) against this suffering. But suffering is a basic condition of dualistic logic. Dualism is conflict. Conflict is suffering. Suffering is ego, the battleground mentality of subject-object departmentalization. Our fascination with suffering drives us to struggle against it. The struggle is the struggle.

• All we have to do, obviously, is to quit.

• Is that *ALL*?

It's very bad wine, and I plagiarize shamelessly. The people from whom I plagiarize probably don't care.

Day 17: world. I resigned from the American Alpine Club today.

Day 18: cabin. A good day, a good supper—why is this evening so bottom-line bad? No one is here, nothing is wrong, and my bummer is demonic and complete.

I didn't know I could do that all by myself.

I've read a Tibetan parable of a monkey locked in a house of his own creation. He experiences various states of despair and elation as he tries to escape his condition, but nothing really works. He's still imprisoned by his hallucinations. The solution, the tale goes, is to recognize and laugh at the hallucinations.

I take comfort in the notion of laughing my way through the hallucinations, but comfort is undoubtedly a cruel hoax. Still, the idea appeals.

Day 18: world. I had an easy and disappointing day today. I showed our client in Michigan his new film depicting the pathetic emptiness—yea, vicariousness—of a tourist's activities in his state. He was pleased by the accuracy of our statement, sure that Michigan's flagging tourism industry would be forever revitalized. I was pleased that this ugly duckling of a film had won his approval.

My related despondency stems from the idea that we pool our mediocrity. My client and me, we're proud to share the total pointlessness of what we say and do because we know that nobody will object. We're just playing the game, and you can tell the best players by their colorful uniforms…

Day 19: cabin. Uppers and downers. It's hard to laugh at the hallucinations. Laugh and the world laughs at you.

Today's hallucination is frustration. Vibrating intimately with frustration is frustrating. Who could laugh except a frustrated cynic? I yearn tragically for a more convenient hallucination—perhaps I'll be a fireman when I grow up.

Day 19: world. I had an easy and disappointing day. I showed our client in Alberta his new film depicting the utter tragedy of marketing his beautiful province to the global letch for uncluttered real estate. He loved it, so we found ourselves in agreement.

We both get paid for what we do.

The reason the monkey is supposed to laugh his way through the hallucinations is that they're his own creations. Why would I make such a mess? How could I be so creative?

Day 20: cabin. Nothing.

Day 20: world. My dog cries outside this shed where I write. She doesn't like to be alone. She has a dream house of her own, and I'm IT.

Day 21: cabin. Pretty boring day. I'm sure getting ready to leave this place.

Day 21: world. Just finished *Tales of Power*. Castaneda makes me laugh and cry more than any other author. At the end, it's all warm tears. "We're all alone, Carlitos," don Genaro said softly, "that's our condition."

Castaneda makes me think longingly of sitting alone on a desert peak, waiting patiently to see the "lines of the Earth."

Day 22: cabin. What the hell is this? I arrived on a Saturday, left three Saturdays later, and it's one day more than three weeks. I've been cheated, detained by order of the hallucinator-general, who can't even count.

In fact, this whole excursion seems like a dream, a hallucination of insight into insightlessness. There have been isolated instances of apparent timelessness, but these too are gone, safely gaffer-taped into continuity.

Day 22: world. I can't say that any one series of days was better, harder, spacier or more instructive than the other. I'm not sure that one even influenced the other. They're just days, without any psychological or moral valence. While I vividly remember occasional yearnings for convenience and companionship while I was gone, I now find myself subject to periodic urges to disappear again.

I'm glad I went away, and I know I'll do it again. I have no idea where I'll go next time, or what I'll set out to accomplish. If I'm lucky, of course, there will be no place to go and nothing to do.

I'm looking forward to it immensely.

—MOUNTAIN GAZETTE, #33

CATHERINE LUTZ

Slouching Toward Simpletopia

BY
GEORGE SIBLEY

Once a rustic, always a rustic—the simple life for me.
But living the simple life these days is a very complicated proposition.
—Ernest C. Steele, 1928

"The simple life" continues to have cachet. In fact, it gets worse all the time. Every year we get a fresh crop of simpletopians here in the mountain valleys: people—young and old, rich and poor (usually either young and poor or old and rich)—wanting to "get back to the simple life," as though it were something we all once had but lost. Simpletopia. On my forays to the urb, which usually include a trip to the Tattered Cover Bookstore just to luxuriate in the abundance, I can peruse whole racks of magazines and long shelves of books about "simplicity." It's not just a yearning; it's an exploitable yearning: a "lifestyle" to be cultivated.

I'm no pot to be calling the kettle black, of course. I was pretty much a simpletopian when I stumbled into the mountains in 1966, one of the young and poor variety. And I did a lot of things back in the 1970s that could have been construed as simplicity-seeking: After most of a decade at the end of the cultural road in Crested Butte, Colorado, my search for low rents and affordable time for writing led me and my family to a caretaking position six miles beyond the end of the plowed road. Even worse—after that, my lack of imagination and desperation for material led me to write a book in the late 1970s about my "life in the woods"; anyone who didn't read it closely might have put it on the "simplicity" shelf. As it happened, hardly anyone read it at all; it probably never even made it onto the Tattered Cover shelves, which spares me the problem now of having to explain very often why my "life in the woods" really wasn't part of the simplicity movement.

But at any rate, my suspect past notwithstanding, I want to take a closer look at the simplicity movement. With the movement increasing in numbers and visibility about as fast as the pace of life in general is increasing, questions arise. Is "simplicity" real? Is it possible? What does it help? What does it offend?

The movement—the revolution, some call it—is most often described as "voluntary simplicity." As opposed, I guess, to the kind of involuntary simplicity imposed on those who have nothing, not even prospects (although such a life is rife with its own complications). "Voluntary simplicity," according to one of the movement's gurus, Linda Breen Pierce, "is not about living on as little as possible or about depriving ourselves. However, it does involve unburdening our lives, living more lightly with fewer distractions."

Pierce—author of *Choosing Simplicity: Real People Finding Peace and Fulfillment in a Complex World*—sees the movement as having two categories, or maybe they are stages. She says in an article in *Simplycity* magazine: "Some people are interested in simplicity for those values that are primarily self-directed—more time, personal freedom, reduced stress, a slower pace, control of money, less stuff, fulfilling work, passion/purpose in life, joyful relationships, deeper spirituality, better health, and a connection with nature."

Others, she says, "are motivated more by other-directed values—protecting the earth's resources, remedying social injustice, serving the community, and caring for others." She observes that "many people first approach simplicity with an interest in self-directed values and later develop other-directed values. An almost magical transformation takes place (as) people...get acquainted with their true selves, and then naturally become aware of their connection to other life forms—people, plants, and animals...They discover that their personal fulfillment is intimately connected with serving others."

Based on the evidence presently stacked on my desk, the chief means chosen for serving others, by the people who have discovered this type of "personal fulfillment," is to write a book or two or three to tell everyone else about the joyous virtues, the virtuous joys, of the simple life. There's Breen Pierce's book, of course, which came out of a three-year "Pierce Simplicity Study" of some 200 people living lives of voluntary simplicity. There's Cecile Andrews' *The Circle of Simplicity: Return to the Good Life*, which has allegedly spawned "simplicity circles" around the nation and a lot of consulting work for Andrews.

There's Sarah Ban Breathnach's growing shelf: *Simple Abundance: A Daybook of Comfort and Joy* and *Something More: Excavating Your Authentic Self*, and bunches of workbooks like *The Simple Abundance Companion: Following Your Authentic Path to Something More*. There's the prolific Elaine St. James, author of *Inner Simplicity, Living the Simple Life, Simplify Your Life with Kids* and *Simplify Your Life: 100 Ways to Slow Down and Enjoy the Things That Really Matter*, and she's also now available on a $60 set of audio cassettes so you can learn how to simplify your life while sitting in traffic jams, or driving a moving van to the mountain village of your choice. Today it seems to be mostly a movement of female gurus, but there's at least one prominent guy: Duane Elgin, author of *Voluntary Simplicity: Toward a Way of Life that is Outwardly Simple, Inwardly Rich*.

There are magazines: a couple of these are *Real Simple*, which puts "life/home/body/soul" under the flag and, for the city cousin, *Simplycity*, which the cover proclaims to be about ideas—style—euphoria—home—frivolities. The more venerable *Mother Earth News* also fits the category in a more woodsy approach. As is the case with most magazines, the real function of these publications is to bring readers together with advertised products that will realize the yearnings the publications feed on, from the serious hardware of the simple life in *Mother Earth News* to the organic lipsticks and "simple shoes" of *Real Simple* and *Simplycity*. One notes that most of the simplicity hardware is priced to keep the dollar flow up: If people are going to be buying less, then it's only fair that they should pay more for it, and that's the reality of simplicity buying and selling.

Beyond the magazines are a host of catalog companies that feed, or feed on, the simplicity movement. At the origin of this species one suspects a May-December coupling of the grumpy old L.L. Bean catalog and the flower children's *Whole Earth Catalog*; the growing progeny from that union of ideas now also runs from serious hardware (*Real Goods* and *Plow & Hearth*) to a host of austere looking catalogs offering simple fashions that require complicated financing for all but the really wealthy.

The extent to which this is a modern growth industry is indicated not just by the product lines it has launched, but by the real and virtual webs of workshops, seminars and information sites spreading across the nation and the globe. You can keep up with the movement, and your manifold options for buying into it, on web sites like www.realsimple.com and www.simplycity.com.

So what's really going on here?

Some of this stuff is so thoroughly precious as to be terminally irritating. Sarah Breathnach, author of the *Simple Abundance* books, holds the pole position at this end of the simplicity spectrum. She writes to and for women of vaporous spirit, women perhaps a little too sensitive for a testosterone-driven world; to such women she offers a sanitized and scented "inner journey" that will, if successful, leave them safely cocooned in a "tapestry of contentment that wraps us in inner peace, well-being, happiness, and a sense of security." Important to this journey is finding one's "authentic self," the "woman we were meant to be," the inner woman who for decades has offered "overtures" that have been ignored: "Wear red…Cut your hair…Study art in Paris…Learn the tango." Et cetera. (Quotes from *Simple Abundance: A Daybook of Comfort and Joy*)

Breathnach's books are for women, so I'm not going to go overboard in turning them into locker room humor. But it's worth noting that, before she latched onto simplicity, her earlier work celebrated "19th-century domestic life," and despite a kind of New Agey tone, her simpletopia is solidly grounded in a Victorian sensibility. I can't dip into her books without coming quickly to think of "Mother," the central woman in E. L. Doctorow's *Ragtime*, the industrialist's wife who had grown up cocooned in a Victorian "tapestry of contentment," and spent most of her time trying to avoid suffocating in its folds, and in her own indulged and somewhat enforced uselessness—especially after she became intimately aware of the extent to which her bourgeoisie tapestry of contentment was woven for the lucky few by the many laboring in industrial hell.

It's no great leap at all from the Breathnach version of simplicity to the simplicity fashion magazines. Yes: wear red, get that haircut. This isn't voluntary simplicity; it's voluntary Victorian superficiality, for those who can afford the time and money. The "real man," who secretly longs for the end of feminism and a return to the time when women were ladies, might try giving each new squeeze a copy of *Simple Abundance*: If she doesn't laugh and throw it at you, you might have yourself a good Victorian keeper that you can get up and out of the way on a pedestal.

Almost as irritating is the ubiquitous helpful hints literature, like Elaine St. James' *Simplify Your Life: 100 Ways to Slow Down and Enjoy the Things That Really Matter*. This is, however, probably the most American part of the genre, because we Americans have always been suckers for someone stating the obvious to us in a numbered list. This represents an almost mystical faith in the power of books: I think we acquire them in the naïve hope that, if we have them on our shelves, or maybe under our pillows as we sleep, we will never have to actually read them; the obvious good sense they represent will just seep into our lives the way a potpourri or a fart permeates our air. I read enough of St. James' book to know that there is nothing there but pretty obvious common sense. "Drop call waiting." "Reduce your needs for goods and services." "Make water your drink of choice." "Do what you want to do." OK. Sure. Makes sense to me.

But does anyone who is seriously moved to change his or her life really need this kind of advice? If you can't figure out your own "hundred ways to slow down and enjoy what really matters," you might as well save your energy and not start, because you're not going to be able to figure out what to with all the time you save when you get there—although probably St. James or someone has a sequel in the works for that stage.

Once past these kinds of irritants, and the didactic, earnest and tendentious zeal that infuses it all, there is some reasonable stuff in the simpletopian literature—mostly because some of the books get past the "self-directed" level of indulgence in bourgeoisie superficiality, and on into the "other-directed" realms in which there is at least some examinable political and socioeconomic purpose driving the urge to simplify. Simplicity for some of its acolytes becomes a way of life in the world, not just an individual's commoditized "lifestyle."

Linda Breen Pierce's book, *Choosing Simplicity: Real People Finding Peace and Fulfillment in a Complex World*, is her effort to find out if all this good formulaic advice actually works for people. "Most of the simplicity books I read told me why or how to simplify my life," she said, "but I found little written about real people who had actually tried it." She was curious about "the thin note of loneliness weaving through many of these stories" of people who had "simplified." The stories in her book seem to be upbeat but generally honest accounts of people fumbling along the path to simpletopia.

She herself admits, "When I look at my own life, I see that I have a long way to go before I can truly walk my talk. Even though I have reduced my dependence on material possessions and cut back on my utilization of earth's resources, I still consume more resources than four-fifths of the world's population...(But) before a child runs, she must walk. Before she walks, she must crawl. I am at the crawling stage." A nice disclaimer that should be more prominent in this literature—but might also leave one wondering if instruction in the early stages of crawling is really going to be worth the price and time.

Cecile Andrews, author of *The Circle of Simplicity: Return to the Good Life*, plows a lot of the same old terrain in the same old earnest way, but her real contribution to the movement is the idea of "simplicity circles," which are a close variation on the Swedish "study circles" that are credited with helping to shape modern Swedish democracy. And the Swedes in turn borrowed the idea from the "learning circles" of the American Chautauqua movement in the late 19th century, a kind of early "distance learning" program that filled a big gap in higher and adult education in the Midwest until it got coopted by the entertainment industry. And the whole idea probably has its richest tradition in the Jewish Torah study groups that gave depth and meaning to the Jewish experience through many a diaspora (for men only, of course).

Essentially, "simplicity circles" are people gathering in living rooms, libraries or bars to discuss their progress in simplifying their lives, the problems they are encountering and new ideas for new efforts. Who knows, all kinds of weird ideas, like democracy and social justice, might emerge out of such homely structures; when the corporate masters of the universe catch on, they will probably outlaw them. But however minute such phenomena might seem in terms of developing meaningful social or political or economic change, they are sure a lot more on track than cocooning oneself into a tapestry of personal comfort.

But a big question lurks: Is this kind of thing really "simplifying" life? Is getting together with a roomful of other people for serious discussion really a "simplification" of anything? Is democracy simple? Duane Elgin, in his book *Voluntary Simplicity: Toward a Way of Life that is Outwardly Simple, Inwardly Rich*, attempts to address this kind of paradox. For Elgin, living "more voluntarily"—more deliberately, intentionally, purposefully, consciously—is as important as living "more simply"; we live more simply mostly to remove the distractions that keep us from living more voluntarily, with "a more direct, unpretentious, and unencumbered relationship with all aspects of our lives: the things that we consume, the work that we do, our relationships with others, our connections with nature and the cosmos, and more."

Nowhere in his book does Elgin make what I think is the mistake of proclaiming that a life of "voluntary simplicity" is a return to anything—Breathnach's bourgeoisie Victorianism, or some kind of earlier, even more innocent Currier and Ives ruralism. He begins in fact by saying he is not talking about living in poverty ("involuntary and debilitating"), not talking about turning away from economic progress (not "no growth" but "new growth"), not talking about embracing rural living ("rural living does not fit the modern reality"), and not talking about denying beauty ("rather than involving a denial of beauty, simplicity liberates the aesthetic sense").

Rather than Andrews' "return to the good life," Elgin calls for a movement—not just a solitary ambling by individuals out to save themselves—toward a way of life that is somewhere we presumably haven't been yet. So doing, he kind of links this end of the simplicity shelf to the shelf of global economic alternative thinking that includes names like E. F. Shumacher, Herman Daly, Mary Clark and Hazel Henderson. He cites frequently from this shelf—drawing, for example, on Dana Meadows' analysis in *Beyond the Limits*: "If the human family sets a goal for itself of achieving a moderate standard of living for everyone, computer projections suggest that the world could reach a sustainable level of economic activity that is roughly equivalent in material comforts to the average level in Europe in 1990."

But what's so "simple" about all this? Elgin is pretty honest about that: "When we combine these two ideas (living more simply and living more voluntarily) for integrating the inner and outer aspects of our lives," Elgin says, "we can describe voluntary simplicity as a manner of living that is outwardly more simple and inwardly more rich, a way of being in which our most authentic and live self is brought into direct and conscious contact with living. This way of life is not a static condition to be achieved, but an ever-changing balance that must be continuously and consciously made real. Simplicity in this sense is not simple."

Ah. Now maybe we're getting down to it.

I want to propose that there is a massive and unfortunate muddiness at the heart of all this "simplicity" discourse. We are, as usual, using words badly, trying to peg down good intuitions with sloppy articulations that we then commit to, as though some god had carved it out in stone for us; and the end result is a lot of misconceptions built on misconceptions that ultimately offend the original unarticulated intuition.

In the particular instance of this "simplicity" thing, I think the problem—in America at least—starts with Thoreau. That old cultural enigma, Henry. All of the simpletopians acknowledge him as the demigod who carved out "simplicity, simplicity, simplicity!" for us all, or found it carved in stone on Mt. Katahdin perhaps. Because we have made him a demigod, there's a general presumption that he must have known what he was talking about, and so we continue to perpetuate what may have just been Henry's massive misinterpretation of all his carefully collected data from nature.

Walden, or, Life in the Woods is one of my favorite books, but I don't like all of it. It's a Jekyll-and-Hyde book, exhibiting two Henrys: one is a classic old-school naturalist "observing Nature," looking for its "sermons in stone, books in the running brooks."

But the other Henry is a preacher, a New World Jeremiah, looking back at a cultural landscape that I don't think he really understood, fit, or much enjoyed. In the long run of the book, I think the naturalist is a stronger and steadier presence than the preacher—but the preacher seems to have captured the low-grade imagination of modern America better.

To see the two Henrys at their close-woven best and worst, consider what may be the most famous passage from *Walden*. When Duane Elgin says "to live voluntarily is to live more deliberately, intentionally, and purposefully—in short, it is to live more consciously," we're just getting a typically earnest and pretty pedestrian paraphrase of the beginning of Henry's famous "simplicity" rant: "I went to the woods because I wished to live deliberately, to front only the essential facts of life, and see if I could learn what it had to teach, and not, when I came to die, discover that I had not lived. I did not wish to live what was not life, living is so dear; nor did I wish to practice resignation, unless it was quite necessary. I wanted to live deep and suck out all the marrow of life, to live so sturdily and Spartanlike as to put to rout all that was not life, to cut a broad swath and shave close, to drive life into a corner, and reduce it to its lowest terms, and, if it proved to be mean, why then to get the whole and genuine meanness of it, and publish its meanness to the world; or if it were sublime, to know it by experience, and be able to give a true account of it in my next excursion."

Well, that's good stuff; no one has ever better expressed that yearning to live "as deliberately as Nature," learning what life had to teach, and when Henry was focusing on that, he was as good as it gets. But Henry's problem was that he could hardly go a page without having to look back over his shoulder at the village of Concord, at which point the naturalist immediately succumbs to the preacher. So he goes directly from the wonderful passage above into the rant that launched a thousand simplicity books:

"Still we live meanly, like ants; though the fable tells us that we were long ago changed into men; like pygmies we fight with cranes; it is error upon error, and clout upon clout, and our best virtue has for its occasion a superfluous and inevitable wretchedness. Our life is frittered away by detail. An honest man has hardly need to count more than his ten fingers, or in extreme cases he may add his ten toes, and lump the rest. Simplicity, simplicity, simplicity!" And he's off and running.

It's not always easy to tell exactly what it was about antebellum Concord that Henry found to be so un-simple, so meanly detailed, so complicated—or why he found it so much more complicated than the profound complexity of the great hardwood forest communities that throve on the south slopes of New England. Henry the preacher was never so descriptive as Henry the naturalist. But I have to say that his diatribes don't fit my own experience.

I started this unfinished argument with Henry quite a few years ago, when I found myself out in the woods for a spell—caretaking a summer biological field station up in the spruce and aspen about six miles north of Crested Butte. I'd like to be able to write as nobly as Henry wrote about my reasons for being there, an hour or four by ski beyond civilization, depending on the weather; but the truth is that I was there for the rent-free living, and the fact that, at a time when I needed a little drying out, it kept me out of the bars except for the weekly ski to town. I was there to see if I could make the big step from journalism to writing and have to admit that the results are still mixed.

I also wasn't there under Henry's condition of solitude; I lived in a 16-by-20 cabin with my wife and son, who was six months old when we went there and a five-year-old when we left—by which time we also had a daughter. So my "life in the woods" wasn't very uncluttered, and it wasn't solitude—except when you stepped outside the door, into the whispering silent realm of all possibility that Henry called Nature.

And was there simplicity there, in "Nature"? (An emphatic separation from "Culture" that bothers me.) Only if you didn't look very closely. Outside the cabin there was momentary relief from the clutter of culture inside the cabin; it was quieter outside my head than it was inside it. But to sit in it for longer than a minute just looking and listening—in our fine south-facing outhouse, or up on the edge of the springbox with the water dipper—was to begin to see and hear (even in the depths of winter) the rustle and rush of life, the grand fractal exfoliation of size and type and texture and color and scent that is nature. Nothing is simple there.

What I brought back from the woods, after four years of watching the ebb and flow of life there in the woods and floodplains of the East River valley—the mix of elbowing and collaborating, competing and cooperating that goes into the annual recycling of a plant community; the play of eagles on the mountain updraft and the badgering of skunks by badgers; the sad beautiful withdrawal of life into itself as the planet tilts toward the long night, and its soggy green explosion as the planet tilts back toward sun—what I brought back from all that was a rough measure of the magnificent complexity of real life, of "nature." And a sense of the extent to which civilization as we know it is just a great (one might say gross) set of oversimplifications against nature.

This over-simplification is undeniable: In the conversion of the wild prairie to vast monocultures of wheat and corn, the conversion of mature multi-storied mixed forests to even-aged pine or Doug fir plantations, the replacement of a natural mix of animals with herds of inbred cows, the conversion of the mountain valleys first to timothy hay and now to monocultural expensive suburbs and golf courses—it's as if our ancestors, and now we ourselves, had looked at that random and rampant multiplication of diversity we call nature, and said: "Simplify, simplify!" Who wants all that motley of life? Who needs it, who can really bear it? How can we assert any economic order if we have to carry along the useless with the useful, the fiscally worth less with the worth more?

Our global civilization is probably different from the swarming of every other successful species—lemmings, army ants, whatever—mostly in the success we've had in extending our swarming phase through our technological adaptation of all earth environments to our needs and desires. We're working on simplifying the whole planet into a set of vast monocultures designed to serve just one single globally ubiquitous species. Even most of our efforts to slow or halt this process—the efforts we call "environmentalism"—have a kind of freeze-frame simplification to them: Let's go back to the moment before us, and make it mandatory.

When we stand back and take a look at this massive project of global simplification humankind is embarked on, solid science confirms what intuition suggests: It's a risky business. The more we simplify the planet in order to supply the swarm's needs, the more vulnerable we become to the kinds of changes that we can't control—the planet's wobble around the sun, the eruptions of tectonic activity in the restless crust of the earth, the subtle vagaries of wind and water in the global climate. Not to mention the human-induced changes we probably could control if we wanted, like the buildup of human-generated greenhouse gases, or our slow poisoning of air, earth and water.

The Mr. Hyde who urged us to "simplify, simplify" was also the Dr. Jekyll who observed that "in wildness is the preservation of life," and therein lies the two-faced paradox of Thoreau's message. If there were a "vast eternal plan" to the evolution of life on earth, which I doubt, then it would be the all-contingency type of plan evident in "wildness": basically, a plan to put out such a proliferation of interwoven, variant and redundant systems of life that there will always be something that will keep the life force churning along no matter what challenges and opportunities universal chance imposes on the planet, short of a direct hit by a bigger lump of galactic matter that knocks us out to somewhere around cold dead Mars.

So in his lauding of wildness, Henry the naturalist saw the need, or at least the incipient rationale, for a diverse complexity of systems. But Henry the preacher didn't seem to pay much attention to Henry the naturalist. So the naturalist's insight gets lost in the preacher's message, because the preacher's message seems to be just to do what we're inclined to do anyway: simplify, simplify our blue-green jewel of a planet down to our own comprehensible little anthropocentric tapestry of contentment.

As usual, irony treads on the heels of earnestness. Look at a couple of our greatest simplifications. The automobile, for example, which we tell ourselves is transportation. And indeed, it does transport us from one place to another: usually, eventually. But the automobile is primarily an instrument of convenience and insulation—simplification. Convenience because it enables us to just get in and go, rather than having to know a bus or subway schedule and having to organize our day around such schedules; and insulation because it enables us to move around without having to interact with all those idiots and assholes out there (except maybe occasionally flipping someone the bird through our protective wind-and-people shield). In short, it simplifies our lives— or seems to, until we start factoring in the replacement cost of fuel, the cost of keeping one running, the cost of paying one off, the time loss in traffic jams, and so on—not to mention the externalized costs of all those by-products going up into the air and the eventual clutter of the landscape with its wornout hulk. When we look at all that stuff, an automobile starts to look a little less convenient. In exchange for a little simplifying of the immediate complexity of dealing with mass transportation (and the masses), we pay through the nose and pile up an incredible clutter of complex problems for our kids to solve. But we allow ourselves to be persuaded by four-color ads that those costs are more than balanced by the gain in convenience and insulation—and we are easily persuaded to buy ever more massive and convenience-laden automobiles that lead to ever more massive traffic jams that make a mockery of "transportation." But transportation is not what the automobile is really about; it's about simplifying.

Consider too the simplification represented by "screens" in our lives. As in movie screens, television screens, computer screens. We are a species that probably survived to the sticking point by telling each other stories. Colorado scientist R. Igor Gamow suggested that "being able to tell a story is perhaps humans' most distinguishing feature...Storytelling was a means of holding early groups together and thus, since this was an advantage, was selected for."

But sitting around in a circle with a bunch of other people, some of whom are hard to like in spite of or on account of their stories: this is...well, it's not simple. Often it's inconvenient. Sometimes embarrassing. So a whole lot of culture has come to be about screening ourselves and those like us from the uncomfortably rich mix of all the rest of us. We go to great and expensive lengths to simplify our contacts; we no longer want to huddle around the fire with the rest of the tribe, telling and singing and dancing our stories, because the tribe is too big; we've swarmed; the stories have gone exponential. So we put more and more "screens" between ourselves and the swarm, screens that only let through the sad sweet songs of humanity in carefully measured doses, professionally filtered doses, with audience-tested stories, in the private comfort of our living rooms—with laugh tracks to tell us what is and isn't funny since we no longer have the rest of the sweating, farting, milling mass all around to clue us. Simplify, simplify.

All of these efforts to simplify our lives are, of course, expensive in a number of ways, and they lead directly to a lot of clutter in our lives and throughout our world—mountains of clutter, vast landfills of it, barges circling the seas looking for new places to pile the clutter. But all of it: just the consequence of a lot of efforts to buy our way out of the complexity of life, into a misbegotten vision of simple, insulated grace.

Some of us do get bored enough, stifled enough, in the clutter of small, screened, insulated tapestries of simplification we've woven around ourselves, to actually want to seek out the complexity of the larger world, and a lot of us instinctively come to the mountains for that, for a couple of reasons: first, because the gross monocultural simplifications of civilization have not taken so well in the more rigorous mountain environments, so the natural complexity of life is still residual. And second, mountain environments have kept the human swarm from arriving in its full strength. A city of millions just isn't possible in a high mountain valley—at least not without levels of organization way beyond our current capabilities.

But once here, most of us do some variation on the Henry dance, imagining that complexity lies behind in the city and simpletopia lies here in these "quaint Victorian mining towns." But it's just the opposite: You've come from a massive but ultimately simple brontosaurus of a culture, a big but pretty straightforward eating machine that consumes the diversity of the earth and converts it to monocultural appetancy—and you've come to a place (if you were lucky enough to get there early enough) where there's little insulation, no cocoons, and the rich yuppie retiree with his 5,000-square-foot menopause manor ends up on the town council with the dreadlocked hippie Marxist freestore saint. And all of us who can afford it of course bring the baggage of the old urbanized simpletopia with us, and eventually recreate a lot of the old insulations and conveniences we thought we wanted to leave behind, because the real complexity of living together in small places, with large mountains watching, is—well, it's complex.

"This (voluntary) way of life is not a static condition to be achieved," said Duane Elgin, "but an ever-changing balance that must be continuously and consciously made real. Simplicity in this sense is not simple." Well, then, let's come up with a better word for it than "simplicity." How about "complexity"? As in "the true complexity of life"? Then we might begin to develop a willingness, however cautious, to grow into the reality of life—to complexify, complexify, complexify! Make life interesting again! Even if it occasionally hurts a little! Throw off those Victorian tapestries and middle-class cocoons!

A quarter century ago, I resolved my one-sided argument with Henry—more or less—by suggesting that what he really meant (this is easier to do when your proponent is a century or two dead) was that we should unclutter our lives of all those expensive simplifications piled up around us, in order to clear the decks of our consciousness to be able to truly embrace the real and important complexity of life that is going to impress itself on us (or our children) in the long run anyway.

In *The Golden Day*, a book about American culture in the 19th century, Lewis Mumford came to about the same conclusion about Henry Thoreau: "Thoreau was not a penurious fanatic, who sought to practice bare living merely as a moral exercise: he wanted to obey Emerson's dictum to save on the low levels and spend on the high ones. It is this that distinguishes him from the tedious people whose whole existence is absorbed in the practice of living on beans, or breathing deeply, or wearing clothes of a vegetable origin: simplification did not lead Thoreau to the cult of simplicity: it led to a higher civilization."

But one has to overlook a lot of Henry the preacher to come to that conclusion—

like Henry's assertion that "in proportion as (one) simplifies (one's) life, the laws of the universe will appear less complex…" Why do we always have to seek this kind of illusion? Why aren't we capable, as Keats begged, "of being in uncertainties, mysteries, doubts," without having to look for 100 easy ways to make it simple? Why can't we turn around in the cave, squint into the sun, and say, Goddamn! How magnificently intricate, interwoven and complex this all is! How can we make ourselves worthy of our limited comprehension of such magnificence?

To just acknowledge ourselves for what we really are would be a first step in answering that question, for of all the complexity that's woven into the tapestry of life on earth, we are probably the most complex thing so far—a form of life blessed, or maybe cursed, with the capability of being aware of it all, and of making choices about how to proceed further in the life project. But weighing the fruits of our consciousness so far, it's a little discouraging: on the one hand, a lot of good science, some good poetry, and a few exemplary lives like Gandhi's, all actively embracing the complexity of life; but on the other hand, hundreds of religions, political ideologies, socioeconomic theories, and cult movements all bent on pushing some particular brand of simpletopia in which 100 simple formulas rule and everyone is promised a tapestry of contentment (in heaven if not on earth) unilluminated by any troubling reminders of our true and truly difficult nature.

But see, I begin to rant on like Henry the preacher, and the first thing you know, I will be making notes for a book to launch a new cult: *100 Easy Ways to Begin Appreciating Life's True Complexity*. Forget it. But also forget simplicity. Go ahead and unclutter your life of some of those things that were supposed to make life simpler now, at whatever cost to the future, because they don't really work anyway—but then, buckle on your cross-country boards, ski up a valley, and try to really look at the fractal history unfolding all around you. Or just go to the local public hearing on the latest subdivision proposal.

Complexity, complexity, complexity! Rejoice in it when you can, suffer it the rest of the time. But what the hell: face up to it, and get beyond all this children's literature about escaping it. Life is our destiny if we'll embrace it; it's us; and it's not simple.

–MOUNTAIN GAZETTE, #79

N.E.D.

Here I am, back from capitalism. For a variety of reasons too incriminating to go into here, it became necessary for me to get some money a short while back. The money that is available in the mountains, concentrated as it is primarily around the selling of recreation or land, wasn't particularly available to me. I had neither to sell even if I could have stomached the idea. So I have been out of the mountains for a while. Spending time out of the mountains probably didn't do a thing for my strength of character—particularly since it was in the pursuit of money—but it did provide some flashy insights about this life that I assume we are all trying to lead.

Selling recreation is a weird business anyway, isn't it? I note that most of the advertisers in this publication are indirectly engaged in same, around some periphery or other. But the people who are directly in the business of selling recreation—ski lift operators primarily, but also the backpack tour guides, mountain-climbing leaders, jeep, mule and raft entrepreneurs, camp operators, etc.—are really trying to sell experience, aren't they? Here, give me your dollars, and I will give you a lift ticket, which is a day of weather, air, exercise, thrills and chills, views and vistas, sensual pleasure, and maybe just a tiny touch of sensual pain for spice. But just a touch. We are selling cold toes, chapped lips, and a nice glowing ache in the musculature afterward. (We're not trying to sell you a broken leg, not at all.)

Since nobody has been able to find a way to keep a day of weather, air, exercise, thrills, etc., from being free, then the experience-sellers have had to learn to embroider the experience. Reduce the effort one must invest (lifts). Reshape the experience to shift the proportion of thrill and pleasure, which means removing the delays and hindrances, or reducing the time one must invest. More skiing-rafting-climbing-camping for the buck. Wipe out the gradual, effortful transition from nonrecreation to recreation. Allow the customer to jet from the heart of commerce and in instant transition be plunged into the experience. That way lies helicopters, of course, machines that finally pose a more dangerous threat to undeveloped country than bulldozers. The highest triumph of the recreation marketeers will be the day they can pick up the business executive from the roof of his office building, let him change into all the tough clothes and powerful gear in the chopper en route, deposit him momentarily on the summit for view-absorbing, deep-breathing and picture-taking, and then whisk him down to the bar in the lodge below so he can talk about it. Except that the motivational researchers will tell the marketeers that somehow they lost the experience along the way, so they'll go the Disneyland route, with clockwork-and-plastic threats and dangers.

A friend of mine escorted a non-English-speaking Italian lady through Disneyland, including the electronic Abraham Lincoln. She broke into hysterical laughter, and my friend had to spirit her away to prevent a riot. The tourists saw Father Abe and heard holy writ; the Italian lady saw a rubber monster and heard gibberish. Emperor's new clothes.

What really pisses me off about the recreation business is that they are trying to sell me my life.

What is even more infuriating than that is that every once in a while they succeed. I buy a little slice of recreation that really turns out prime and am left reflecting on what a twisted and incompetent manager of my days I am, that some goddam lift-stringer can enrich my time.

I wish there were something new and relevant to say about land, and the sale thereof, but I haven't been able to think about it enough. Gurney Norman has been thinking about it, in a way. Norman was the Whole Earth guy whose novel, *Divine Right's Trip*, was published along the margins of the last volume of that seminal publishing milestone. He is writing another book, and notes for it, under the title "*The New Provincials*," appeared recently in a magazine called *Place*; I think he's trying to define who we are, but some of the notes have so much to do with land that they are worth quoting in this context.

> A community's history is a story. Its residents are the characters, what they do is the plot, the theme, the action. There is no way that participation in a community story can be contrived. You can't buy your way in. You can only live it.
>
> A story has a setting. A setting is a place.
>
> And this is a question: Does jumping around from place to place, American mobility, produce chaos in the mind because the individual's story-setting is never still? Localis interuptus; premature withdrawal; abandonment; loss; death?
>
> Could it be that sanity is simply knowing what story you are in?
>
> The tragedy of old people fading away in nursing homes is that they have been kicked out of the family story. They have lost their roles, they have lost their dialogue, they're isolated from the setting.
>
> Part of our national dilemma is that the American people are in danger of losing the setting of their national story. That was the

tragedy of the American Indians—we took their setting away—and now we are in danger of taking the setting away from ourselves, too. The physical landscape is being altered, acre by acre, river by river. In places like Appalachia and Black Mesa, whole settings are being bulldozed away. That's a common fact that all ecologically-minded people know, but is there a deeper way to know it?

If wise men say, "Be here now," then where are the people who are ripping the continent apart?

Absent.

They are absent.

It's an interesting little meditation to roll around in one's brain the next time you see the half-acre lots, $99 down. Or the haste with which someone is building a condominium in a nice place that will make it stop being a nice place, thereby dissolving the only reason for building a condominium there in the first place, but by which time it will be too late. Feeding the slick-fingered American lust to Get There First. Emergency blood plasma for the politics of scarcity.

I wish I could tell you that my expedition into capitalism involved one huge successful dope deal. That's the new American mythology: I understand that within another six months, every farm in the continental U.S. of under 500 acres will have been bought up by young freaks who have pulled off that one big dope deal, half a ton from Mexico by ITT submarine. Who took their accumulated pile (and half their imported product) and put it into an isolated farm and a piece of land, free and clear, and are sitting there now smoking, grinning and learning the banjo.

It's a nice version of the American Dream, a kind of 1970s version of Joe Kennedy running in all that scotch during Prohibition and founding an economic and political dynasty that we haven't seen the end of yet. Can you imagine what great American fortunes of the future will turn out to have been founded on three kilos stashed in the headliner of a Volkswagen bus? Dear dead Hashbury as the birthplace of a line of politicians and captains of industry?

Anyway, there was nothing so romantic and socially useful about my own voyage on the seas of commerce. I just quietly went off and went to work—bathed, shaved, wore clean clothes, the whole nine yards. Me, N.E.D., out there selling, imagine the mortification. Selling stuff that no one needed for five times what it was worth, taking my cut and paying my debt. There is no way I can possibly justify any of it in my soul, except that the things I was selling were so abysmally silly that no one could conceivably buy them if he had any other use for the money, and I chose to believe that would indicate that nobody was being actively hurt by my naked greed.

So for the past 18 months or so, I have been out there in the nonmountainous commercial world. I have tasted and smelled and observed—senses no doubt sharpened by my years of retreat—and I propose to tell you about America in what I presume we will always refer to as the immediate pre-Watergate years. The whole sick scene, motels and martinis, charbroiled styrofoam neon high-compression hair-sprayed commercial America, the keen-eyed journalistic operator skimming the essence of truth off the detergent foam spinning in the eddies of the waterways of finance. Bringing the Truth back to the mountains that we may all guide our lives beyond the errors of our countrymen.

On second thought, I think I'll go see R.J. and Amy Simple, and try to forget the whole thing. The money? Well, I paid my debt and had a little left over. Bought some land with it. Shit, there really isn't any way out of this, is there?

—*MOUNTAIN GAZETTE, #10*

Mountain Gazette 25
SEPTEMBER 1974 / 60 CENTS

GALEN ROWELL

Hanging Around

BY
DAVID ROBERTS

"This is all very different from the South Col!" you can remark crisply, as you watch bikini-clad girls swarming over the rocks like chameleons.

—Tom Patey,
The Art of Climbing Down Gracefully

One cringes at belaboring the obvious. But every once in a while the need for such an article as this arises. Traditions grow fuzzy, and the legends of one generation need footnoting in the next. In my recent rambles around the climbing world, a melancholy fact has gradually nagged its way into my consciousness.

The fact is this: There is emerging upon us right now a whole new crop—one would be tempted to call them a "school," but for their woeful ignorance—of young, dedicated climbers who have got the most central tenet of mountaineering ass-backwards. For their benefit, and at the risk of sounding pontifical, I here reaffirm that tenet. Namely, that the very object and essence of mountaineering, its be-all and end-all, is Hanging Around. All else is peripheral, a distraction.

The youngsters, it grieves me to report, can be seen everywhere these days obtusely infringing upon the long-established convention of Hanging Around. Some of them, for instance, behave in a climbing store as if it existed in order to sell them equipment to go out and abuse the cliffs with—not as a hanging-around scene for its own sake, sufficient unto itself. Others can be observed leaving a bar at 10 p.m. in order to get "an early start" in the morning. Some actually come to slide shows to get ideas for expeditions. I do not exaggerate. In general, there is a disturbing tendency among the young generation to take climbing itself altogether too seriously—which cannot but lead to a deplorable loss of finesse and subtlety when it comes to Hanging Around.

For their sake, then, and for the sake of posterity, which seems to have a knack for losing track of the most obvious facts of the past—it takes *My Secret Life*, for example, to remind us that English women wore no underpants before the 1870s—I shall set down here in cold print the hows and whys of Hanging Around, as it was practiced in the golden heyday of mountaineering, by the best climbers in every civilized nation, some few of whom I was privileged to loiter and vegetate with in their best days.

The Climbing Store. It is important to realize that the climbing store exists not to sell equipment but as a mutual sniffing ground in which mountaineers establish their credentials. The essence of proper behavior, as in so many climbing scenes, is not to try too hard. The climbing store is particularly effective in allowing the local talent to size up the visiting outsider, and vice versa. The game is harder to play for the outsider, since indifference and disdain come automatically to the proprietors, who are sick to death of selling kiddie-packs and freeze-dried lasagna. There are a few tricks, however.

For instance, in Southern California you can usually get away with talking tents. A successful dialogue might proceed as follows:

Customer (musing over the store's own design, prominently pitched in the front foyer): Very pretty. Nice colors.

Owner: Notice the extra-long sleeve door. And optional mosquito netting.

Customer (With a chuckle): Ah, yes, mosquito netting. What won't they think of? I was wondering if you have something with a ridge pole, you know, a little sturdier? (At this point it doesn't hurt to push skeptically against the tent wall.)

Owner: We do sell the Glacier Designs four-man. It's very popular with Sierra Club outings.

Customer: I was thinking, really, of a two-man. With snow flaps.

Owner: I see. For winter use...

Customer: Well, yes, of course, but really for next summer. Up North. I hate to spend so much money, but my old Alp Sport's in tatters.

And so on. In the East, talk Jumars vs. Gibbs ascenders. In Boulder, cagoules. In Jasper, bivouac hammocks.

In the climbing store, of course, it is essential not to hang around too long. Twelve to eighteen minutes is about right. Longer brands you an equipment freak. And, of course, one does not browse among the books (implying unfamiliarity with the few good ones), but only among the magazines (implying that you would never subscribe to one).

Another gambit is to dash in and ask abruptly for a very specific item. Half a dozen nuts of the same size, say. If questioned, it is legitimate to mumble something about "taking the aid out of Steppenwolf"—but never volunteer this information.

It goes without saying that proper attire is vital. The sort of fellow who shows up in a climbing store with a nylon runner for a belt has blown it irrevocably. Likewise with worn climbing shoes, frayed knickers, or too conspicuous cuts and scratches on the backs of the hands. Greasy down jackets, however, are OK.

The Local Climbing Area. Here the important thing is not to get conned into climbing anything hard. A once-clever gimmick, now ruined by over-use, is to show up in sandals. (Future guidebooks will list first sandalled ascents; the standards are already tough.) However, fresh variations on the nonchalant just-happened-to-drop-by approach still work. A friend of mine recently scored valuable points at Carderock (near Washington, D.C.) by becoming the object of awe-struck whisperings: "You know ____? He doesn't even use chalk."

In the old days, a favorite way of beefing up local prestige was to take the visiting celebrity out to the local area, ostensibly "Just for a little climbing," but really to see what the hot-shot could handle. It's sad to reflect that even champions like Terray and Robbins fell for this trick, and struggled valiantly on 5.10s some local wizard had spent the last half-year perfecting. By now, of course, the visitor knows that all the hardest local climbs are hopelessly beyond him, and that the way to play the game is with studied magnanimity. Thus, the locals are in danger of blowing it themselves—as happened here a year or two ago, when, following a visit by Chouinard and Frost to the nearby climbing area, the local store wore its hero-worship on its sleeve, immortalizing the visit in a poster-board display (photos of Chouinard's shoe poised exquisitely on a foothold) and repeating fervently to all who would listen Frost's assessment of a 15-foot aid climb, "Yessiree, we don't do much in the valley that's that hard."

The local climbing area is an ideal place to show up with an ulterior motive: say, hustling a touch football game, or (although this is a trickier tack to take) soliciting information about the big climb planned for the next weekend at The Gunks. "Anybody know about the top pitch of Carcinoma? How's the protection?" The danger, of course, is that someone there will have done Carcinoma, or that someone else will remember two weeks later that you were going to do it last weekend.

If forced to climb, it is best to wander absent-mindedly along traverses five feet off the ground. If you fall off, you can usually imply that you were trying it one-handed. Never bring a rope. The reason for not soloing climbs, naturally, is for fear of encouraging beginners who may not know any better. Groups of beginners have other uses, too: You can often claim that you showed up (and can persuade others to join you) just to watch the University Outing Club practicing prussiking.

The AAC Board of Directors Meeting. A very strange scene, one which, for better or worse, most mountaineers will never be privy to. The weekend of a directors meeting begins with a Friday evening cocktail party. The principle of the disguise, developed originally for use in climbing stores, seems to have reached macabre perfection here. Most of the directors show up in business suits, wearing shiny black shoes, carrying attaché cases, talking committee reports. At the party, after hearty back-slapping, they drink—not beer or cheap wine, as in campgrounds the world over—but real cocktails, scotch-on-the-rocks in hefty belts.

The next day, during the business meeting, the directors continue the masterly charade. All of them display an astonishing familiarity with parliamentary procedure, and they sling around with Congressional pomp phrases like, "If the chair can prevail upon the head of the Rocky Mountain Section to instruct his membership that..." They seem to know about budgets.

There used to be a dangerous pressure at directors meetings to knock off Sunday and go climbing together. Fortunately, the business of the club has grown to such proportions that it can be counted on to spill generously over into Sunday afternoon. Thus the Board need not be haunted by the specter of an aging director securing the token Yosemite youngster with a classic European over-the-shoulder standing belay, or of a boozy ex-hard man developing sewing machine leg on a 5.4 move the day after a hard night arguing membership qualification standards for the AAC.

The Slide Show. Slide shows used to be fun, easy-going. No longer. There was a time when you could get away with sunsets. Tilted climbing shots. Under-exposed bad-weather shots. There used to be a gentle give-and-take; the audience was on your side. In fact, I can remember the hoary days when you could get a chuckle out of old chestnuts like, "The belay I gave Joe was a strictly psychological one," or, "Ten minutes after we climbed this slope it was swept by an avalanche."

Nowadays a slide show audience is hostile in proportion to its sophistication. This makes it very hard to give slide shows. Good photos are suspect; if you have time and the weather to take a beautiful picture, obviously it was posed, or at least you weren't spending all the effort you should have getting on with the desperate business at hand. Bad photos (poorly centered, out of focus, dusty, over-exposed) have their integrity. In fact, the perfect slide show, like Apollinaire's perfect poem, would have no pictures in it.

The right tone is extremely difficult to strike. One must imply that he is somewhat bewildered by all the fuss being made about some climb he and his buddies happened to blunder up last summer, that he is quite astounded to find anyone interested in the fact that he woke up one morning last January to discover himself in Patagonia at the base of Cerro Torre. The "what am I doing here?" tone is best, which modulates easily into, "Wouldn't it be more fun if we all went out to the local bar?" tone, which lends itself admirably to a transition to:

The Climbers' Bar. The bar, it is understood, is the place one goes to commiserate with other climbers when, darn it all, the awful weather has spoiled the climbing. Like the climbing store, it also serves as a mutual sniffing ground. Here the game is to figure out what route the other guys have up their sleeves, or, failing that, to imply that you have their much-coveted route up your own sleeve. I recall a conversation in a brightly lit bar in Banff (all Canadian bars are brightly lit) a few years ago with the resident master, who was quite suspicious of our plans. After the ritual pleasantries, he said:

"Well, what brings you chaps up here?"

"Oh, nothing much," we answered cheerfully. "Some of the classic routes. Robson, maybe."

"Lovely mountain. We had a great weekend there in May."

Thus we had established that we were both after the same new route. Now the sniffing got sharp. After half an hour's banter, we managed to compliment the old hand's route of a few years before on our intended mountain's north face. "Must have been a fine climb," we said.

"Nothing special."

"That route up the middle of the face would be something else. Scary, I'll bet."

His eyes glinted. "Suicide, I suspect. When we were on our route, we saw this huge bloody rockfall sweep the whole thing."

He had scored there. We must have betrayed a little panic. "That's what I figured," one of our group managed lamely. "Anyway, after all this weather, a thing like that wouldn't be in shape till August. If then."

"Maybe not too hard a climb," the master said as he got up to play shuffleboard. "Just suicidal."

That we were out the next day, nursing our hangovers on the hike in, was irrelevant. The essential battle had been fought the night before. The master had won, by insinuating that we were crazy if we tried the route and chicken if we didn't.

Bars, too, serve as scenes for epic gross-outs and obscene displays. These are a specialty of the British. A rule of thumb to follow is, if you can't be spectacularly offensive, then get quietly drunk in your corner. Anything in between is bad form and smacks of seeking attention.

Corollaries: Make sure you show up in the bar on marginal days—defined as days with any clouds in the sky. Only perfect weather, you want to imply, will lure you into attempting a scheme of the boldness of the one you and your cronies have been hatching. And you need a few days of sun to get "the summit snowfields" in shape. There is a whole style of slouching, of brooding pensively, that reeks of hoarding strength and of building up psychologically. Cultivate it.

Bush-pilot Hangars. A scene requiring very adroit one-upsmanship. The politics of hangar-waiting in Alaska, for example, during the mandatory three-day storm after arrival, are fairly subtle. One tries to imply an old friendship with the pilot, leading to the automatic assumption that he will take your party in first as soon as the weather clears, no matter when you showed up. Meanwhile, an air of calm confidence attends your perusal of the other groups' gear. The Japanese in Alaska always used to provide a few laughs. "Very interesting pickets," you would say, grinning, as you dubiously flexed their hopelessly flimsy stakes. "Pickets, yes," the Japanese would grin back. "Much snow McKinley."

In Don Sheldon's hangar there were archaeological layers of leftover rations from past expeditions. There, behind the cartons of Pepsi, were Terray's lemon drops; and, yes, under them, could those be some of Cassin's meat bars? One liked to strike a pose of being willing to donate, say, an extra loaf of logan bread to that food museum.

Relations with the pilot himself are carefully ritualized. Self-evidently gauche is any palaver about how-soon-do-you-think-you-can-take-us-in—in a category of crudeness, really, only with haggling over rates. On those rare occasions when the man himself comes in view, one hallos out a hearty greeting, as if you are quite surprised to find him there, in the middle of your summer's food-boxing and equipment-puttering. Most coveted of all: the special invitation to have dinner in the pilot's own house, there beside the air strip. The green looks of envy on the others' faces, as they labor over the Sveas, are worth a whole expedition.

Finally:

The International Scene. Having never climbed in the Alps, I consulted a friend just back from three years in Europe about the international scene. He had done some of his best hanging around in Chamonix. I was curious to know whether European climbers understood the traditions of the activity as well as Americans do. He was reassuring.

"Basically, in Chamonix," he told me, "the scene centers around two bars: the 'Nash,' or Bar National, and Le Drug Store. Only the English hang out at the Nash, which is a pretty raunchy place, small and seedy, with a foosball game and a one-eyed waiter named Maurice. At Le Drug Store, you find all nationalities. That's where the English go when they're looking for a fight."

"A fight?" Never in my climbing days had I seen a fight. American climbers are pacifists.

"Yeah. They hate the French. They get drunk and stand up on the tables and sing songs like, 'If you've never fucked a Liverpool man, you've never fucked at all.' One day I was walking to this climb with some English guys, and they saw two other climbers in the distance. Naturally, they assume they're French. So they shout out, in inimitable Cockney, 'S'enculer!'"

"Which means?"

"Roughly translated, 'Up yours!' There's a pause. Back comes a voice in inimitable French, 'Sheeet!'"

"Why do they hate each other?"

"God knows. They're English. It's English to hate the French. When I was there, I helped them heist a table from Le Drug Store to take out to their campground."

"A whole table?"

"Yeah. They camp on this private land owned by Snell's, the local climbing store. Snell's has a tacit agreement with the English: They can camp on the company's land, and in return they won't shoplift from the store."

"Weird. What about the others? Say, the Germans?"

"Oh, the Germans. They all camp in another campground, tents in perfect rows, neat and tidy. And the climbers are all very hardy looking, neatly dressed."

"But they hang out at Le Drug Store."

"Yeah. All except the Japanese, who are far too serious for the bars. The Japanese only know two words in English: 'North Wall.'"

"Not like the old Japanese in Alaska."

"Nope. A different generation."

"How about Americans? Besides yourself."

"Oh, there's a typical kid from Washington who's done all the 5.9s in the Seattle area. He's come over to do the Bonatti Pillar. The day after he gets there he hears two people are killed in the approach couloir. He manages to sprain his ankle falling off the boulder in the campground parking lot."

"I see why you associated with the English."

"Yeah. Hey, did I ever tell you the story of how MacInnes inherited Terray's down jacket after he fell off the statue and broke his leg?"

At least the English, I concluded, have the proper respect for the grand traditions of Hanging Around. While we were talking, a 15-year-old kid had just made the crux move on Amanita. Two 17-year-olds beside us were planning their expedition to Nuptse. An eight-year-old girl in tennis shoes was perusing a well-thumbed guidebook. We shook our heads, finished our peanut-butter sandwiches and headed down to Ramon's for a morning beer.

—MOUNTAIN GAZETTE, #19

Crooked Road to the Far North

For Ajila, whose name means life.

**BY
LITO TEJADA-FLORES**

At a beer garden in Berkeley, putting down a pint with my old friend, Chris Jones, the subject of summer climbing trips came up, and that's how I got back on the road...Chris had planned a mini-expedition to the Devil's Thumb, a redoubtable granite spire in Southeastern Alaska, with two Salt Lake City climbers, George and Jeff Lowe. Something had come up, Jeff couldn't go. Did I want to come? Sure.

The juke box was blaring out Elton John, the air warm with summer and tasty with pizza smell, students and freaks and half-naked chicks swirled around us; through the open patio, the popcorn machine was spitting and the avant-garde cinema beside us disgorged its Fellini-eyed crowd into the night...Not exactly an atmosphere for reflection but perhaps "up there" would be more real than "down here." Sure. We left La Val's drunk and enthusiastic. Chris and George would drive most of the way, via Salt Lake and Canada, as far as Prince Rupert. I would go by bus to Seattle, then ferryboat to Petersburg, Alaska, and meet them there. We'd fly into our mountain, air-drop our gear, and climb it, even if it took us a month.

We were laughing and joking, and it wasn't really too clear just what I was getting into. (On a last minute practice climb before leaving, I discovered how out of shape I was, and a real anxiety about the climb began to build up inside me.) But one thing was sure: Before I knew it, I'd be traveling again, stepping out, in the grip of strange currents again, and it felt good.

> *So it's starting again:*
> *Once more, the mad*
> *last-minute dash,*
> *hastily packed bags,*
> *smokey white tiles of*
> *Greyhound waiting rooms.*
> *Once more, the motives*
> *are uncertain & the means*

*confused with the ends. Only
the need to go (but where?)
to do (but what?) to feel
again (but why?) is real.
So it's starting again:
Once more, my mind overflows
with debris from the future,
with scraps from the past:
sleepless nights, narrow seats,
wide country, endless roads...
Somewhere in the far north
a mountain will do for a goal.
And perhaps this time
the crooked road
will lead me straight,
instead of sending me off
again in search of myself,
starting again & again...*

Early in the morning in Seattle, bleary-eyed and loaded-down with ragged old duffel bags, I find my way down to Pier 48. False front of wood and plastic, giant Indian totem designs, yellow and blue Alaska Marine Highway signs. Right away, I've got a problem: The list of walk-on passengers is full, closed. No, we don't make reservations, but all these folks came in yesterday; first time ever; we'll put your name on the stand-by list; yes, we'll know around three or four this afternoon. A helluva note. Only one boat a week from Seattle, and my friends up there waiting. Screwed. Well, what can I do but trust my Karma, again and always? Leave my pack and bags at the feet of a giant stuffed Kodiak bear in the waiting room and go out to see Seattle.

After an endless walk in cowboy boots, I decide to eat lunch at The Prague, a gallery/restaurant in a waterfront district of run-down brick buildings, slowly being remodeled into a funky, posh shopping area. I know it's overpriced but shrug my shoulders: There won't be anything like this up north...

*Prague or Seattle, it's all the same,
only the mind travels...only the feeling
that something is about to happen (or
just has or is, right now, around
the corner) counts. Why search for
Gothic images here? Steinbruke or Golem?*

> *A skid-row panhandler meets me at the door*
> *& dust-mote light pours down green vines,*
> *shines the wooden floor, dusts old brick walls,*
> *starkly hung with post-Klee prints:*
> *this is post-Kafka Prague, post-Dubcek too.*
> *Only the mind travels...Beside my soup*
> *spotlights hit exploded heads, magic wheels,*
> *spikey suns & transcendental paddle boats*
> *gliding through intaglio seas of scratchy green:*
> *horses & priests, life & death & sex,*
> *safely under glass frames from West Germany.*
> *Waitresses without breasts shuffle by*
> *On three-inch cork heels...Only the mind travels!*

Lunch is good: watermelon and fruit, cold meat and cheese, a mysterious central European soup to go with the tangled images on the wall. It's two o'clock. Time to say goodbye to Kafka and Klee, Telemann and Bach on the stereo, time to say goodbye to Prague and Seattle and, somehow, get on that boat to Alaska.

Back in the ferry terminal, the situation has deteriorated: There are now some 90 frustrated, confused people on the stand-by list (and a few really angry ones). Their story is the same: We telephoned from Tucson, drove all the way from L.A.; no plane till Wednesday; they promised; I told my husband, planned this trip for three months; they'll tell us at 4:00, no, at 5:00...

In the middle of this displaced-persons atmosphere is a lovely slender girl, tallish, in faded jeans and a big, loose Levi jacket. How do we begin talking? An unimportant, impersonal remark addressed at random to the milling crowd. A minute later she is saying, with a smile: Want to hitch-hike north with me if we don't get on? Inside me a small voice is already shouting yes, yes! I am surprised at myself. She has a little child and an enormous duffel bag. We compare luggage, miles, laugh at the impossibility of it. Don't worry, we'll get on!...A long conversation begins. Something else has begun.

Her daughter's name is Ajila: a lovely smiling face, a snub turned-up nose, short blond pigtails. She sits between us on the high ticket counter and draws with a ball-point pen on application blanks for MasterCharge cards...We exchange names, fascinating bits of information that unfold and unfold: Kathleen, unlike her daughter, has dark wavy hair pulled way back, a pale oval face with only two spots of color on her cheeks, prominent without being high-boned. A tiny gold dot in a pierced nostril makes me think of gypsies, central Europe, faraway places. She is beautiful without being beautiful. She doesn't sparkle, she glows. But she's real, she's tired, has real problems, a real mixed-up past, her own crooked road leading her north. She's going to Ketchikan to meet her old man, they'll travel, work, she's not sure, maybe she'll wind up working as a nurse as she did in Crescent City...Her dreams are close to the surface: She talks about going to South America someday, adopting a lot of kids...Ajila? It's an Arabic name, her father studied Arabic, no someone else, she married him to keep him out of the Army...

Around us people fester and complain. Behind the counter, the harassed clerks with their gold and blue ALASKA HOST pins pretend not to notice the people on the other side. But eventually the purser's list arrives with 96 free places: room for everyone, I think. My own name is third on the list. I buy my ticket and stagger to the gangplank under my enormous bags.

The M.V. *Malaspina* is so big that its levels, decks and passageways seem, at first, a labyrinth. I lug my duffel bags in relays to the solarium on the top rear deck of the big blue and white boat. This is home. What next? Look for Kathleen, of course. I meet her at the top of the stairway from the car deck; her bags are down there, so we go down again and I carry them up on deck. We're still amazed at having got on board at all. And we sit down, out of breath, and stare at Seattle, rising up the hill behind the waterfront in grey tiers of freeways and office buildings, a grey city under a cloudy sky. It looks like rain.

We relax, the three of us, on a wooden box-seat full of life preservers, peeling oranges while Kathleen makes cheese sandwiches from the food left in her old carpet bag. A bushy-whiskered, prophet-like figure of a man walks by (prospector? recluse? hermit?) dressed in Army fatigue pants and an old brown sweatshirt. Are you hungry? Would you like some? Kathleen knows, offers, dispenses, smiles. Strange easy-going vibes are all around her, all around us.

Paul is, indeed, an old recluse with a philosophical bent, going up to Wrangell to "work in the woods." He takes me down to the deck below to show me a part of the ship he has "captured," hanging a large blue and orange tarp across a corner of the covered walkway. Beneath it are his incredibly worn-looking possessions and his pride and joy, a big black iron pot. He pulls two beers from a paper bag. One for your wife. No, she's only my friend. A strange rush of emotion that will be explained later, or never.

When we get back to the solarium, the ship is just casting off; silently, imperceptibly at first, the long wharves slide away from us, gathering momentum as the whole panorama of Seattle distorts, expands, recedes...It takes a long time to lose the city astern, but already we're in a new space. The North is already more real, our day-to-day lives already half-forgotten. Under threatening skies we enter another world:

> *Bluegreen water*
>   *beside*
> *Greygreen forests*
>   *beneath*
> *Yellowgrey clouds*

> *thru*
> *The sundeck roof*
> *where*
> *Raindrops dance*
> *on the grey*
> *Plexiglass windows*
> *& the grey*
> *Velvet fabric*
> *of dreams.*

    The solarium is full of backpackers, young people, freaks. Kelty packs and down sleeping bags are everywhere. Paul spreads a 100-year-old, hand-embroidered quilt on a kind of raised dais, like a legendary bearded pasha out of the Arabian Nights. Down inside the boat, there is a second scramble for the remaining staterooms. The other walk-on passengers, those of a certain age, or a certain lifestyle, will be spending the night stiffly upright in airplane-type lounge chairs. They don't look very comfortable, or very happy...

    The evening is forever. Already the northern latitude gives us more daylight than we're used to. It's a late long twilight, the lamps are on, Kathleen and Ajila are tucked under a forest-green sleeping bag. I lie on my stomach on a deck chair beside her, and we talk, ask, answer, tell: What kind of women do you get involved with? A funny question, women like you. (My words surprise me, the feeling doesn't.)

    Our talk takes us back to our other lives, takes us forward to the edge of the Far North, the edge of our own dreams. We surprise each other by talking of death, finding that we've both met it, thought about it, made a temporary truce with it. Kathleen has put a degree of order into a confused life. She talks of her "life plan," a good one if slightly impossible, as anything must be that makes sense. She asks me questions that stop my standard answers cold. Her face is full of possible answers. Her beauty is as hard to understand as my reasons for taking this trip, or wanting to climb that mountain. I fall asleep beside her with an open heart.

    Tomorrow we'll wake up to the same gentle motion, the same heavy clouds, anonymous forest channels moving mysteriously past.

> *The boat glides on and on*
> *shedding an outworn skin of miles*
> *behind it, while I too*
> *wriggle painfully forward*
> *out of my own past,*
> *leaving the transparent scales*

> *of a hot California summer*
> *shimmering in the wake...*
> *And so we drift North*
> *through the fog,*
> *toward a second summer*
> *and a new skin.*

The next day was long and lazy, monotonously beautiful, and at the same time full of a quiet excitement that had nothing to do with the scenery—the low forested hills sliding by on either side, the stark rocky inlets, isolated homesteads, tiny lighthouses, lonely channel markers on a lost spit of granite, sudden waterfalls cascading out of the clouds into the inky blue of the strait.

We walked and read, and even ventured into the high-priced world of the ship's cafeteria for coffee and hot chocolate for Ajila. We met our fellow vagabonds on the rear deck, and talked about their trips, and their scenes. The real landscape of people and faces began to take shape. There were climbing boots to grease and free hot showers to enjoy, and the lazy quiet flow of water on every side to pace us through the day. In the evening I wrote a small poem for Kathleen and gave it to her:

> *Frontiers are places so beautiful,*
> *and so empty, that men*
> *have to fill them with dreams.*
> *Frontier women, too, have*
> *calm deep faces that*
> *make men dream...*
> *It's good to know that*
> *both still exist,*
> *and that you're*
> *one.*

Her smiles went through me like knives. Wherever she went on board, the air would ripple around her. I enjoyed watching her random movement on our deckside world: finding her and losing her, smiling, exchanging private glances, watching her disappear around a bulkhead, spotting her through a window, noticing how other people were attracted into her orbit, coming up to her to offer their smiles, their gifts, listening to Ajila play with the other kids on board, only smiling when I noticed that her mother wasn't really as beautiful as she seemed to be...Her beauty was beneath and behind beauty. I was in love with a gentle dark-haired puzzle in faded blue jeans.

Everyone in the solarium that evening looked hungry. In any case, no one could afford to eat in the cafeteria, much less the dining room. The prices were unbelievable. Someone, I think one of the kids on a bicycle trip from Seattle, suggested pooling our food for a community dinner. (Some hadn't thought to go shopping before our departure, and others, caught in the stand-by list, hadn't had time.) It was a huge success, a feast, not a dinner. Food appeared from everywhere: bread, cheese, sardines, celery, fruit, peanut butter, cold meats, cookies…I bought some Rainier beer and Jay, the neuro-psychologist whose daughter played with Ajila, contributed some wine. We were already drunk without it. Serious bearded faces, young hairless ones, homely girls beginning to look pretty because they were having such a good time. Paul was there, beaming like a prophet; and the art-school teacher from the East Coast who had sketched Kathleen resting against her duffel bag; two teenaged boys from Maine who looked as if they hadn't eaten for days; pint-sized touselheaded Ray, a diminutive chain-smoking 16-year-old whose dad was a steward on the ferry, on his way back to Alaska after a year in an "institution" for some adolescent craziness, smiling and stuffing his face like the rest of us. Ajila was kneeling at her mother's side. Incongruously, older "straight" people were drawn into the warm circle of our picnic on the floating bank of this endless winding ocean highway.

After dinner we borrowed the ship's vacuum for our crumbs, then played charades till midnight, laughing, jumping up, crying out, losing track again and again, still drunk with each other, coming down slowly, slowly, like the long pale northern evening, reluctant to give up the last light in the sky, or in each other's faces…

Fatigue finally triumphed. The kids were already asleep. Ajila was a blond Moslem angel under her green nylon sleeping bag. Lounge chairs were pulled out flat to sleep on. Kathleen and I headed below decks for a midnight drink in the ship's bar. A perfect day, perfect evening. I wanted to stretch it out, talk to her until the words dried up, until there was nothing left to say—knowing that in a life you hardly begin, that one more evening wasn't even time to begin…

In the bar we talked, drank Scotch because neither of us could think of anything else to order, listened to a guy in the booth across the way thump out a bluegrass polka on his banjo, laughed when two of the cyclists from Washington, brother and sister, started to dance crazily up and down the narrow aisle. Out of breath from dancing, Mike came over to sit with us: his hair sticking out in all directions, his chambray shirt pulled out, his thick smiling lips covered with fever blisters (beautiful people, we learn, don't have to be too beautiful). And out of the blue, he delivered a crazy, moving, totally disarming speech about Kathleen and me, about having watched us on the ship, about the way we stayed together without grabbing onto each other, about watching the way Kathleen treated her little girl with such respect, letting her choose what to do next, what clothes to wear in the morning…And going on to talk about himself, his efforts to find himself, not to be possessive with girlfriends, with people…And he said a lot more, but what moved us was how he said it. Letting the barrier between himself, his ego and his words become eggshell-thin, exposing himself in a strange trusting way to talk to us like that, so that we learned more from where the words were coming than from the words themselves. At any rate, we blushed when he talked about the two of us, but he was right: Our bond in the present was so real we hadn't even begun to hold onto each other for an imaginary tomorrow.

There you are. Happiness, desire, perfection. Where, if not inside you? Who, if not us now? When, if not here now? We fell asleep, warm on the cold deck, arms outside our sleeping bags, hands clasped.

Reborn under dazzling blue skies. They stole the clouds during the night. Nothing to do. A million things to do. Time rushes forward out of control. Before we can adjust, Ketchikan is swimming into view like a postcard of some far Norwegian village. The dark blue water is full of pale white jellyfish. At the railing, Ajila has a tearful moment, imagining that we'll have to swim ashore. No, it's not like that at all. Minutes later I'm carrying their duffel bag downstairs and across the ramp to Ketchikan. Farewell is a little picnic on the rocky bank, a few words, Ajila crying out: Oh, mommy, you kissed him! And an incredible knot of emotions in my stomach. Kathleen's old man, the fellow she's been living with for two years, should be arriving on the afternoon ferry from the north, and the three of them will have to begin the business of making a new life. All my concern, my good wishes, for her, for them, seem superfluous; of course, it will work.

Back on board, I remember my Solzhenitsyn novel, *The Cancer Ward*, that she was reading, find it in my pack, and manage to run ashore at the very last minute to give it to her. An extra farewell, stolen kisses. Kathleen and Ajila running out to the dock's end as the *Malaspina* pulls slowly away. My eyes are full of tears. I've just lived through something unbelievable. Paul is standing beside me at the rail. Whatever he is saying seems to make sense, with such a long grey beard he must have lived through all this, too…The knot in my stomach starts to untie itself, we're still moving north.

Back on top deck I find my friends the bicycle-campers and a few others sitting around in shorts and cut-offs, having (in the simplest possible way, and so unselfconsciously) a kind of Quaker-like Sunday-morning communion, sharing a giant hunk of rye-crisp, taking a few moments to think about it and share their thoughts.

Breaking bread
with brothers and sisters
sitting in sunshine
sharing white wine
sharing our weakness
warm in the sunshine

> *feeling our strength*
> *not yours nor mine*
> *quiet communion*
> *on a grey steel deck*
> *taking our turns*
> *talking of love*
> *listening by turns*
> *to brother and sister*
> *sitting in sunshine*
> *breaking bread.*

Everything that's happened so far has made us all high, and that's the way it stays, all day. Paul leaves us at Wrangell, the next port, with his incredible collection of surplus equipment and his beautiful antique quilt. Hot-rodding teenagers are driving motor boats through the pilings of the pier, and one runs headlong into a cement footing, flips 20 feet into the water and emerges unhurt. A carnival atmosphere with everything but flags fluttering in the breeze. But the magnet keeps pulling us, we keep on moving North.

> *Perhaps this is the Far North:*
> *The water opens and islands*
> *pull back their forest tongues*
> *for us to pass...Overhead*
> *black and white clouds are fighting*
> *their ancient Taoist battle*
> *(summer's victory a fragile truce).*
> *In the distance now, rain streaks*
> *are staining the pale sky with rust.*
> *White mountains rear up, like*
> *welcoming ghosts or new friends*
> *on the far edge of our dreams.*
> *Wrangell Narrows swallows our boat.*
> *Rainbows welcome us to the Far North.*

Time to repack my bags. We're almost to Petersburg. The northern sunset has just begun in a high-contrast battle of burnished gold and inky blue. Someone says: Look, your mountain! Yes, there it is all right, even though it isn't mine. The incredible Devil's Thumb, even this far away, it's overwhelming! I wonder how we'll ever find the courage to climb it, but I know we will. It always feels like this. Kevin, one of the cyclists, gives me a small card, a poem that a friend had written for him a long time ago. Here, take this, for the summit...

> *"It's the time you have*
> *wasted for your rose*
> *that makes it so important"*

    Thank you. Thanks for everything. It seems impossible that I'll ever forget all this: today, this evening, these people, Kathleen and Ajila, reading poetry on the rear deck, sharing our food, such perfect uncomplicated love. I didn't come North to find this, I have almost forgotten why I came, what was behind it all, luring me up here.

> *Behind snowy coastal ranges,*
> *behind the cobalt blue*
> *of Fredrick Sound evening,*
> *the crackerbox waterfront*
> *of smalltown Petersburg,*
> *behind all this there's only*
> *a granite dream at sunset,*
> *too icy perfect to believe in*
> *but just real enough*
> *to pull me off this boat...*
> *And all our island friendship*
> *and ferryboat love*
> *becomes one last shout*
> *from ship to shore*
> *and back...*
> *I gather up the echoes*
> *in my rucksack*
> *and promise myself*
> *to spread them around,*
> *promise to pass them on...*

II

    In literature things end. Stories end. Poems end. But in real life (as they say) things and people and events go on and on, world without end, forever and ever, amen. In real life you never reach the end of the crooked road. In real life there is always a part two.

    For, of course, getting there is only a part of the story. There one is, there you are, here I am, but yet not completely here. I'm still traveling, or else a part of me has already moved on, and ultimately I'll look back to realize that I left without ever completely arriving…And the Far North is not different. The Far North: dream image, typewriter cliché, dimestore poetry, five-and-dime metaphysics.

But I did get here. What now? What next? Now that I'm back in charge of my own life, at least temporarily, at least partially…The present is Petersburg. The immediate future is the mountain, lonely and frightening on the horizon, really a thumb-like Thumb—poking up into an 11 p.m. sunset or hidden in rain and clouds, but still waiting. And the far and future future? Out there somewhere, behind the Devil's Thumb and those long snowy ridges, invisible guesswork.

Things change, time is no longer a slow crystal river.
But there are still new friends, events, poems:
This is the calm before the storm:
an extra day, waiting for friends,
down by the public float, where
spikey boats thrust long trolling poles
up into the low clouds. The seiners
are coming home tonight, the tenders
will be here in an hour, and the girls
who work in the egg room
are having one last cup of coffee
before reporting to their Japanese foremen.

Alaskan summer: reflections in green water
a simple life (here too, brothers and sisters)
the circus scene of snow and trees and boats.
Hippies migrating north for cannery jobs
have come to civilize this hard wet land
with their gentle talk and long wavy hair.

Even the clouds look friendly now,
and the next cloudburst, when it comes,
will only lay more dust in my heart.
A float-plane takes off with a roar,
buzz saws whine, below my feet
an old hull is being scraped on the grid,
overhead: eagles, gulls, giant ravens
circle together, scream, fight and wait.
The sunshine is supercharged with mist

> and the clouds are shot full of holes,
> torn by sundrop, scattered by rainbeam.
>
> I'm soaked to the skin and don't care.
> Surely, this is the storm before the calm.
> Afterwards will be time enough
> to grab a paintbrush, tear fish from nets,
> finish this life or start a new one.
> This must really be the Far North,
> and from here the only way is down,
> all roads lead back to the world,
> lead south and home...Yes, surely,
> this is the storm before the calm.

It passes quickly. It takes forever. The people of Petersburg are my clock. It's always raining, it's always cloudy, the mountain has disappeared for good. I am happy here and at peace.

The telegram reads: CAR PROBLEM ARRIVE FRIDAY EVENING PLEASE GET BATTERY CHRIS CALGARY ALBERTA. So I have an extra week to wait. I can wait. Secretly I hope they never get here, I know they will. And sure enough, when they do, a whole lifetime has passed, the rain stops, the clouds lift enough for us to fly in, drop our gear, land at the lake. And we're right on schedule again, the real schedule.

The weather keeps improving, luring us on in two days over a high pass toward the Devil's Thumb. Chris Jones, my old mate from Fitz Roy, lean and cunning, witty and optimistic. George Lowe, old friend but new companion, smiling tousle-headed physicist, full of power and quiet strength. Me and my doubts, still glad to be here, glad that it's started, that it's happening to me, this incredible scene: the ice and granite battlefield of the Witches' Cauldron, the spire-like satellite peaks, the overwhelming bulk of the Thumb itself, this mad adventure unfolding.

The weather gets better and better. Two days of rough packing to Base Camp. Collecting our fluorescent orange air-drop boxes, strung out across the glacier. Dazzling white snow, dazzling blue sky, hard-edge sunshine. There will be no rest day tomorrow. Such weather is too good and too rare to waste. We have to start tomorrow and scramble to get ready. The Thumb towers overhead.

> Magic mountain—
> in all probability
> we are enchanted
> & not the mountain.

> *Let's hope this magic*
> *pulling us up there*
> *will bring us down*
> *again.*

    The rock begins tricky, stays tricky: slantwise traversing pitches, awkward leading, awkward hauling, awkward following. George and Chris are full of fire and energy, but still they move up oh-so-slowly. At the bottom of such a great granite face, I feel—as I must have known I would—overwhelmed by the situation. A voice inside me is saying this is no place to get in shape for hard climbing and today I manage to lead only one pitch; it will be my only real lead of the climb. Uncomfortable about contributing so little to the pointed end of the rope, I knock myself out trying to clean the pitches fast and efficiently. We do six or seven pitches gaining the shattered diagonal ramp system that leads up across the face. A square turret-like buttress does for a bivouac platform. We're in the clouds now. The Thumb is claiming us for its own.

    The next day takes us on up into the clouds, into the upper face, even into the long final summit dihedral, but also into the rain. Only a drizzle at first, but it doesn't feel healthy. It's almost midnight when Chris rappels back down out of the mist after fixing two more ropelengths above our ledge. We have an impossible time wriggling into our sleeping bags in the dark, everything wet, boots still on, eating a small snack, shouting across to George on a ledge 30 feet away, trying to sleep sitting up on our miniature ledge, fumbling with our anchor slings, waiting for dawn.

> *Ice cliffs crack & groan:*
> *huddled on a tiny ledge,*
> *damp climbers moan & dream*
> *of warmer places, of*
> *lovers' faces that seem just*
> *out of reach, & each one*
> *wonders why he came*
> *& how long it will last.*
> *In the night*
> *enormous sheets of mist*
> *& rain blow slowly past.*

    New morning: drizzling first, then rain, then snow, whipping along in the wind, fat wet snowflakes. George and Chris are cold and wet, but still optimistic. They look up, push on back up last night's fixed ropes.

    This is the low point of my climb: Alone with the hauling bag and a giant wet pack, I feel sick, weak, lost. It's really snowing now. The rock is turning white. My friends are out of sight, somewhere above. For a minute I find the voice to question our judgment in going on under such deteriorating conditions, but George calls down that everything's OK. What can I do? Push off the ledge, pendulum across and start prusiking up the long thin nine-millimeter rope. Part way up, I feel my tail rope jammed behind me, back across the traverse. I have to jumar down and back across to free the rope. I want to cry or curse or hit something out of frustration, but save my strength for going back up…

An incredible day: Everything goes slowly, awkwardly. Our dihedral world, our long vertical rock corner is half white with snow, wind begins to whistle up the slot. Near disasters follow each other like desperate warnings shouted in vain at the deaf and dumb.

I reach a tiny belay ledge and reach out to steady myself against our big hauling bag. It starts to topple slowly off the ledge—my god! The knot in the tie-in sling has come undone. I only just barely catch it. All our food, sleeping bags, everything, a few more inches…don't think of it!

I'm slowly cleaning the pins from a steep pitch at the top of the dihedral, but below me the rope jams again: I rappel down to free it and start back up on stirrups and jumar ascenders. Sudden rumble/explosion: the rope has dislodged a cluster of big flakey blocks, they cascade down on my head. By the time I can duck it's over, I'm only scratched—but no, don't move! There is one more block, a big one, teetering some fifteen feet above me, pinned, held in place by the rope I'm standing on. I hold my breath. If it falls and cuts the rope, well…George is just above on a small snow-covered stance; he sees, understands, throws me the end of another rope. That way they can't both be cut if the rock comes down. I tie on, still holding my breath, then slowly shift my weight to the other stirrup. Ouf—the block tumbles, hits me in the face, falls into my arms. I manage to hold on to it, turn in my slings and heave it off, safely beyond our ropes. My gloves are red with blood, but it's only a cut lip. On we go.

Just below the summit, standing in a kind of notch out of the wind, roaring up the dihedral at gale force. We're soaked and shivering and we have to take our sweaters off to wring out the sleeves. We lower George into a gully, he disappears around the corner and ultimately, the hauling line leads up at a cockeyed angle over an immense overhang. I'm too tired and impatient to be careful and when I swing out on it, I find myself hanging five-feet from the wall and screwed up higher than Hogan's goat: I've rigged everything wrong, my slings are so short I can't move, my safety loops are somehow clipped into each other and not the rope…It takes a while to get everything straightened out, but I do, and now moving slowly up this yellow-green thread into the sky I get another nasty surprise. The rope is icing up and time and again my jumar clamps suddenly come loose, dropping me with a thump, so that only the extra Gibbs ascender clipped on top of everything just in case keeps me from dropping to the end of the rope every few feet. It's too much, I'm really getting psyched out and call up for a top rope just in case. Once more George saves the day, although it's probably all psychological.

Chris has been waiting in the wind through all this, and now he's too cold to lead on, so George takes the last pitch, disappears over the black, hoar-frosted summit ridge. We're up. I'm the last one off the face. The rope is an icy white cable. As I clean the pitch there is a momentary clearing, the mist parts below me. A surrealistic vision: snow-plastered slabs dropping into the Witches' Cauldron, the giant twin Cat's Ears spires rising out of the gloom under my feet, so cold, so hostile, so beautiful.

The clouds close back in. On the other side of the summit ridge, George and Chris are shoveling out a bivouac ledge like demons, clouds of spindrift pour over them, a snowslope shelves steeply off into the greyness. The summit is up there in the mist but it looks easy, we're here, we're up, and tomorrow...

> No summit
>> no mountain
>> no earth
>> only
>> three shadows
>> walking
>> on top of
>> mother of pearl
>> clouds
>> no climbers
>> no climb
>> no victory
>> no defeat

Afterward came another wet cold bivouac, another day, another bivouac, another day. Our fiberfilled sleeping bags, soaked and frozen, somehow keeping us warm and alive. Our soggy feet and hands carry us, lowering us, carrying us again, over the top and down, and down, and down.

A few moments snatched out of the mist: the top or thereabouts (a series of bumps on a long thin ridge, who knows which?) where we spent an hour or two moving above the clouds. Our shadows accompanied us on the clouds some few hundred feet below, each inside its own rainbow halo, Brocken's specter.

There was our happiness at finding two of Fred Becky's old rappel rings from his first ascent of the mountain in '48. Just to think that somebody else...And then at last the glacier, an out-of-focus world of subtle shades of grey:

Cloudwalkers,
or fallen angels,
we stumble forward
across the
uncertain interface
of snow and sky.
Why escape?
We may already
have left the earth
far below us
to keep company
with invisible gods,
tramping silent circles
thru the infinite
white on white
of endless clouds.

It was whiteout city, but we just kept on trucking. And sure enough there was an end to it. The edge of the high Stikine plateau, the escape route into and through the ice fall; and four-thousand feet lower down, after we'd run like madmen under the last ice cliff, jumped the last crevasse, our beautiful blue tent was still waiting for us.

After six long days, we took off our rope, threw off our packs on the moraine, kicked off our wet boots and smiled at each other and the world. This time the crooked road had led us straight. We'd been to the Far North and back. The journey was over and no longer mattered. That too was OK, and still is.

Thank you lord for rest days:
for sun on the boulders,
for camp in disorder
with drying ropes and
clothes and bags and all
our rainbow colored junk
spread out around us.

Thank you for this lonely place
for these empty miles
of cracked glacier tongues,
for these stark grey walls
towering into the clouds,

*for letting us be here
where we don't belong.*

*Thank you for these safe sounds:
the rain on our tent fly,
and not on our faces
under soaked bivy sheets,
the roar of ice cliffs
collapsing high above
and far away.*

*Thank you for our mountain
which frightened us
but didn't kill us,
for a safe route down,
a world to return to,
friends and women to love,
for today and tomorrow.*

—MOUNTAIN GAZETTE, #15

GALEN ROWELL

Alaska: Journey by Land

BY
GALEN ROWELL

H e was walking down the road in a drizzle, a few miles from Watson Lake in the Yukon. An ageless Indian. Maybe 50; more like 30. We couldn't tell, and it made no difference; he put out his thumb and we stopped. Now there were five of us plus a thousand pounds of gear in the station wagon. It wallowed around curves like a waterbed on roller skates and bounced over chuckholes with the shock of a steel ball rolling down stairs.

The Indian was quiet and oblivious, eyes focused at infinity. His twisted Asian face seemed incongruous above Western cowboy clothing. His nose had been broken, and when he turned my direction it still pointed at ten o'clock. We asked him questions and he answered in a guttural, barely intelligible monotone. What did he do? Worked in a motel part-time and lived off the land. Trapping, shooting game. Suddenly the dam opened and information poured forth, almost too predictably. Many brothers. Hard winters on the trap lines. Constant referrals to boozing. All the stereotypes of the Northern Indians. We began to wonder, was this all for our benefit? Was he just telling us what we wanted to hear, guarding his private life in a last vestige of dignity? Or was he for real?

We asked him his name. He looked at me intently: "Just call me the Black-Haired Yukon Kid."

We took the Kid to the bar of his choice in town. It smelled of beer and piss. An imaginary Mason-Dixon line separated the Indians and the Whites. Ours was the only integrated table. The Kid was on edge. Conversation flowed, but not in a stream. Rather, it resembled a canal system with locks that ended abruptly before a change to a different level. He didn't tell stories, he dropped fragments, and unlike a skilled politician, he made no attempt to tie them together.

Realizing the window of his soul was closed to us, we began to talk about our trip. Halfway up the Alaska Highway. Six hundred miles through the dirt and 600 more to the border. We talked about glaciers, wildlife and all the things we expected to see in Alaska. Our palms sweated when we talked about climbing; each of us was unsure how we would perform on 5,000 feet of granite in the Alaska Range in May.

The Kid was drunk. Something roared up inside of him and he yelled at us. Screamed. Bellowed. Just as quickly his anger subsided and he began to sing a weird improvisation of primitive sounds and modern jazz. He sang of his family and his mother and just as suddenly he stopped. It wasn't embarrassment. Maybe pride and anger for having given us an inch of his soul. The window closed; he clamed up for good. Communication dwindled to awkward tense stares. We gulped our beers and left him sitting there.

We drove two blocks from the highway in the center of the town of 500 people, parked the car, walked 50 feet and slept in boreal forest. It was twilight at midnight and sunny at three.

We were back on the highway at six a.m. When passing cars we kept our windows rolled tight and the heater fan on high to prevent dust from sifting through every nook and cranny. Trucks were a different matter. At the beginning of the highway, near Dawson Creek, a passing truck had unleashed a rock that hit our windshield like a bullet. It was the first of seven breaks. But the dust was more dangerous.

Winnebagos and Aristocrats crammed with American geriatrics crawled along the road at 30 mph—pathetic products of the Affluent Society whose only touch with the environment was an occasional forage for food, gas or souvenirs. But the truckers drove the road at 60 to 70, leaving a wake of dust and gravel that defies the imagination, until you try to pass one.

Fifty miles from Whitehorse we were cruising at a comfortable 65 mph, raising an opaque cloud behind us. Wisps of dust, at first barely noticeable, began to appear on the road in front of us, like cirrus streamers before a storm. Soon, thick dust surrounded us on all sides, pierced only by the bouncing taillights of a charter bus, glowing an eerie maroon through the murk. We followed it for miles. I couldn't decide whether to pass, stop and wait, or hang behind in the dusty pall. The bus was traveling a consistent 60 mph and would be much more difficult to pass than the shorter, slower houses-on-wheels. I made up my mind to pass after I chugged behind the bus at 20 mph as it climbed out of a ravine. When a half-mile straightaway appeared, I pulled out to pass and found myself going 75 to get around the accelerating bus. Gravel from its tires rattled off the station wagon like machine-gun fire. Forward vision was totally obscured when we came abreast of the wheels. I determined my position on the road only by watching the side of the bus. Suddenly the strafing ceased, vision returned, dust began to settle in the car. We breathed a sigh of relief. But the dusty wisps continued on the road in front of us. A minute later we encountered another set of maroon taillights in another murky cloud. The passing scene repeated itself, but to our horror we were behind a third tour bus. The view through the windshield was a continuing explosion, and the rear-view mirror was filled with the front of the second bus. Smog and the Manhattan rush hour seemed placid by comparison.

In Whitehorse we replaced our third tire. All three steel-corded radials had burst in about the same place on the sidewalls, while a rayon tire rolled along just fine. We would never know why. It was like the story of the plastic Jesus still standing upright on the dashboard of a demolished car.

Whitehorse is on the railroad, the highway and the Yukon River. It is the transportation and tourist center as well as the capital of the Yukon. Winnebagos and clicking shutters surround dead hulks of old sternwheelers, beached near the middle of the city. Neon and Mounties and Pavement.

The Alaska Highway was built across Canada by the U.S. Army Corps of Engineers during World War II. After Pearl Harbor, the Pentagon boys decided that having the Japanese in the Aleutian Islands without a supply road to Alaska was not a good thing. So they told the Corps to get with it. Nothing beats Hard Work and American Dollars, so tens of thousands of people and dollars were sent to the North along with a suitable number of bulldozers. The effort began in March 1942, with crews working from both ends toward the middle. On September 2, 1942, the bulldozers met at Contact Creek in the Yukon near Watson Lake. Never before had so many bulldozed so far so quickly—1,200 miles in less than six months.

Unlike more habitable places—such as Southern California—the Yukon has not been quickly populated in the wake of the road builders. Even today, the entire population of the Yukon is only 18,000 in a land larger than California. A boom may be coming, however. Mineral exploration has tripled in the past five years. Two new national parks will bring in more tourists, but only in the summer. A record temperature of -81° Fahrenheit was recorded in the town of Snag in 1946.

Near Whitehorse we visited the Yukon Game Farm, which advertised "Wild Animals of the Yukon in Their Natural Setting." For two bucks we drove our car past some sickly Dall sheep and caged predators. A golden eagle tried to flap its wings in a cage the size of a closet made of cyclone fence. A wolverine slobbered and grunted on a half-chewed piece of plywood inside a similar enclosure.

We imagined how a family might tour the farm. Mom and Dad would gaze through dusty windows, commenting ecstatically at real wildlife, while sitting in the rear of the mobile home, the children might possibly view a scene closer to reality by watching "Wild Kingdom" on Whitehorse TV.

A hundred miles past Whitehorse we reached the shore of Kluane Lake, more than 75 miles long at the foot of the St. Elias Mountains. The big peaks were hidden from view by a front range of mountains under 10,000 feet. Even so, the monotony of the relentless boreal forest was broken by views of glaciers, green fell fields and the giant, still-frozen lake. At the lake's inlet, white dust clouds blew across the flood plains. The dust was glacial milk, but it gave the place the impression of an alkaline desert.

We were on the edge of Kluane National Park, second largest in the Western

Hemisphere. We talked to one of the two park wardens and found out that the entire population of the park at that time, including tourists, was probably less than ten. There are no roads and no facilities in the park. More than half of its 8,500 square miles lie under ice. Mt. Logan, Canada's highest peak, rises to nearly 20,000 feet in a remote section of the park. When it was first ascended in 1925, the party spent 70 days installing supply caches over 130 miles of the route in winter, because the terrain was too rough to pass with pack animals in summer. When horses could go no further, they guided dog teams over the snow-covered ice fields in 50-below-zero weather. We felt pretty insignificant complaining about the dust and the chips in the windshield.

At the edge of the lake, a sign next to the highway proclaimed "Sheep Mountain." It was a beautiful setting. Sun, wind, ice, green hillsides and white Dall sheep. But here, in a land where the human population was less than one person for every ten square miles, the sheep were adversely affected by people in many ways. As the sun and the temperature dropped on a May afternoon, we were not alone scrambling on the hillsides for a closer view. Many tourists stop at the sign, and the area is mentioned in most travel guides. The sheep's normal predators, wolves, bears, eagles, were greatly reduced by hunting and trapping. Gradually the sheep lost much of their fear of the approach of large mammals, developed over eons of time when life hung on the thread of seeing and escaping enemies without being seen themselves. They have little use for their powerful telephoto vision, often likened to 6X binoculars. A chink was missing from their boldness, and they lay on the hillsides like bundles of inanimate white wool, moving only when I approached them very closely. I might have had stunning photographs except for the fact that some self-centered biologist had hung collars on many of the animals. The old ram with huge horns wore the latest in wide natural leather, while an adolescent yearling was attired in a Day-Glo pink. I could not explain why I found the collars such a flagrant affront to my senses. I have not felt that way about tags on the ears of campground bears in national parks. It is more like the hatred I would have if I visited New Guinea and found the tribesmen primitive except for Sears Roebuck tennis shoes —an unnecessary, degrading intrusion of the modern world into one of the last strongholds of wildness. I wondered why the study of wildlife is so often pushed so far that it robs the very wildness it seeks to comprehend.

After hours on the hillsides, we squeezed into the car and drove along the lake. We had traveled more than a thousand miles on the dirt, and it began to seem like home. We now expected dust and rumbles under the car. The road is perhaps the best unpaved highway in the world, carefully maintained by large crews that constantly repair the frost-buckling, chuckholes and wash-outs.

The original course of the road was not intended for modern tourists and truckers. Quite the opposite. It was purposely twisting and winding to safeguard military convoys from aerial gunfire. Even across flatlands, the highway meanders. Many of the winding sections are being slowly replaced with modern straight roads and wide right-of-ways. Pavement is one step closer.

The dirt road—almost continuous for 1,200 miles except for small sections through towns—was something we originally dreaded. We thought to ourselves how easy and pleasurable our journey might be if those miles were only paved.

We drove through the long hours of subarctic twilight. After midnight we reached the Alaskan border. The U.S.A. and asphalt roads beckoned on the other side of the customs building. A small but determined customs agent searched our car, sure that our youth, laughter and long hair meant illegal drugs. He looked serious, dedicated and definitely unhappy when his search was fruitless. Like big-game trophies on a hunter's walls, the waiting room was adorned with drug-oriented spoils: elaborate pipes, bags, bottles, etc., mounted in locked glass cases.

We rolled out on pavement. The end of the dirt and dust. But the road was worse! More curves. Poor banking. Harder to drive. Frost-buckled pavement caused us to hit more bumps in the first five miles than on all the miles of dirt. The bumps were severe and unpredictable; but the worst thing about the pavement was insidious. We sensed it but could not express it. An element was gone from the Alaska Highway mystique. Finally, after an hour of winding through the mountains next to double yellow lines, someone said what we all felt, "It doesn't seem like wilderness anymore. It's just like any other paved road in any other mountains."

After a long day on asphalt, we reached the end of the drive. Talkeetna, Alaska. Fifty-three people, two airstrips, hundreds of dogs, uncounted drunken Eskimos and two bush pilots who refused to speak to each other. It was truly the Last Frontier. We watched two young men in buckskins walk through town, carrying rifles and knives. Ever-present Winnebagos occasionally rolled past, turned around at the dead end, and rolled out again. Grotesque, dusty, myopic eyes of our times, unable to focus at one point longer than the time required to stick a decal onto a window. Behind the first row of houses a man dressed totally in leather negotiated with a bush pilot to be airlifted, with his dog team and canoe, to a very remote lake. He talked only about one-way terms; no mention was made of coming out.

GOLD
GOLD
GOLD   Alaska is
GOLD   the place to go!

I found this inscription on a men's room wall in a Winnemucca casino, but in Alaska I saw the other side of the coin. Rising gold prices have brought would-be prospectors back to the North. On the airstrip a man with tomorrow in his eyes thrust a chunk of rusty metamorphic rock in my hand. "Gold," he said simply. "It assays at over $200 a ton, but I've got to fly it out. I'd be rich if only there was a road." That night he was dead drunk in the Fairview Inn.

At ten in the morning the main street looked like a typical western town in the 1950s. Two-lane road. Dirt shoulders. Neon yet to come. Outside the Fairview sat two young men—one wearing Levis, a sweatshirt and a crewcut; the other with a Hell's Angels style vest, cowboy hat and gun belt. Rattling the doorknob of the closed tavern was a wizened, almost blind Eskimo, leaning heavily on his cane. Age, booze and 20 hours of daylight had obscured his awareness of time.

At the corner of the building, a chattering group of Japanese, wearing double boots and bright parkas, busily crammed equipment into the rear of a brand new van. In the distance, far beyond the railroad, the river and the spruce forests, the Alaska Range loomed above the horizon. A 40-minute plane ride would transmute us into an Ice-Age scenario. Like a shabby time warp in a cheap movie, we would find ourselves in a primeval, uninhabited land, staring alternately at unclimbed, unnamed mountains and at our mound of tents, skis, beer, freeze-dried food, paperbacks and week-old newspapers. We had arrived.

—MOUNTAIN GAZETTE, #17

E. LACHAPELLE

Growing Up High

BY
RANDY LACHAPELLE

During the International Geophysical Year (1957–1958) my father initiated a research project on the Blue Glacier. The Blue Glacier lies on Mount Olympus, part of Olympic National Park in Washington State. A 16x24-foot field station was erected on a rock outcropping adjacent to the Snow Dome. This was to be my summer home from age five until I was sixteen.

That time spent living on the mountain has never left me. I find myself driven back to those memories. The experience looms inside of me, a psychic presence of great power. I return to cleanse my past and dwell in peace with my mountain parentage.

Each summer morning I would awaken and walk outside to see the mountain. On clear days I would walk the horizon on a radiance of light. The glacier was sharp as the shale beneath my bare feet. The air was unburdened, carrying only the cold and light. It would pass as a breath of wind, barely noticed, leaving the mountain to touch me unhindered. On cloudy mornings the island of rock around the cabin was the extent of my world. Dimly felt shapes would press against the clouds, hinting at the world I knew to be there. The focal point of the mountain constantly changed. I have grown a whole mythology of remembrance around the view from the glacier. Vast distances and oppressive nearness were the gods of my life.

I used to climb to the top of a small peak behind the cabin and look out over the valleys below. The horizon extended from the Cascades in the east to the Pacific in the west. To the north lay Vancouver Island and the peaks of the Coast Range in British Columbia. I had the entire Puget Basin as the crucible for my imagination. I have difficulty filling the vacancy such space has left me in life—it has been a Promethean wound.

Often I would sit on the peak and stare out into the valleys below, from which the wind carried the smells of the forest upward to my resting place. Unexpected openings in the currents of air revealed whole meadows. The smells were over-rich in the mountain air. I could see one particular valley from the peak, which fascinated me. It would usually melt out by the middle of July. Four small lakes lay on the floor of the valley. I was the hermit of the valley and had responsibility for the adventures that took place there. I peopled the valley with my fantasies. I have often intended to visit the valley in person, but never have. It remains sacred to the visitations of a child.

My day was at the mercy of the mountain. Drifting patches of clouds raced their shadows across the snowfields and enveloped me. The sun wandered across the sky and my body simultaneously. I was overly sensitive to the lighting and mood of the glacier.

I was content to live in a child's eternity, with the peaks as my companions. No brothers or sisters, no playmates to seduce me from the mountain. It was an innocence that had its own ecology. The research station and its debris were as natural as any other part of my glacier life. I suppose now I could find the rationale of protest. I could disapprove of such wilderness contaminations. But it would be a lie. My emotions cling to the unchangeable unity of my childhood.

My earliest memories come from when I was two years old. At that time we lived in tents on the moraine of the lower Blue Glacier. The tents were strung out along a thin rocky top. Heather and a few wildflowers grew among the glacial debris. I remember it as being always foggy. The glacier and the fog intermingled, creating a uniform world of white. The dim forms of trees could be seen downslope from the moraine. Looking upward the snow blended into the sky. Out of that mingling of heaven and earth, climbers would occasionally appear. I remember having an unquenchable awe for the bundled men who went high above my world.

The peak-shaped, six-man army tents, filled with the incense of Coleman stove gas and musty canvas, were my personal temples. They took on a substantiality that seemed to magnify their interiors beyond all reason. All the space in the world could fit, together with my sense of security, inside those tents. My staple food in those days was peanut butter and logan bread. I burnt out on both. Even now peanut butter has little hold over me. I grew up being the kid who didn't like peanut butter—a stigma which would set me apart. One of the many ways the mountain separated me from my age. I was timeless, with no need to grow up, or change, or be any different than I was. That eternal time sense lingers on, subtly conditioning my world even now.

I had few toys then. A red ball was my prized possession. One day I lost it, the ball rolling down into the glacier and then disappearing into the icefield. I still distrust that icefield. I feel cheated by the convolutions that robbed me of my toy. I can watch that ball disappear endlessly in the inattentiveness of my life. The glacier is my silent answer, the judge of my willfulness to find my lost possession and pick up the pieces of my craft.

I was two when I made the first walk up the Hoh River Trail to the glacier. My only memory of that particular hike is of my father coming to meet my mother and me three-quarters of the way up the trail. He had come down from the camp on the moraine. I remember him bouncing along the trail, tanned and full of energy. His presence was overwhelming. I was captivated by my father, transformed as he was into an emissary of the mountain.

The bridge where we met was the demarcation point between the valley floor and the climb upward. The area of the bridge has always transfixed me. It spans the Hoh River as it meets Glacier Creek. The Hoh runs in a deep chasm at this point. The boiling water in the narrow canyon was mesmerizing. The water welled up in massive spirals. A deep green movement muted with the glacial dust. An overpowering urge to jump would strike me like a disease. It was with little surprise that I learned of a suicide from the bridge a few years ago.

The trail upward reads clearly in my memory. The slow places, the first view of the mountain, avenues of moss and the lake all appear when called upon. The personality that emerges along the trail is as complete as any other part of my glacier life. The steady fall of footsteps measured off the thoughts and feelings of the journey. Always there was the image of the mountain that displaced the monotony. The clear crunch of snow underfoot echoed along the hot, mosquito-paved way. The other worlds of snow and ice haunted the footfalls of the forest.

Such comparative spaces are the legacy of the mountain. To have lived in the highlands has rendered the lowlands incomplete. My intellect rebels at such thoughts, but in my heart I feel it to be true. I am inflated by the mountain. Tendrils of perfection reach out from my past, usurping the present.

Time is elastic. It stretched to break a child's patience, riding the eternal presence of the mountain. The coming and going of the sun was the heartbeat of glacial time. During the long periods of storm, the days would lengthen into boredom. The four walls of the cabin were the measure of my sanity. Among my refuges were cases of army survival rations that had been airdropped at the start of the project. They were intended as a stock of emergency food; instead they became part of my personal domain. Opening all of the carefully packaged food, uncovering treasure after treasure, had its own delight. A six-year-old can of fruit cocktail may not seem luxuriant, but entwined in the ritual of my childhood, it was pure ambrosia.

Tiring of the usual cabin occupations, I would sit and watch the fog blow past the window. Cunning wisps of cloud streamed from the wet rocks. It was as if the sea had risen and was seething in phantom shapes, turning the valleys into cauldrons. The presence of something just beyond reach often came to me. There was a pregnant depth to the storm that fascinated me. Even though it might be sleeting or snowing, there was an indefinite heat that seemed to flow from the greys and whites of the storm. It was the incarnation of storm demons, and my fascinated stare struggled to uncover their birth.

More often than not, a child's boredom quenched the mystic in me, and I would turn to the cabin for amusement once again. The opiate of card games, books and food shut out the storm.

During the earlier years of the project I often flew into the glacier on the supply plane. It was a small ski plane with a payload of three hundred pounds. It seems incredible to me that one small plane supplied the vast machinery of the project. Yet I know the vastness I conceived is a product of a child's vision. There was a self-contained quality to the station that assumed the weight of my whole world. In actuality the project was run, after an initial investment of $60,000, on $15,000 to $20,000 dollars a year (with the majority of that going to salaries). A paltry sum for financing a whole world.

I remember spending hours at the Clallam County Airport waiting for the supplies to be organized for the flight in. The airport was an old World War II fighter strip. It was absurdly large for the small aircraft that flew out of it. I filled some of the space by learning to ride a bicycle on the acres of unused pavement. Those times at the airport were gilded with the edge of anticipation. The subtle warmth of knowing I was on the edge of another way of life.

For the flight in, I was seat-belted and then surrounded with the rest of the baggage. I would wrap the plane around me like a chrysalis and wait to emerge on the mountain. Invariably, I was confounded by the terrain on the way in. As the plane climbed, a snow massif would appear that looked deceptively like the Blue Glacier. I would be amazed at the speed of the flight, but a little disappointed in the shabbiness of the vision before me. This mountain was spectacular, but did not have the aura of the glacier. I would resign myself to the counterfeit and then the plane would pass beyond, revealing the real thing. The Blue Glacier quickly filled the gaps in my memory with the ease of recalled landmarks. I was navigating again in the security of the past. The plane would land and I would stagger out, ducking to avoid the blow of light. But there was no escaping the radiance of the mountain; I was home.

The safety record of the station underscores the sense of security I had about the place. In the 17 years of its existence, the only injuries there were a sprained knee and a mild case of food poisoning. Often I played at the cliff's edge, cautious but unafraid. I had an inner freedom that betrayed an unsaid bond between the mountain and me. There was trust between us.

I learned a subtle secret of movement in those early years of scrambling over the rotten Olympic rock. The secret was to trust the mountain and move as low in my body weight as I could. There is a touch that sustains you when you abandon yourself to a mountain. Some days I could move unencumbered within my body—an effortless flow from somewhere around my waist. There was no separation between who I was and where I wanted to be. I travel in the mountains now and am amazed at how some people move through the land. They appear to walk from somewhere in their throat or chest and have to constantly readjust their balance as a result. "Come down," I want to tell them, "relax and walk with the mountain."

The early ease of the glacier has left me in a naïvete about mountains in general. I could easily disappear into strange peaks with nothing more than a shirt, pants and shoes. My conditioning draws me to high places, but leaves me unprepared—a traitorous beguilement.

Watching the sun set was sacred to the glacier ritual. I doubt if I have forgotten any of the thousands of nightfalls I have seen. As the air stilled countless times on the point of the setting sun, I traveled the length of the Puget Basin in the stillness. The lights of Victoria and Vancouver would gradually emerge. The Pacific coastline sometimes broke free of fog, the Tatoosh Island echoing the constancy of my sunset awareness. My vision could be entirely different each time, and yet the same mood would flow over me—the sense of being on the edge of eternal time.

I felt a peculiar attraction to the cities at night. The cold of the evenings would come up fast, biting through my parka, but often I stayed and lingered in the after-colors of the sun. And I would gaze out toward the cities. I treasured the presence of other people. The lonely place in me that was an only child on the glacier took comfort in the distant lights. Vancouver and Victoria have always been my favorite cities.

The only times I can remember feeling a sense of acute danger on the glacier is when I journeyed into a crevasse field. The really large crevasses, the ones that could swallow several buildings easily, petrified me. I would creep downhill to the edge and look in. I could never dismiss the glacier when it opened itself with such massive wounds. The crevasses always chilled me. It could be a hot day, and I would still get a cold shiver looking into the ice shadows. On either side of the prominence where the cabin was situated, there were large crevasses—catchment basins, should I ever slip and try for the valley floor. They were the watchdogs of my world. In my mind I have fallen many times, the result of these musings being hours of imaginative attempts to extricate myself from the glacier. Candy for a child's mind.

As the summer deepened, the sun cups on the glacier surface would grow more profuse. If they became too large they would endanger the landing of the ski plane. I can remember many times trudging out after dinner to do battle with the sun cups. It was glacial housekeeping. We would carefully shovel the top layer of snow away from the landing zone, leaving a strip of cleaner, less-contorted surface behind. It always seemed futile to me, rearranging a small patch of the glacier in hopes of fooling the sun for a few more days. It was like scratching the back of an enormous beast so our gnat of an airplane could land.

The ice worms came out in profusion during the evening hours. Every shovel of snow was peopled by sun cups and little black worms. I learned much about the tenacity of life by digging down through layers of algae and ice worms, uncovering the thin layer of life that the glacier begrudgingly sustained. Throwing ice worms into the sun, I would watch the day end.

The glacier was an ocean. Whenever I stepped off the rocks and headed onto the snowfield, it was an entrance into another medium. In the morning the crust would be unyielding, demanding a tight, controlled step to navigate between cups—like walking on a thin membrane of tension, protecting me from the deeper currents of the icefield. I was always amazed at the distances I could cover in the early morning. As the sun rose higher the glacier gradually sucked at my feet. Footholds had an ephemeral existence as the soft snow gave way under my weight. Distances would drag out the glare and heat of the snow. The glacier could be a desert, an ocean and an icefield, all within the change of a few degrees of the sun. The sun molded the substance of my world—shaping it and giving it color and tone. The ocean of snow was extremely sensitive to the slightest change in the environment. It was a resonant membrane that suffered us to walk on it.

Scientifically, much of the early research conducted on the mountain was to determine the energy exchange of the glacier. Instruments were set up to monitor the incoming and outgoing radiation. Thermocouples monitored the air temperature over the snow surface. Anemometers measured the wind flow. The microclimate of the glacier was constantly logged. The living movement of the glacier was transposed into dots and lines of yards of recorder paper.

There was an interesting progression of research over the years. The initial experiments studied the energy exchange on the surface of the glacier. The surface was mapped and survey points were established. Sites were picked for ablation measurements. The glacier was monitored for extremely subtle changes. As the years passed, there was more concern with the internal structure. Hot-point drills wormed their way down to measure the depth. Pits and tunnels grew larger and deeper. The hardware associated with the research transformed; electronics gave way to air compressors and jack hammers. The permutations of gravity as it played within the ice animated the glacier. The life of the glacier was communicated in plastic flow. The mechanism of that flow was the focus of attention for the research.

Science was a ritual that I took for granted. It was something automatically done when confronted with something as mysterious as a glacier. I knew that all the instruments and all the experiments would never really unmask the mountain, but the ritual was important. There was a subterranean purpose behind the daily tasks of the project. Even now, I have only a fleeting glimpse of that purpose in my mind, enough of a view to know that science really was a sacred approach to the mountain—a way of expressing the connection of man with the mountain.

The last year I spent on the glacier was the year of the great tunnel dig. An ice tunnel was bored into the western ice fall. The effort involved fourteen people. An unheard-of population for the station. The silence of the mountain was pushed into the background with the intensity of the activity.

The tunnel, when completed, reached to bedrock. One section of the tunnel deformed so quickly that it repeatedly had to be dug out. The tunnel bisected several small crevasses. It afforded the interesting comfort of being able to look up a crevasse while secure within it. The end of the tunnel revealed the soul of the glacier. The ice was exposed in its massive response to gravity. Long feathers of ice peeled away from the ice-rock contact. There actually was considerable space between the two mediums at various points. I always felt, standing at the tunnel's end, that I was intruding on some private relationship between the glacier and the earth. I would catch the movement of the glacier in the transitory light of a flashlight. An intrusion of human perception that violated the time sense of the glacier.

The psychological space of the project was far different than during any previous year. The drama of living together with fourteen people in a small mountain cabin became more interesting than the physical surroundings. An example of the change was the transformation of one of my favorite scrambling areas into an arena of competition. I joined with the others in trying to make it up impossible routes. The old ease with the mountain was gone. The freedom of scrambling was caught in personality structures. I was growing up, into the world of adults. It was fascinating, but it was also drowning out my connection with the mountain.

My descent into adulthood has been arduous. No climb in any of my glacier years can match the shock of having to grow up. I no longer have the psychological edge of the mountain. I am stripped of my uniqueness, my mountain overrun with hundreds of climbers. I have been forced to let go of the inviolate sanctuary of my youth. The memory vault has opened, and everything the mountain was pours out, demanding a new reconciliation. The power of my connection with it emerges despite the saccharine memories I cloak it with. Going back to these memories has liberated the real mountain of my youth. This dual vision, of wishful memories and real power, is my legacy of childhood in the peaks.

—*MOUNTAIN GAZETTE, #40*

GALEN ROWELL

Fear

BY
DAVID ROBERTS

Buddhists know that the *duhkha* of this world—the suffering, discontent, anxiety—outweighs the joy in it. And I knew it, too, during those early June days in 1963, as I watched the mud and dust of the Alaska Highway reel past from the back of our VW bus. I was afraid of Alaska, where I had never been; I was afraid of McKinley, a mountain, I was sure, too formidable for my three short years of climbing. But in 1963 I would not have said so. Then I might have seen in the gloom of the spruce trees plodding past our window only an uneasy regret that I wasn't spending my summer as I had the previous one: working construction in my home town and dating old high school girlfriends. Where were the grassy parks and blue lakes of my native Colorado? The highway was not even mountainous. Instead, a gauntlet of gravel-lot gas stations, expensive stale hamburgers, generators chugging all night; in the cafe kitchens, prim Canadian radio voices through a fog of oatmeal and bacon. And between oases, those endless stands of shaggy spruce, with hardly a glimpse of a view.

But it was not the scenery I was afraid of, not nights without girls and beer. In the threat of those somber woods, of those dismal gas stations, there was something all tied up with our plans for the Wickersham Wall. I was afraid—like all climbers, though I didn't know it then—of getting killed. I had never seen an avalanche, never stepped across a crevasse. We were too brash, I thought, seven college kids blundering into water over our heads.

On the way up, as we drove along, I'd argued with Hank Abrons about danger. Mouthing Rébuffat, I'd contended that we would enjoy climbing all the better if it were perfectly safe; we seek out difficulty, I quoted, not danger. Too naïve, said Hank. Without risk, there'd be nothing to it: Climbing would be reduced to a game. I was sure at the time that he was wrong. I wanted him to be wrong.

Like brash college kids, we climbed our route gaily and fast without a real mishap. When you got down to it, when you stared an avalanche slope or a crevasse field in the face, it lost much of its terror. Notice, we would say, that the slope only slides after 10:00 a.m.; in this part of the glacier, we observed, the crevasses are pretty regular. The only close calls—a few falling rocks and a river-crossing—were over before we could work up a healthy fear; and bravado, we found, was a more effective weapon against them. At McKinley Park Hotel, we were received as conquering heroes, and so we began to believe we were.

Yet eleven years later, I remember my depression before and during that expedition better than I do the feeling of triumph afterwards. It claimed me in the wee hours as we traversed the Denali Highway, when I shivered even in the driver's seat, at a mere 2,000 feet, and stared at the unsympathetic plains of tundra to the left and right. It emerged while we waited, for four days, most of them sunny, for Sheldon to deliver our air-drop at base camp, 35 miles in. We heard imaginary planes, played cards and dice and read out loud to kill the time, cutting back at last to two meals a day, the "what-if-he-doesn't..." never quite spoken. During storms, too: my first taste of them, of the insidious, clammy waiting for three days in a sleeping bag, watching the tent walls frost, sponging the floor. Twinges, twenty days in, at the thought of fresh bread and salad and oranges. The gloom depended on standing still, or lying still, waiting and wondering. Once in motion, I was safe.

What sticks in my memory about McKinley is that first perception—which I'd never had in Colorado—of a natural world to which I did not belong. There was too little order to those highway forests, and the dreary lives led in those well-spaced truck stops somehow reflected that fact. There was a hugeness on both sides in which mere human travel seemed inconsequential. The mountain made more sense than the tundra; the hike in was coherent because it led to the climb. All around the edges of the mountains lay a world we specialists would not define. Why did the river wander there? Where did the silt in the sandbar come from? Better hidden, but there just as surely, the sense of the alien lurked on the mountain itself, to be glimpsed in those quiet, depressed moments of waiting. It made no difference to the wind if it blew for a day or a week. The snow turned to ice randomly, not for our step-kicking convenience. The mountain knew nothing about routes.

Six years later, on the north face of Mt. Temple in the Canadian Rockies, I thought there was a fairly good chance of getting killed. I was with Hank again, and Denny Eberl, whom I'd never climbed with before. It was about four in the afternoon of a late June day; we were three-quarters of the way up the 4,500-foot face. We'd gone up impatiently the day after a long storm, climbed the lower wall fast, and now, with simple logic, the snow was beginning to slide in the afternoon sun. There was no ledge to hide on, no going down.

I'd seen avalanches, I thought—from the safety of base camp below McKinley, on adjacent walls on Deborah. But not up close, like this. An hour before they had begun, little sloughing trickles of wet snow presaged by runnels of powder. A deterioration in the texture underfoot. Now they were in earnest, and fear and bravado seemed equally impotent to cope with them. I watched as a big one piled over the short cliff we crouched under, only a hundred feet to our left; tons of wet snow dumping noisily into view, gathering force and size downward, for all the world like the cement I watched as a kid pouring out of mixers down a wooden trough.

We did what climbers must always do when their real urge is panic: climbed as fast as we could upward, yelling advice, skipping belays, moving desperately. Over the cliff, up into the bright sun on the steep, long slope where all the danger was focused. Like cement, I kept imaging, and us with no anchor to hold us on, no roof to keep ourselves dry. The grace and naturalness of avalanches seen from the valley floor reduced to a squat, heavy, gouging ugliness up close. Here, too, was proof that we did not belong.

Denny was leading. I went last. In the middle of the slides, starting his own little ones, he plowed through the deep snow upward, as fast as he could move, at a snail's pace. With time to look around, I spotted an escape route left. Five minutes of extreme danger, then safety. I shouted up; Denny yelled back "no." Maniacal, he drove himself straight up, toward steeper snow and an ice cliff. I knew my way was better. I yelled again, told Denny to wait. The three of us assembled, stood on a rotten snow platform, and harangued each other. I thought of unroping and deserting Denny. Hank tried to mediate. All three of us pictured the absurd possibility of being swept off while we stood there arguing.

At last Hank voted with me; Denny grudgingly gave in. I led the dangerous traverse; in ten minutes, perhaps, we were safe. The rest of the way up to the summit, I walked in a euphoric stupor, prickling with apology toward Denny, yet glad—oh, how glad—to be alive. Hours later I realized, with a chill, that what must have driven Denny upward was the abstract image of our route, the straight line on the photo. We had traversed off, cheated, lost the true first ascent. At the edge of easy and obvious death, the game had still mattered to him, the line of the route, the human imposition of purpose on a place that mocked purpose. Was he the fanatic, or was I the coward?

A year ago I went to Arizona, alone, during March. It was a deliberately structured trip: eight days alone in the unfamiliar terrain of sandstone buttes and arches and pinnacles. A calculated way of forcing a new experience upon myself: I didn't know the desert, in fourteen years of climbing I had never spent more than two days in a row alone. To upset my habits further, I would define for myself no goal, no arbitrary point to reach; I would simply wander. And I would leave my camera at home.

It was, I knew, a safe experiment (that too was calculated). I would do no climbing, only hiking: I had gear to stay warm in a Colorado winter, plenty of food, good maps. I worried more about snakebite than frostbite, Navajo laws than safety rules. To be sure, I could slip, break an ankle and die before anyone found me. But I was safer in the desert than driving there.

It snowed the whole time. The country seemed strange, depressingly so, illogical to my mountain-trained mind. But traveling in it was so like winter camping that I found myself oppressed with familiarity. Juniper, it was true, burned smokier than sub-alpine fir; north was less obvious than in a timberline cirque. But lighting the stove, cooking the usual glop, pitching my tent, finding water: all the rituals of camping seemed deadeningly habitual.

Only the solitude was new, and I was surprised to discover how it frightened me. I could rationalize that the discomfort of the storm made life miserable; but it was not so much a damp sleeping bag that disheartened me as the poverty of myself, the emptiness of waiting and being alone. I turned for solace to my habits. The watch on my wrist became my most valuable piece of gear, for it measured out the pieces of time that I could sanely squander on each necessary task. Time: too much of it, the nights too long, eight days an eternity. If I started to cook breakfast at 7:30, it might last till 8:30, then packing up would take half an hour. Two and a half hours of hiking to an early lunch; by 12:15, chilliness would drive me on. I was tired by 3:00, but it seemed too soon to camp. Perhaps I could eke out the day till 4:00 by angling up this side canyon. The tent up by 4:30, dinner over at 5:30—depressingly early. My diary wasted an hour, then reading by candlelight until 8:00. But I could seldom sleep before 10:00.

My habits were my only companions. I leaned on them, taking a meager comfort from the just-so placing of my boots beside my sleeping bag, the efficiency of Jell-O cooling while the glop was on the stove. I missed my camera, paid less attention than ever to my surroundings. As long as there was a trivial chore to do, I fought off the fear: even gathering wood or sweeping out the tent. But if I tried to sit and reflect and hear the silence and watch the snowflakes fall, I recovered my misery. Even writing in my diary, where I tried to explain the fear, kept the fear away: It was the purposeful act of moving the pen, filling the pages, accounting for my time, that gave the time meaning.

The country had made sense beforehand. There were jumbled chaoses of contour lines I had longed to investigate, claustrophobic canyons to explore. But in the deep snow, most such goals were too far away. I confined myself to a timid circle, and decided, on the fifth day, to go out.

It felt like cheating, there was no denying that, no matter how I rationalized my decision. On the last day, it cleared suddenly, confronting me all the more squarely with my defection. It was truly beautiful here, the red sandstone hissing dry in the dazzling sun; but beautiful also was the prospect of car and town and newspapers. I edged past the smoking hogans and hostile dogs of the Navajo, whom I never saw, and unlocked my car with a sense of relief.

As I drove into Gallup, the thought occurred to me that, in a deliberate attempt to upset my habits, to force myself into new ways of knowing the wilderness, I had only reinforced the old ways, discovered how much I needed habit and ritual. And I knew disquietingly that already the comforts I had longed so to get to were dwindling in value: that however familiar the sliding of tent pole in nylon sleeve was to me, the clink of a quarter in a Coke machine was more so.

In the summer of 1974 I found, with a kind of pleasure, that the fear was still there, as naïve and presumptuous as ever. A few miles south of McKinley, Galen Rowell and Ed Ward and I were climbing on our second day on a 5,000-foot face of Mt. Dickey. Part of the fear came from knowing that this was, simply, the hardest climb of my life. The mist had closed in at dawn; now it seemed certain that a storm would follow. The trick was to get high enough to be okay when the storm came. But, of course, the higher we got the harder it would be to back off. Already a labyrinth of potential rappels lay below us, invisible in the mist.

The granite had to be read in code: orange meant good, white bad. The pitches twisted, traversed deviously; rope drag was a personifiable enemy. All through the day ran the motive thread of fear—subsumed by our skill, tamed by precaution. I grew to hate the feel of carabiner and jumar handle. I kept my mind on a mental abstraction of our route that I could collate with the Washburn photo in my pocket.

Then there would come one of those moments for which the only apt word is *Augenblick*—the blink of the eyes signalling the absurdity of the self-evident perception, *yes, here, I am indeed here, exactly here, here rather than any other place*. Absurd, because it was as clear to me in this twelfth straight year of climbing in Alaska as it had been on the drive to McKinley—that here we do not belong.

Safety: the imaginary line leading up through the bad white rock; the point of the climb, the definition of the game. Against its purposefulness, those existential *Augenblicke* that catch me at a loss to understand what I am doing, and a running commentary of pious fears, like half-remembered nursery rhymes: "And if this rope should break, or if that rock should fall..." "Down will come..."

Now, five days later, the route successfully behind us, I sit alone on the Ruth Glacier, waiting for Sheldon to fly me out, and begin to write this essay. Galen and Ed got out yesterday; this morning, when it was to be my turn, there was a thin drizzle, and Sheldon did not appear. For all those 12 years of Alaska, waiting is as intolerable as ever, and I wonder why. We are safe now; the white rock cannot collapse above us. I have no duties but to cook myself three meals a day, and wait. But I lose interest in the books I pick up, I eat too much, and the rest of the time I catch myself listening for that distant, lovely whine of the plane.

What puzzle is working itself out in me? Which me does the fear define? Am I the same person here, on Mt. Dickey, as I am in an armchair at home? I think now that I am not. But as long as I continue to climb, I cannot separate the various strands of my ego, or figure out how to splice them together.

I know now that Hank was right. Without fear it would be nothing, or little—another game. I have always loved games, loved winning, and I know that much of my urge to climb comes from that: For whenever I see the kind of wilderness that defies route-finding, I see it with a heavy heart. The unclimbable exfoliated north wall just opposite Mt. Dickey, for instance, or some of those aimless canyons in Arizona. But why does the game, if that is all it is, continue to vitalize me? I know by now how good or bad I am at other games; I know winning is a dead end. But at the top of our wall on Dickey there was all the violent relief and joy there ever had been, the victory (or escape) as gratifying as ever.

Then why do I so wish, right now, that I were off this glacier? Why does the thought of a beer in the Fairview Inn or a magazine stand in Anchorage hold out all the false promise of civilization, even though I should know better?

The fear connects with, is implicit with, the waiting. An allegory occurs to me: This waiting for salvation by airplane is, in its temporary way, not unlike the waiting of our years for that archangelical moment of death—so much of life is spent impatiently, waiting in lines, for vacations, for quitting time, for that favorite theme in the last movement. The fear I felt on Temple, on Dickey, however, had the power to transform 15 hours of hard work into an almost timeless, unified day, a day when there was no urge to look at the watch, no need to eat or even sip from the water bottle. And the ironic moment of the *Augenblick*—does it not give the illusion of time stopped, seized, photographed?

Yes, to all that; but still, why me? If I need the fear, why don't others? Why aren't other lives barren or tawdry for the absence of it? Could I change? Learn to live less mechanistically with others (for climbing is mechanistic), sensitize myself to weakness and love as I have to strength and passion? Could I learn that Buddhist calm that would forever make climbing unnecessary?

But it is almost time for dinner. The clouds are lifting over the Southeast Spur, yet it looks foggy down-glacier. I think he'll fly in tomorrow morning. I really think he will.

—*MOUNTAIN GAZETTE, #25*

Flat Mountain

BY
CHARLES BOWDEN

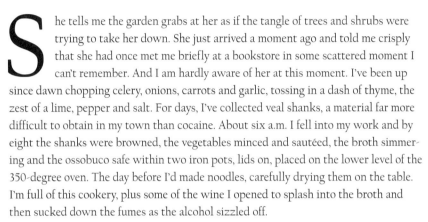

She tells me the garden grabs at her as if the tangle of trees and shrubs were trying to take her down. She just arrived a moment ago and told me crisply that she had once met me briefly at a bookstore in some scattered moment I can't remember. And I am hardly aware of her at this moment. I've been up since dawn chopping celery, onions, carrots and garlic, tossing in a dash of thyme, the zest of a lime, pepper and salt. For days, I've collected veal shanks, a material far more difficult to obtain in my town than cocaine. About six a.m. I fell into my work and by eight the shanks were browned, the vegetables minced and sautéed, the broth simmering and the ossobuco safe within two iron pots, lids on, placed on the lower level of the 350-degree oven. The day before I'd made noodles, carefully drying them on the table. I'm full of this cookery, plus some of the wine I opened to splash into the broth and then sucked down the fumes as the alcohol sizzled off.

The woman is the waif that we beckon to our tables on holidays, an acquaintance of my mother's who has hit a bad patch. She enters like a wind, takes in the glass wall that forms the south of the house, and suddenly she is in the tatters of the garden and walking down the path. She moves with girlish speed. She is in her late seventies and in a few months she will be dead. She was told this a few weeks ago. Her face is lined, browned from the sun, and her thinning hair bristles from its short cropping. She wears sensible slacks and shoes. She has that assurance the old earn when all the fires are banked and gone cold and nothing really matters anymore. Her body is disintegrating as she wanders the garden with the thorny trees grabbing at her clothes. The garden has no straight paths. I do not trust straight things.

We sit in the living room sipping glasses of wine, there are about a half dozen of us. The talk is brittle and to no point, remarks about the mild weather, comments on the fire. Around us the city is silent, dead with the languor of the holiday. The mountains framing the valley hold in the smog. They stand mute and dumb and pocked with pines and patches of snow. I used to climb them until I learned that the top led to nowhere. Now I stay in the flat pan of the desert where I can't look down on things. I am tired of cooking and anxious to eat. For hours, I have thought of the veal slowly leaking its rich marrow into the sauce. I resent the wait, the talk, the sputtering fire, the sparkling wine in the flute in my hand. The food is ready and I want to savor it at its peak. I hate the holiday and everything about it. This whole dinner happens with my resentment. I want to scream leave me alone, I'm sick of people. I want to simply devour the meal and have silence return. The aged woman with one foot in the grave is a nuisance whom I barely acknowledge or listen to.

She begins to cough and then this racking comes from deep within her. She rises, excuses herself, and slips out the front door. My mother, eighty-five and recovering from a stroke, says, with disapproval, "She is going to sneak a cigarette."

I nod.

Finally, she returns and we sit down to eat.

The meal flows with platters of ossobuco, noodles, a salad, and a Norwegian sweet bread the woman has baked from some ancient family recipe. The wine glasses fill and refill. A numbness descends from the food and idle chatter, the great calm of feasts that have been observed for centuries and from this long march of gluttony have created their own rhythms and silences. I am counting the minutes until this will all end. Light floods in from the garden, and the only movement there comes from dozens of birds darting on and off the feeders.

Would you like some more wine?

The old woman considers and nods. I fill the glass like a robot.

"I remember when my husband retired from the military and the party we threw," she says suddenly. "It was the best hangover of my life."

The light pouring in through the glass wall catches those outlaw hairs that sprout on the chins of old women.

She speaks in a soft voice with a sharp undertone, a vinegar lacing through the soft petals of her words. She is back in Wisconsin and she is coming home from college in the Fox River Valley. A huge storm sweeps off the plains. She and the other passengers spent the night snowbound in the rail car and then when she gets to her town, she must walk a mile to her parents' house through the drifts. She is wearing high heels and nylons, it is just after the Second World War, and the cold goes to her bones. And my God, when she finally makes it home, her parents are without food, the storm, you know, so she changes and puts on her skis and glides into the village market. The thing that keeps her going then is theater, the arts and she loves the theater so much that when people listen to her they think she must be part of a company, but it is all really in her dreams. She saves and in '48 takes a vacation in Europe. And, she laughs, she does not come home for three years because she meets a soldier boy, and then come the three children, the moving from base to base, that wild retirement party, my God, when she came down the following morning with that huge hangover, the living room was a shambles and after she passed out the night before, the drunken revelers blowing out the candles had gotten red wax everywhere, even on the walls. The marriage lasted forty-eight years, and she still drives his big Chrysler, he insisted on solid and large automobiles.

I can smell the whiskey and cigarettes of those barracks days. The aroma of the ossobuco has filled the house for hours, the rich scent of bone marrow and meat and wine clogging my senses. But now the whiskey and the smoke knife through this aroma, and there is lipstick on the tumblers, red lipstick smiling from the rim of the glasses where the ice slowly melts and slowly lightens the amber glow of the drink. And I don't ask about the three children, there is nothing to ask since she is alone on a holiday and eating at a stranger's house and dying at a rapid clip. I can see the snow clinging to her young strong legs, clinging to the nylons with a dark seam etched along her leg. I can see her dreams where she is painting sets and cueing actors and dancing on table tops after openings in New York where the war hangs like smoke and is slowly blown away by the post-war boom. She is cross-country skiing across continents and oceans, tossing down a drink now and then, peeling her nylons off as she sits on the bed with only a low light glowing in a strange and hopeful hotel room. And she is not alone in that room and the heat rises off her.

She begins to cough again, that bad cough that says I am not a cold, I am not bronchitis. The cough that leads to a rattle in the throat and then to nothing at all. She excuses herself and says, "I must go out to the car for my medicine."

As soon as the door closes behind her, the rest of the table drifts into the mutterings about the cough, about concerns, compassion, how brave she is, all the small notes we make to avoid the short word: death. The mutterings all say: we are sorry. And the mutterings never say what we are sorry about. Because that would ruin all the meals.

And then, she is in her chair, as if she silently glided back to the table on skis schussing through perfect snow.

It is time for coffee. We are full. The old woman's eyes seem to float over the walls and the walls are cluttered with photographs, paintings, masks, fetishes, amulets, prayer sticks, a visual noise that makes me calm. I cannot abide orderly walls or any form of symmetry. Order terrifies me, like a hand choking me to death. Suddenly, her eyes alight on a Dorothea Lange print, a shot of an Okie mother in a California camp of the Depression, an image that has come to be seen as the Migrant Mother, a kind of Madonna of that dark and deep economic trough.

She says, "I love that photograph, that is my favorite image."

Her voice rises, there is vivacity, the guileless energy and appetite of a girl in a college classroom, a girl fresh off some farm, who has just had the hand of this thing called art touch her breast for the very first time. I look at her, the wrinkled face, the hairs sprouting on the chin, I sputter something about the history of the photo, how it was taken at the end of a long day, how Lange almost did not go down the road to the camp, how she was tired and it was late in the day and she saw the turnoff and said the hell with it and then twenty miles later felt this nagging, this sense of missing out on something, and wheeled around, went off the side road, got out of her car, and only spent ten minutes, took but five shots, how she stumbled on the woman being devoured by her children like a sow in a farrowing pen, how the contact sheets reveal the way Lange slowly zeroed in on the famous image, the shots distant, then closer, the baby sucking at the white tit, then the camera waiting for the tiny mouth to release the nipple, the blouse sheltering the breast again, the children flowing around the mother like a flood of flesh, and then click, this icon imprinted in our minds that says hunger and fatigue and the flesh willing us to march on and damn the reason why. And it almost didn't happen, just a happenstance, a trip taken with surprises.

"I love it," she says simply.

We all sag from eating too much. The words float away, their enthusiasm alien to our stupor. The meal begins to break apart. My mother full of years must go home, the dark hours will come soon and she does not drive at night. And since the stroke, mornings are her best time and the afternoons are more of a struggle against fatigue. The old woman too must drive and beat the setting of the sun. And there is the cough to constantly rein in, the medicine to swallow. I'm tired also, I've been over my pots and kettles for hours, and there is a table to clear, china and crystal to put away. I want everyone gone, just the fire flickering and then I'll drink all the open bottles until empty, drink them dry and hear the glass clink as I toss them one by one into the trash.

I walk the old woman out. She gets into her big Chrysler, talks about how she knows she should trade it in and get a smaller car, but you know, she continues, they're not safe, nothing but damn little shoe boxes. So she'll stick with his car even though it is an ugly brown, she couldn't talk him out of the color. I watch her drive away.

Then I empty the bottles. The wine tastes so rich I think a carpet of just stomped pulp flows across my tongue. The bitterness lingers like a gift as I swirl the cabernet in my mouth. I am through with the peaks. I tell you it is all a flat mountain. The peaks are a lie, or at best a fantasy. The skis glide, the smells hit the face, the hotel rooms glow with weak light, the lipstick smears the rim of the whiskey tumbler. A hand is on the body, the buttons come undone and the skis keep gliding, silently gliding, and the peaks are never apparent until long after they are passed. The seam of the nylon snaking down the creamy skin of young fresh legs, breath hanging in the frozen air and blinding light. And the mother, that cracker face, staring straight at the camera, unsmiling, as the children take her down in the undertow. Drove by and almost did not come back. Just ten minutes for five quick shots.

The best hangover of my life, the red wax even on the walls.

—*MOUNTAIN GAZETTE, #79*

GREG WRIGHT

The Impsons, Ed & Ma'am

BY
JOHN PETERS, M.D.

Christmas Eve, 1961: a holiday at home for most folks, but it was the night Lydia Impson walked five miles through a raging blizzard to get help for her dying husband, Ed.

Five miles in that savage storm would have been an ordeal for a trained mountaineer in the prime of life. But imagine an 81-year-old woman, crippled with arthritis and almost blind. It sounds impossible, but then, you didn't know Lydia Impson. She may have been nearly unable to walk, and well-nigh sightless, but her spirit was as strong as the mountains.

Ed had "taken sick" about two weeks before Christmas and had gotten progressively worse. He was having trouble breathing, his legs were swollen, he had a high fever, and he was passing blood and pus in his urine. To make things worse, the Impsons didn't have a phone, and their nearest neighbors were five miles away, down a dirt road blocked in winter by drifting snow. And their only transportation was Sally, an ancient brown mare that hadn't been ridden in years.

Ed stoically took stock of his situation and told Lydia it was no use, that there was nothing she could do. No one could help. "Ceptin' me," she said later, proud but matter-of-fact. Over Ed's protests she put on her "shawl and greatcoat," as she put it, and headed out into the storm. She went to the barn and got Sally out of her stall. "I couldn't sit on no horse, 'cause of the arthritis in my hip," she said. "So I just grabbed onto Sally's tail, and headed her down the road toward Bob's place. I knowed Ed was a-dyin', and I knowed there was a purty good chance I was goin' to die too. I've been livin' in these mountains for nigh onto 75 years, an' I've helped bury more than a few young, strong folks who got themselves caught out in storms like this 'un. But I figgered, if I don't go, Ed's gonna die. And if he dies, ain't much use in me goin' on either.

"Ain't neither of us afraid a dyin'," she smiled a tranquil smile. "It's close to our time anyways."

So the old horse and the old woman plodded on through the storm. Sally forced her way through two-, three-foot high drifts for five miles that night, along that old river road, avoiding the steep dropoff on one side. One misstep would have sent them both plunging into the icy river below...but their steps were sure.

No one knows how long they were out in that storm. As crippled as Lydia was, it must have taken them eight, ten hours. "That cold wind was a-bitin' right through me. And these old hips and knees, they felt like they was gonna freeze up solid," Lydia smiled. "I could only go a little ways, and then I'd pull on ol' Sally's tail, and we'd stop and rest for awhile, and then we'd go on. We kept doin' it till we got there."

Around midnight, Bob Magnus, the game warden, heard a sound on his front porch. He went outside and found Lydia collapsed on his doorstep, nearly dead from exhaustion and hypothermia. The faithful Sally stood a few feet away, watching over her. Bob carried Lydia inside and phoned me.

"Hey, Doc, I've got Mrs. Impson down here at my place. She's been out in this storm, and she's in pretty bad shape. My wife and I are trying to get her warmed up, but all she keeps saying is, 'Don't bother about me. Ed's back up at the house, and I think he's dying. You've got to help.' If you can get down here right away, we'll take my Jeep and try to make it back up to the Impson place."

"I'll be right there."

It took me almost half an hour to make the seven-mile trip. The storm hadn't slacked off a bit; if anything, it was coming in harder. The snow was still falling and the wind was up, causing whiteouts that reduced visibility to a foot in front of the windshield. After many a white-knuckled moment I pulled up in front of Bob's.

The first thing I did was give Lydia a quick exam, over her protests. "Don't you worry about me," she scolded. "It's Ed who's the sick one, Doc. He's got the fever and the cough, and he's peein' blood. He's real short of breath, and he's swole up like a poisoned pup. He's probably dead by now—it took me so long ta get here," she sniffled.

"It's a miracle you made it here at all," I said, patting her on the shoulder. "Listen, Bob and I are going to head up right now and check on Ed. Don't worry."

"I'm a-goin' with you," she said determinedly.

"Lydia, you almost froze to death tonight. Wouldn't it be better if you stayed here with Mrs. Magnus and rested up a little more?"

"No sir. My place is with my Ed. If he's still alive, he'll be a-needin' me. He's blind, see, and can't do for himself. Just pack an extra blanket or two and I'll ride in back, out a the way of you menfolk. But I ain't takin' no for an answer."

Well, that was that.

It took us nearly an hour to get back up to the Impson place, in four-wheel-drive with chains on all four wheels. How an old horse and an 81-year-old woman ever did it on foot is beyond comprehension. I guess love does have unseen powers.

The Impson house was an old-fashioned earthen-walled place with a sod roof. As we stepped in the roughhewn wooden door, I could immediately see Ed was in bad shape. He was sitting up in an overstuffed chair, puffing for air with waterlogged lungs. His legs and feet were horribly swollen, like the limbs of an agonized giant. He was coughing up blood. There was a milk bottle on the floor next to him, half full of an unholy mixture of urine, puss and blood. Ed had a deadly combination of heart failure, pneumonia and a severe kidney infection. Thrice-deadly, really: any of the three was enough to kill a man in his 80s. Ed really needed to be in a hospital, in intensive care, and even there his chances of survival would be poor.

"Ed, I'm going to give you a shot of penicillin and something else to strengthen your heart and get rid of the fluid in your lungs. Then we're going to get you out of here and down to the hospital."

I began to prepare the injection. As I did, he reached over and squeezed my arm gently. "Doc, I sure do appreciate you comin' out in this weather, and I don't mean to be disrespectful, but I really don't want to go to no hospital. I won't." All this was said with painful effort, like a man pushing a boulder up a steep mountainside. Before I could say anything, he continued. "Soon as I leave here, somebody's gonna steal my ranch."

The Impsons' "ranch" consisted of a few acres of hardscrabble mountain terrain, with hardly enough flatland to make a pool table out of, buried in snow seven months out of the year; but to an old-time Westerner like Ed Impson, it was "his spread." And beyond price.

"Who's going to steal your ranch?" I asked gently.

"Don't matter who. Someone," he rasped. "And I ain't leavin'."

"You may die if you don't go to the hospital, Ed."

"Then that's what I'll do."

"You mean go to the hospital?"

"No, I mean I'll die."

Just then I felt a tug on my shirtsleeve. It was Lydia. She drew me aside and whispered, "If Ed says he ain't a-goin', he ain't a-goin'. And don't be thinkin' of tryin' to make him go, neither. He may be 86 years old and blind as a Pharisee, but he's still got that ol' Colt 44 six-gun in that coat hangin' on his chair."

"Thanks for the advice," I said. "But don't worry. I don't ever try to make anyone do anything they don't want to. But I'll tell you, I'll do everything I can for him right here, but I don't think he has much of a chance if he stays."

I was, as you will see, wrong...

Bob Magnus and I stayed with Ed and Lydia that stormy night. I gave Ed a shot of penicillin every four hours, plus heart stimulants and diuretics. In between, we talked with the tireless Lydia, who insisted on staying up with us. I'd noticed a big scar on Ed's right groin, and the fact that his right testicle was missing. I asked her if he'd lost it in some kind of accident.

"No, Doc," she said matter-of-factly. "He lost that nut o' his back in 1898. We'd been married about a year, when Ed went and got himself a bad case of the mumps. Well sir, the mumps went right down him, the way they'll do, and they settled right in that nut o' his.

"We were livin' up on Lizard Head Pass at the time. I loaded him in the buckboard and took him down to the hospital in Telluride. He was in that hospital for 'bout four days and that nut just kept gettin' bigger and bigger. It was hurtin' him something awful. So Ed called me over to his bed. 'Ma'am,' he said—he always called me Ma'am—'Go hitch up the team and bring the buckboard around. Take me on home. I asked the doc here to cut the damn thing off, but he won't do it. Says it'll get better by itself. But it's plain to see, it just keeps gettin' bigger and bigger, and I'm afraid it's gonna burst. I'm gonna have you take care of it.'

"Now I was gettin' a might nervous. 'I don't know if I can do it, Ed,' I told him. 'Sure you can, Ma'am,' he said. 'You watched them cowboys castratin' them steers last year. Just do the same thing to me.'

"So that's what I did. We went home. Ed sharpened up his hunting knife, and then fired it up till it was redhot. Said it wouldn't bleed so bad if it got seared while it got cut. Then he took the Bible to bite on whilst I cut off that bad nut."

Thank God the story was over! I was sweating just thinking about someone performing a semi-castration on me with a hot knife and no anesthetic. I looked over at Magnus; his face was a sickly shade of green. Lydia smiled beautifully at us.

Well, ol' Ed went through a crisis that stormy night, but by morning he had made it through and was beginning to make a miraculous recovery. Bob Magnus and I checked on him every day for the next three weeks. He was completely well in a month. Like I said, I was wrong, thank God.

There was just one more installment in Ed's medicinal story. On one of my visits, he beckoned me to come close, and confided in me in a low voice: "Doc, Ma'am told me you talked 'bout me losin' that right nut o' mine. I guess my left kept workin' purty good, cause she gave me a baby boy two years after that. And it kept workin' good till about a year ago. Maybe it's 'cause I'm gettin' a might old, but me and Ma'am can only make love about twicet a week now. It's beginnin' to worry both of us. There was a time when twicet a day weren't enough. And now, here we are, down to twicet a week," he said mournfully.

"Ma'am works with me most every night, but I can only get hard enough to please her a coupla times a week," he went on. "I'm beginnin' to feel like I'm only half a man, Doc. Next time you come down, can you bring one of them testyrone shots I heard about?" I did what he asked, and, placebo effect or not, it sure seemed to work. He asked me to keep giving him the "testyrone" every time I came. By that time I was only seeing Ed every few days, and if I was a day or two late Ed would be beside himself with concern. "Thank God ya got here in time, Doc! We were afraid that testyrone was a-fixin' to wear off!"

Several years later, Ma'am broke her hip. It never really healed right. She never was able to walk again, and went into a nursing home. Ed tried to stay home by himself; he hired a nephew to stay with him, but it didn't work out. Within two days he was out of his head, screaming, hallucinating, threatening to kill himself and everyone else within range of his Colt 44. He just missed Ma'am so much he couldn't take it. We had to arrange for him to stay in the nursing home with Ma'am. As soon as he heard her voice, he was fine. Love conquers, and cures, all. They had a rule back then that males and females had to stay on different floors in nursing homes, even if they were married. Soon after he arrived the nurse on duty called me. She was upset, to put it mildly: "I can't keep Mr. Impson out of Mrs. Impson's room! And he's always trying to have—" I could actually hear her blushing over the phone—"intercourse with her!"

I tried not to chuckle as I replied. "Mr. and Mrs. Impson have been making love at least twice a week for over 70 years. I guess they're just luckier than you or me and our spouses. Now listen, I want you to tell your administrators to change the rules for the Impsons. Put them in the same room, and keep the door closed if real lovemaking bothers you. And tell your administrator if he doesn't do it, I'm gonna come down there and when I'm through with him he won't be able to have intercourse for six months. If that doesn't work, I'll move the Impsons to another nursing home, and he can kiss that income goodbye."

The nurse laughed nervously. "You're joking, aren't you, Doctor?"

"Of course," I said. "But don't tell him that, okay?"

"Okay. There's just one more thing."

"What's that?"

"Well, Mr. Impson has some kind of padded belt on. He won't let us take it off him, even when he's bathing. Could you talk to him about it?"

"I'll be down there in a couple of days, and I'll see what I can do."

The padded belt turned out to be Ed's money belt; and it was stuffed with $32,000 in medium to large bills. I persuaded him to let an attorney put it in the bank for him.

Ed then told me he had another $120,000 in cash, that he and Ma'am had buried in coffee cans under their old house: this was "the ranch" he was so worried about somebody stealing. I immediately called the San Miguel County Sheriff, and he went right out to the Impsons' place. Ed's fears proved correct. The floor was ripped up, and there were empty coffee cans strewn around...

Ed died at the ripe old age of 101; Lydia died a few weeks before he did, 96 years old.

Ed probably could have lived forever, but when his wife died, he decided to die, too. Yep, decided: he just stopped eating and drinking, there was nothing anyone could or should have done about it, till he quietly passed away.

And the thieves who stole their money? Well, in an intimate place like San Miguel County, we had a pretty good idea who they were. And frontier justice had a way of making things right in those days. But that's a story for another time—when the Statute of Limitations on certain "crimes" runs out...

—*MOUNTAIN GAZETTE, #79*

Lady with a Baby

BY
GILBERT PRESTON

Dear Editor:

That devil's advocate, fire-and-brimstone-type article you suggested I do for the *Gazette*, re: Where Are We All Going and What Does It All Mean, Environmentwise, is a hell of a lot tougher to do than first appearances and delightful telephone consultations might suggest. And then, there was the unforeseen question of The Lady with a Baby, which we never got to, but which, as these things go, finally pulled me down. As you recall, I was hot to truck on down to Livingston, Montana, to listen at the feet of important environmental gurus, and Ed Abbey, for latest hot goodies on how to save wilderness, trees, good scenery, et al., here in the Big Sky Country, and to use same, and notes therefrom, aforementioned telephone talk with yourself, dear editor, as platform for a most extraordinary and precise, not to say surgical, analysis of implications of chosen environmental positions *vis á vis* central government intervention in the life of ordinary guys, energy development and subsequent economic rates of growth in western industrial societies, sprinkled liberally and subtly with kernels of nutritious wisdom on the meaning of life as gathered in the waiting room of one of the smallest medical practices in the Intermountain West, etc., etc.

I was all prepared, you recall, to discuss, along with whatever happened or did not happen in Livingston, the article by William Tucker in the December issue of *Harper's*, wherein William Tucker, a freelance reporter in New York, documents the destructive role the environmental movement played in preventing the construction of a hydroelectric power plant at Storm King Mountain on the Hudson River from 1962 to the present; yea and verily, a destructive role carried out in the name of nothing more sacred than the pristine view of the river enjoyed by several estates of the landed gentry across the river and into the trees of Duchess County, a view enjoyed also by wandering folksingers and their merry bands of otherwise stoned followers, which view, if truth be told, meant exactly jack shit to those of the megalopolis who would not get to see said view, but who needed, desperately as it turned out, the power the plant might have provided, when in 1965, and again in 1977, the blackouts, of which we are now all informed, occurred, and which did little for the urban scenery themselves. Well, I was going to say, is that the kind of shit we truly wanted to stand behind? Hell no! I would rush to add, answering my own rhetoric, as writers (sic) so often take the liberty of doing, it is, if nothing else, nonsensical, and that at least; but moreso, destructive to the entire notion of environmentally aware and concerned citizens standing up for what they believe is right in the face of Philistines of the running dog establishment tearing it all down.

I was going to write, see, a little dialogue, imaginary of course, wherein conservationist ecology groups are faced off by overfed, cigar-chomping, Cadillac-humping, wife-beating—you should pardon the expression—industrialists, giving the industrialists the best part of the course, so as to make the point (here, borrowing the techniques of pulp fiction): "Oh Yeah," the running dog says, spitting a little plug of cigar end out onto the polished conference table, "Youse are the dumb fuckers what blocked the power plant upstate because you don't want poor people to have enough electricity to give birth by, youse are practically genocidal, you know dat?" Big guffaw, "and now youse are tryin' to stop a stinkin' little strip mine what ain't gonna do nutthin' to nobody but ship a little piece of Wyoming and tiny slice of North Dakota off to Italy! Shame on youse." And like that. Oh what a piece it would've been! Dear editor, a piece among pieces!

After I opened with a summary of the Tucker article, I was going to sail into, and perhaps this might have been seen as something of a polemical backwater, unnecessary to the mainstream of profound thought I had planned for the piece, nevertheless, I had it all planned then to launch into the tiniest bit of liberalism, wherein I would attack the central government, and by association, those who do things to hand over more power to the aforementioned entity—asking my readers whether or not they truly wanted to hand over more of our precious, and already nearly hemorrhagic personal liberties to the same bastards who gave us Amtrack, and the Post Office, and Watergate (thank God for Watergate), etc., etc. Listen, I was going to write that more and more the government in this society grows like a junkie's monkey, more and more we can't live without it, and we can't peel it off our necks, and then, I would have said, the only way to get free is to shoot for it, Cold Turkey, sweat it out the hard way before the monkey comes to own us.

Shit, I'd trade off a view of Wyoming sagebrush for 200 bureaucrats out of work any day—don't you see I was going to write that these bastards have jumped onto what to some of us are spiritual concerns, and are enlarging their empire by seeming to stand behind us—here, dear editor, I was going to paint yet another of my humble little word pictures—the scene would have been the penthouse office of the beloved minister of Health, Education and Warfare, Environmental Hygiene and Right Thinking—the scene is set—this guy is five-eight and weighs 240 pounds; he hires a filipino male prostitute to cook breakfast, lunch, and supper for him, little goodies like baked lizard pineal gland, sauteed camel lungs, fine wines, and things like that, you get the picture, and here he reads this article in *Mountain Gazette* see, they read these things to get new ideas, somehow they grasp ideas as important, they don't quite know why, but empirically have intuited that the guy with the right ideas gets to hire Filipino faggots to cook lunch and play with their thighs, see, and here's this lovely little piece, unspeakably sincere and from the heart: "Entelechy," see, about a guy who spends more time than most of us have to make doodoo, running and eating bean sprouts and twigs and feeling so utterly fucking *clean* inside, he makes you want to run into the bathroom and Waterpick your pancreas or something; but when the minister reads this idea piece, he takes it as his own see, and the very next week (this happened) this jowly bastard stands up in front of the Congress of the United States, a body not noted over the past 200 years for its personal devotion to Zen asceticism, and he tells these old farts that half the problem with Americans is that they eat too much, and don't get enough exercise, and the Congress bleats and cheers, and gives this guy five trillion more pesos, and another Filipino faggot.

These were the kinds of things I had in mind, dear editor, and merely a beginning at that, so stimulated was I by the power of your editorial direction and my own forgetfulness in missing my daily dose of lithium salts. I know, I know, the piece needed a strong title, and preferably quotes from several of the wise men we all know and love, and I was ready with that, believe me there. I had a title, THE WORLD HAS GONE STONE BLIND, and my first quote, you see, was going to be from Billy Shakespeare, a long-time favorite, something like:

There are more things in Heaven and Earth, Horatio, Than are dreamt of in your philosophy.

I was going to use that when I got to talking about Wendell Berry and his charming little notions about our failure of culture: See, I was reading one of those throwaway journals doctors' offices are littered with, and while I was thumbing through it, waiting for my personal physician to arrive with the lithium, or a Lady with a Baby, I'm no longer certain which, I found this pearl in amongst the recipes for what to do with all that protein left over at the end of surgical procedures and how the administrator of a big city hospital turned his losing budget all the way around by recycling meat that had theretofore been wasted. After that article was the little quote by Bill S.

So where was I? Oh yes, I was going to explain how all fired up I was after our extraordinary discussion that day, and about the Lady with a Baby, and why I never did actually arrive at Livingston, nor did I at that time hear Abbey's speech, but whose mellifluent baritone I have so often truly enjoyed in the old days, while sipping the wonderful Wild Turkey and smelling the good clean smell of one of Ed's piñon wood fires, and sharing too the company of the lovely Renée and her fine and simple cooking—which is not, of course, to any particular point, but which does put me in good company so that perhaps my bitter medicine might be swallowed somewhat more readily. Look, let's be honest dear editor, if one were to write something like—"Shit, me and J. Paul Jr. were sitting up late the other evening, sharing the fair jar of Dom Perignon and a half a box of pre-Castro Habana Habanas, when it occurred to us that too often those environmentalists are wide of the mark..." well you get the point don't you? I mean one doesn't run these rapids blind and chanting poems, and it was *my* dime that began this madness. After all, the notion of the most important mountaineering and environmental journal in North America, the one with the most perspicacious editorial staff and most brilliant writers (I mean, nobody reads *Audubon*, do they?—We all take it for the Eliot Porter pictures) running an article or two on the ways in which the environmental movement might be, in fact, destructive, might be, at given times, a major factor in the Balkanization of our society, to the ultimate demise thereof, might be antilibertarian, and, even, antienvironmental in any final analysis; I mean these old notions on your part were nothing if not brilliant, and although the dime was mine, I stood ready, as you remember, to fly down the interstate, eager despite the raging storms and ice and snow on the roads, yea eager to follow your merest whim, notebook aflutter, cover the Montana Wilderness Association meeting, interview Abbey after a fashion, and to use all of that as a platform as it were, for the aforementioned antimetaphysical detritus. I was truly ready to *do* it! I mean, these questions had been burning in my mind for ages and the roof of my mouth was beginning to ache with the weight of ideas bursting through the bottom of my frontal lobes. My hapless and shifty partner, an excommunicated Apache shaman, would cover my meager practice (after all, who would patronize a medical man who sits and reads the throwaway journals in his own waiting room?). But alas and alark, such was not to be my karma. Enter, stage center, *sotto voce*, The Lady with a Baby.

    Here I must digress yet again, dear editor, for what is known as shallow background—when I was but a seed in the Big Apple's core (*???*), and a punk of one at that, it was usual and customary practice to fight one's way onto the crowded subway train by yelling at the top of one's lungs, *LADY WITH A BABY, LADY WITH A BABY, HEY, GET THE HELL OUTA' THE WAY!*—somehow the idea of a frail and helpless mother caught in that seething human maelstrom folded ordinary people at the knees, and as they folded, one ripped off their seats. Something about a lady with a baby at her breast you see, was like CIA steam to a sealed envelope, and here, therefore, I most humbly invoke that selfsame notion in fondest hopes of arousing your editorial sympathy. As I say, after our extraordinary dialogue concerning where some of our environmentalist postures might be forcing us, and your own careful and brilliant discussion of the Tucker article, and oh so many other items, I was, to say the least, all fired up! So many brilliant notions flashed through my mind, stimulated, no doubt, by your own scintillating observations on "progress" and what in the hell it all means. How can I forget you so clearly stating, "What the fuck is progress, and don't we have enough of it?" I mean, you said, "Isn't that the entire point of the so-called ecology movement, the metaphorical, and, if necessary at times, literal act of pouring sticky Kayro syrup down the intake valves of a boring and decadent social engine, a thermite birthday cake, every so often *n'est pas, mon ami?*" isn't that what you said, playing devil's devil's advocate, as it were? I admit that at that moment I was truly at a loss for words stunned in fact in the face of such volcanic thought, and could utter nothing but weak and defensive, yes but uhmmm, while at the same time quickly flipping through the well-worn pages of my always handy *College Pocket Thesaurus* for the *mot juste* in reply—a reply you left me no time to offer, busy as you are, announcing, just as the words slipped to the tip of my tongue, that important people were anxiously awaiting their turn to kiss your ring, and slamming the receiver in my ear.

What I was about to say when you so rudely cut me off, no doubt to meet your *consigliere* and *capo de capos*, and what I would have in fact written did not the Lady with a Baby command so obviously just then, stage center, was this (or something quite close to it, depending, among other factors, whether or not I took, and how close it was to my next prescribed dose of lithium salts): There seems to live in some of us this unquenchable need to find order within chaos, the lumpers as opposed to the splitters, if you will, and from time to time this need swells to such immense and unhealthy proportions that great masses of human beings, like suicidal lemmings, are driven to murder their collective birthrights, their freedom, their liberty, their right to sautee camel lungs as they see fit, for the illusory security order affords. Simultaneously and always there are those among us with an inveterate need to destroy, gratuitously, whatever order may from time to time appear within our universe. So that, we feed upon ourselves, the one unable to exist in absence of the other, like warring tribes of the New Guinea highlands, or like the pulsations of the universe itself, expanding within potential cosmic cleft until, energy expended, it collapses down and down to the infinite density of the universe as black hole, only to bigbang and lightspeed off again and again recreating creation; order from chaos, chaos from order, and always, these silly little atomic accidents, also known as us, struggling so comically, yet desperately to prevent it all from happening again as if such a thing were at all in our power. On such yearning notions are romantic philosophies armatured. The paradox of life on this planet is, I think, that we must use it in order to survive as men—we have as much right to be here as the trees or the stars, the saying goes, and by force of using this world, we diminish it, and so ourselves, because, whatever silly little notions, ideological systems or mantras we choose to chant so as to reduce chaos, the fact remains that we will be here until we are gone, until there is no more, and no matter what we do, that won't change—we'll use it up and then we'll die, and maybe some quark on Alpha Centuri will worry about it then in his/her? own way. And if all of that is true, and I see it all that way, then the only way to run it is hands off, oh Ministers of Right Thinking and Environmental Hygiene. But, one of the many prices mankind has paid for the gift of upright posture, along with hemorrhoids, and sciatica, is politics and politicians, and if we are destined to suffer that ignominious abomination (I found my thesaurus), then I suppose we must also suffer, somewhat more gently, the Wendell Berrys and Ed Abbeys among us.

And that was where we might have stood at the end of my dime, if you had not been drawn off to discuss, with your associates, your until now well-hidden plans for a Frisbee condominium and massage parlor on the west approach to the McKinley summit. Despite my rude treatment at the hands of so rich and powerful an editor such as yourself, I was, at the close of our conversation, feeling truly wonderful. It had been perhaps 12 hours since my last dose of lithium salts which so cruelly tie one to earthly concerns, and I was feeling nothing if not high. Then the Lady with a Baby blew it all to hell. I mean, the lyric poet with whom I share my simple pallet, and who is kind enough to boil my tea and green twigs each morning before I walk the stony road down the hill to my office, there to heal the halt and lame, wipe away the pain and deliver up of child (MasterCharge, Medicare, Medicaid OK) was off the ground for hours, and no Maharishi to been seen! "Oh what good news" she cried into the cold Montana night. "An assignment from a real editor!" Then she pedaled down to the K-Mart and bought me one of those little lined steno pads you always see journalists carrying, and patched moth holes in my old ivy league trench coat. I would wear the trench coat over my Levis, Woolrich shirt and Galibier Peuterey's, so everyone could easily identify the camps to which I belonged. I don't smoke, or she certainly would have purchased a pack of Camels for me.

But when the lovely and loving Antimony arrived back at our little cabin in the snow with the aforementioned journalistic accoutrements, it was already too late. The Lady with a Baby had found her way down out of the mountains, holding the tail of her pinto pony, and I was gone to the hospital. It seems as if the Lady's bag of water burst while she was feeding hay to her stock, and by the time she reached the hospital, the labor pains and cramps were rushing her like one of those insane Yankee fans, really grabbing hold of her and pulling her all to pieces with pain; then abruptly they let go, just like that!, and that, dear editor, is where the assignment fell all to hell. One minute those pains were kicking the living shit out of her, and I was all primed to make the catch—green pajamas, Ben Casey cap and mask—clip the kid's cord, and take off for that important meeting of the country's top environmental people, take notes, interview Abbey, rush back down the interstate (as you recall, there was no expense account for this, and a poor shaman such as myself can hardly afford the rich life of Livingston on his own) and write one hell of an article so brilliant, that one half of the readership would go stone blind, and the rest would send in subscriptions for life. Then the goddamn labor pains stopped, and the whole business went to hell in a handbasket as they say. Babies are like that, they control the entire affair, as you know. Labor begins when the kid is good and goddamned ready, and not a microsecond sooner. The kid's little adrenal glands squirt some kind of special speed into the mother's bloodstream, hits momma's pituitary like a hot poker, makes the old thing jump and shout and squirt some of its own stuff down the womb via the bloodstream, and the next thing you know, the whole damn thing is doing the Bosa Nova, later you get a baby. But sometimes, heh heh heh, nature goes awry! Which is where doctors come in, and that's how we got Porsches, and thoroughbred horses, and tax shelters in Costa Rica, etc., etc.

Here you see the subtle interdependency of human and natural institutions. Now I can see it, maybe that's where we made our first mistake, huh, inventing doctors? I mean ladies were having babies maybe two, three million years before Hippocrates invented Harvard Medical School, and things went pretty fine before this failure of culture, didn't they Wendell dear? (I think he is so cute! I just *love* his picture in the *New York Times Book Review*, the one they show in the advertisements for his little book of philosophies, the one where Ed says he, Berry, is our own Isaiah, and so forth; he has this great pompadour all blown apart by the wind and feeding hogs down there in Kentucky, and this boyish grin with the space between his two front teeth—I would buy almost anything from a guy with a face like that, except maybe, peanut butter.) As I was saying, all the while those millions of ladies were having babies all alone under the trees, out in the fields, or maybe for all I know somewhere in the back of the cave, thousands of 'em, maybe millions were dying—oh, nothing too serious, a little blue in the face now and then, a tad of heart failure, a kick, a whimper and dead.

And while they were all so many of the dying, I imagine the ecology (hats off here, cover your breast) was really in fine condition. The Carrying Capacity of The Biosphere was unsullied and indifferent, if anything, to all those little deaths; how else could it be?, and even more, I mean what is truly wonderful about all those deaths is that they were not in any way wasted you see, the amino acids, carbohydrates, vitamins and minerals were so nicely returned to the earth mother of us all—oh the Yin and Yang of it! The utter consistency it provides for our notion of Taoistic Thermodynamics, and all like that! Alas, something went awry—a little failure of culture, an inappropriate intervention on the part of the technocrats, and all those who failed to recognize the organic complexities of life on this planet, and who were naïve enough to suppose that Life could be made to speak in the one zero vocabulary of the digital computer and its managers. Oh Hippocrates, you fool! See what your meddling has wrought! Medicine, once invented, like agriculture (please forgive me) would simply not go away, and ever since, increasing numbers of the little bastards have plopped from the womb and then have had the audacity to take their own sweet time about returning to earth mother, paying back the cosmic soil as it were, a trifle later than the time before. And what's worse, ecologywise, all the while the little fuckers are pussyfooting it down toward death's door, they insist on making all these *demands* on our Limited Resources—which, to close the wheel of life, gives birth to *Mountain Gazette*, and my own feeble efforts. *Mea Culpa!* I now and for once truly understand that what it is I do so as to pay the rent and alimony money necessitates at times my contributing to the earth's overpopulation. And if all I do is write, I kill trees. Shit! There's no way out of this maze of guilt. (Maybe that is why, like Hunter Thompson, I stash a loaded .44 magnum under my desk as I write.)

So, here I stand, Lady with a Baby completely shut down, her husband looking at me like I sewed her knees together, and my first true assignment on the block. Well, I figure there's nothing to do but wait, these things are bound to work out in their own sweet time, as they have for eons and eons I tell myself. The Montana Wilderness Association meeting is two days long, and at least Abbey doesn't speak until the second day. Well, dear editor, it was not meant to be. The baby never did come out until Saturday morning, like waiting for an elk on opening day, huh? And when the blessed event finally arrived, it was blue everywhere, and the lungs were funky and slack. After that things went further to hell. The little fucker stopped breathing, and then his heart quit, and somehow, what with his mother and father watching and all, I just couldn't bring myself to let him return to the earth just then. I know I was wrong, ecologywise, I know this poor planet has suffered enough, resources gone to pot and what not, but to tell you the Godshonest truth, at that precise moment there was nothing I would do but breathe in the little guy's mouth, and pump his tiny heart, and dear editor, once one deigns to begin one of these affairs, there's no giving up the ghost unless and until the Holy Ghost beats you to the punch. That chewed up Saturday, quicker, really, than my dog gulps bacon. By the time the helicopter from the University of Utah flew over the mountains and into Montana to carry babe back to the intensive care unit in Salt Lake, my entire assignment was shot to shit. Just like they say it happened in Nam, you call for a medevac, and the bastard gets here tomorrow when you were hoping for yesterday, or the day before that. And that is what truly happened. The final ignominy is we used all that aviation fuel in the deal, and now it looks like at least three score and ten before the kid pays back mutha. I don't know what to think. What philosophy covers all this? Albert Schweitzer said philosophy resembles a pretty apartment where pets are not allowed. All I know is I sure don't know when I am ever going to capture an assignment now: I know I am unreliable and cannot be counted upon to keep appointments and honor a deadline, but I must humbly beg your leave, my Liege, and pray for one more opportunity to prove my usefulness to your most Worthy kingdom. May God have mercy on us all...

—MOUNTAIN GAZETTE, #68

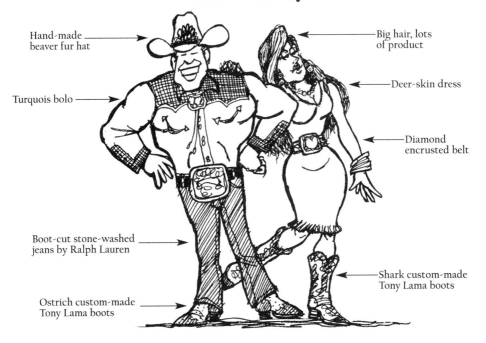

On the Frontier

BY
STEVE WISHART

"Does ropin' really beat dopin'? I wonder."
Luckily Shady's mother decided not to make the trip.
I had been trying to get someone to go to Cheyenne Frontier Days with me for weeks. And while everybody flushed enthusiastic at first, they all tended to have second thoughts about going up there for six days of ritual red-necking in the end. While there were more than enough wild stories about the rodeo, they usually ended with someone in jail or on the wrong side of some cowboy's idea of what constitutes good behavior.

I was beginning to grasp at straws for companionship in what my paranoid mind had escalated into a fearful, sweaty excursion into the World Out There.

I knew those people were different from me, and I was worried. But Shady Lane's mother? The idea just didn't set right, especially after he'd told me that she hadn't said anything all winter.

It was while he went around telling whoever would listen that I had refused to drive his mother to Cheyenne that it first occurred to me that Shady's mother was...ah, no longer among the living. All along he had wanted me to escort a small box of ashes back to Friendly Soil, the family's ancestral home, Cheyenne.

But sober morning thought convinced Shady that a beat-up pickup, loaded down with the refuse of my manner of traveling, was not the suitable vehicle for such a delicate mission, and I was spared the problem of having to explain just why I was toting around somebody else's mother's ashes. I mean, what if the memorial park had decided not to take them? ("Listen, buddy, no telling who's in that box. And besides, you don't look like you belong in Cheyenne, anyway. WHO ARE YOU?")

No, I was going to have to do this one alone.

Cheyenne Frontier Days—The Big Daddy of Them All. Packing them in since 1897. The best riders on the best animals at the best rodeo, competing for huge amounts of prize money. And all of this was about to be covered by the least experienced Rodeo Reporter in the western United States.

My past trips into the world of rodeo had been limited to the Hotel Jerome Bar in Aspen, listening to one of my fellow bartenders run on about his roping horse, Jack. But I knew the feeling ran deep, because he was housing his family in a log cabin, and his idea of a vacation was an all-night drive to Muckstick, Nebraska, for three days of roping school...with Jack.

New Boy in the Press Room. After negotiating myself up to Cheyenne and into the only available motel room—and judging from the price of my accommodations, the management was only too aware of this fact—I called Press Central.

Oh yes, they had received the letter of introduction but, uh, blanket press passes were very tight and if I would see them tomorrow, they would see what they could do for me.

I had been told very seriously before I left that if I wanted to "get along in Cheyenne" I would need "a pair of boots and a good straw." I had the boots, holdovers from my cowboy days in college in New York City, but decided my old Lyndon Baines Johnson white felt would have to do for a head-piece.

I could have used the straw.

I could also have used one of those shirts with the pearly buttons. And a four-pound, silver bucking-bronc belt buckle. And something other than my octagonal purple sunglasses.

Understand, it wasn't outright hostility that greeted me when I showed up in the press room for my credentials. Rather, it was a certain wariness. A silly reaction on their part, I thought. Hell, I didn't go up there to savage their cowboys. On the contrary, I was afraid those cowboys were going to savage me.

After a little fencing with good ol' boy Tracy, my main contact at the press office, it was decided that I would be given a daily pass to the VIP press box. This cat-bird seat was tucked under the eaves of the grandstand across from the chutes. While it lacked the immediacy of the old press box adjacent to the chutes, it did provide the best all-around view of the spectacle, and that's what I was seeking—the entire, majestic sweep.

(An aside on the press relations: toward the end of the week, a friend, attractively female, showed up in Cheyenne and when she accompanied me to the press room, my stock rose sharply. The next morning, when I showed up with two lovely ladies—the second one our photographer—it absolutely skyrocketed. One of the previously unavailable magic blue "blanket" press buttons appeared for the photographer, and I finally was admitted to the old press box next to the chutes. As in most places, tits and ass are still coin of the realm in Cheyenne.)

Back to Day One. With press pass in hand, I made my way up and up into the grandstand, stopping only for beer and a little grease for sustenance...up past the Burns boys and out onto the catwalk toward the press box. Suddenly it dawned on me why I had long ago given up wearing cowboy boots (except when I visited the West End Bar): Cowboy boots hurt when you walk. Which makes sense, because they were designed for wearing astride a horse. But why were all these other people wearing them? Wobbling along, I wondered if they all felt as conspicuous as I did.

I finally made it to an empty seat, set out my afternoon's food and beer, stacked up a couple of chairs to lounge on and got ready to enjoy the show.

On the first day, the show began, as it would every day, with the grand opening ceremony—enter all the dandies of the "Daddy of Them All." Below me, astride good-looking horses, I saw about forty ladies and as many different shades of matching pastel hats, coats, pants, boots and sashes—the latter proclaiming things like "Miss Routt County Whoopie Days." Lots of flags and every name in the parade announced. And bringing up the rear, the Ogalala Sioux of the Teton-Dakota Nation. The only walkers.

An interesting show, one time out, but after that first day, I took to sitting in one of the beer tents on the Midway, downing a brew and reading the paper until I heard the call for contestants' events.

I Have to Admit that I Developed a Taste for Certain Parts of the Rodeo. This, in spite of some of the cruelty and violence. Or, maybe, because of it—but we don't have to get into that...do we, Herr Doctor Krafft-Ebing?

Take the brahma bull event. Actually, bull-riding doesn't sound too hard to take at first: just some guy trying to stay on top of a very large animal for eight seconds. But how angry this large animal can get quickly becomes apparent. And an angry bull rarely is satisfied merely to rid himself of his baggage. He often wants to put his rider on his horns, as well.

Since the bulls seem pretty docile in their pens, it's hard to figure the hostility one of these 1,500-pound beasts will exhibit in the arena toward the mere fly of a man on its back...unless you know about the flank strap, that is. (In bull- and bronc-riding events, a leather strap is noosed around the lower stomach of the animal and cinched sharply just as rider and mount exit from the chute. A veritable pinching of the balls. It makes all the animosity understandable.)

Following the brahma-bull event, I got my first taste of steer-roping—and I found it a bit gut-wrenching. It goes something like this: A steer is released from a pen at one end of the arena and given a 30-foot lead before a lone rider chases him down, puts a rope cleanly around both horns and then rides by at full gallop. The animal usually is lifted entirely off its feet as the rope goes taut between horse and steer.

Seeing that first steer ripped off its feet sent a chill up my spine. (Anyone who has ever suffered a serious neck injury probably would recoil.) The chill returned a while later when one steer got up with its head cranked crookedly toward the sky, trying to run away, yet veering again and again into the arena wall.

But after the initial shock, the cruelty doesn't seem to bother the sensitivities much. Along with that one broken neck, I saw perhaps four animals break a leg. I also saw lots of flanks bloodied by spurs. But it was all tolerable in context: Rodeo, after all, is a direct link to the working cowboy, who had behind him the whole tradition of honest labor. The cruelty of making the whole thing a sport is thus softened. It's not at all like bear-baiting or cock-fighting—or pro football. There are, in fact, several events in which the animals have a fair chance of winning. About half the cowboys in the bucking events gimped back to the chutes, and several had to be helped off the ground.

After steer-roping came steer-wrestling—definitely the brass-balls event. If you've never seen it, picture a large beast flying out of a chute, followed seconds later by a rider armed only with his bare hands. The cowboy runs down the steer, leaps off his mount onto the animal's horns, stops the charging brute, and if all goes well, throws him on his back. It is a little life-and-death tableau that lasts perhaps 25 seconds and was repeated before my eyes about 20 times every afternoon. No doubt because of my own days as a wrestler, I was completely caught up in this event. I felt like Norman Mailer vicariously enjoying a boxing match. No—I felt like I was *out there*. If the steer put up a solid resistance, I was fighting him, too, knocking over chairs in the press box, digging my heels into the floor. On a good toss, my head would snap to the scoreboard at one with the cowboy's. Yeah! Together we were delighting in a little one-to-one combat of man against beast.

Except for steer-wrestling, my enthusiasm for most of the events was pretty general. I enjoyed watching them, but I never got so deeply into any of them that I followed the standings. It's the kind of pick-and-choose entertainment where you can appreciate a good performance here and then move on to something interesting over there. A rodeo, in fact, is just like a circus.

The Analogy of The Circus Runs Throughout. All the events are structured so that there is always something going on. Your attention is constantly courted. If things start to slow down, on come the clowns or in march the Indians.

Some of the make-up events were classics. Like the Dinner-Bell Derby. I prefer to call it the Tit Race. Visualize a half-dozen mares with their unweaned colts: mother and child are separated from each other and moved to opposite ends of the grandstand. The colts are released. Well...you get the picture. The winner, of course, is the first colt to hit the right tit. I'm not sure what bothered me about this particular event. Was it demeaning to the horses? Latently sexist? In the end, it just seemed stupid.

After the first day, I became acutely aware of the repetitions. All the jokes were old and the gags the same: "She must have been a welder's daughter...she had a set of lean legs..."

But, happily, there were also some rare moments of unplanned humor—like the afternoon Monte Montana was out on the track, doing his grand finale, roping five riders at once with a great loop. He's spread the loop—some 20 feet in diameter—across the track and was earnestly heading toward the riders when he was brought up short by a track official caught up in the other end of the rope. Just as the five riders thundered by. Monte was yanked from his saddle and did a flip over his end of the loop while the sport-coated official at the other end went flying. It was pure Keystone. The move won the biggest hand of the day.

The Indians, On the Other Hand, Were Pretty Depressing. I know they've performed at rodeos for years. They're a stock act, brought in to evoke memories of the cowboys and Indians of the Old West. But it all seems so wrong, especially in this day. At Cheyenne, they trooped out to do a few steps in the afternoons, walked in the parades, danced at night and starred in hundreds of snapshots. They were the trained bears of the rodeo.

Women aren't treated much differently. I suppose it should come as no great surprise that the Great Awakening hasn't hit Cheyenne with full force just yet. (N.O.W. was represented with a float in the parade, but carried a decidedly mild message about equal rights under law.)

No, the rodeo is a man's world, and it has enough tradition behind it to cause change to come slowly. So the rodeo woman finds her role pretty clearly staked out: barrel-racing, pony bronc-riding and, of course, beauty queening. This year's Miss Frontier Days, Beth Murry, had ridden in the pony bucking contest previously, but to continue as part of the rodeo world, she eventually had to make the shift from participant to object.

This is also reflected in the fashions in Cheyenne. That chic thinness that passes for beauty in all the right places hasn't made many inroads in Wyoming. Those men still like their stock corn-fed. Shapely, yes...but substantial. Most of the older women were on the large side and relatively unself-conscious about how they looked. In fact, it seemed the men were much more into their image than the women: It wasn't uncommon to find a slim gent with his hat cocked just-so and his pants razor-creased, sitting next to a woman in curlers and pedal pushers. I didn't see too many halter tops in Cheyenne, either. Rather, those pointy bras that aim a little too high and sensible shoes were the style. (A substantially built cowgirl at full gallop probably would look a little sloppy without a bra.) But I digress.

Every afternoon the rodeo ended with a little bit of insanity sponsored by *The Denver Post*, called the Wild Horse race. Ten teams of three men each were given a haltered, unbroken horse with instructions to saddle it, get a man in the saddle and then get both man and animal around the track. An event for the crazies. It wasn't necessarily a team's own horse that caused the greatest problems, however...it was those nine others. Usually, just as one team succeeded in quieting its steed and stood numbly arranging the saddle, another horse inevitably careened toward them, dragging with it some fool with visions of money impressed too strongly in his head to let go. That kicked off a reaction in the first horse, soon sending cowboys flying in all directions. A race by fiat. Aside from the delicious confusion, I liked this event because the horses were decidedly in control.

But It Does Make You Wonder Why They Do It All. Why were those people out there competing in the first place? There is some money involved, to be sure. But according to the Rodeo Cowboys Association, only 100 of the professionals win a pile of it in a season.

At Cheyenne, more than 1,000 contestants paid $75 or $100 apiece to enter each event. With a misdirected flick of the wrist they could throw away that money in mere seconds in an event like the calf-roping contest. Of course, if they hit, they stood to make a lot of money quickly. And we all love a gamble. But the risks involved are very real, and to most of us, they would hardly seem worth the relatively small return.

I don't know why they do it. Maybe there are mystical reasons. A group of outlaws on the fringe, living a free life that always has been associated with the best in the American Character. With the mechanization of ranching today, our stereotyped cowboy is fast becoming obsolete. But he can live on at the rodeo.

As a nation we seem to need a stiff-lipped, independent, macho-type as a part of our collective psyche. Look at all those cigarettes Marlboro sells. And why not? It is a romantic life, a cut above the one most of us lead. I need those fantasies myself to help get me through the week. I can do without the lumps and bruises. But I'll buy a ticket and all the fantasies that come with it.

One Afternoon I Was Sitting in the Press Box Watching Quail Dobbs Shoot Wilbur Plaguer in the Ass with a Shotgun, just like he did every afternoon about that time, when an official-looking type came up and asked me if I was "working press." Paranoid, I reached into my pocket for the piece of paper that attested to that fact, when he said, "No, no. I mean, do you want to come to a press conference for Danny Davis?"

Seeing my duty clearly, I nodded, and after the rodeo, headed for the Rouge Room at the Little America Motel and Country Club to meet my first Cowboy. Or so I thought.

After we media people had settled around the table, a "cowboy"—in matching robin's-egg-blue shirt and pants—walked in, with 15 pounds of turquoise and silver around his neck. As the conference progressed, it became apparent that Danny Davis was from Massachusetts, by way of Nashville, was heavily into the Music Business and hadn't had cow shit on his shoes for quite a while. Danny Davis and his group, Nashville Brass, had played hundreds of fairs and rodeos. The music they played for the four big night shows at the arena was much like the people who attended them: straight, if a little country, easy to understand and essentially harmless. Doc Severinsen and the Now Generation Brass, The Loretta Lynn Show with the Coal Miners, Hank Thompson, Jr. and the Brazos Valley Boys—and Danny Davis and the Nashville Brass. Even the names didn't threaten—unlike Traffic, Chicago, War, and Blood, Sweat and Tears. But the music, if a little bland, was very professional, as it should have been. Because playing fairs and rodeos is Big Business.

Danny Davis is a case in point. The Nashville Brass has recorded 16 albums, each one of which has sold *at least* 100,000 records. Each one of those records is a catalogue seller, which means that the longer it's out, the more records it sells. And now about all the group does is make records and play at fairs.

Danny seems to have a sense of what he's doing. The Nashville Brass is solidly musical (of the nine musicians, five have doctorates and two have masters' degrees, and Davis himself is the only one without a degree); but it is also something of a novelty act. Because they do have a limited audience appeal they push hard. They make all those shows in their own plane, with a traveling group of 17. The plane has two bars and originally was designed for 54 people, so I guess being on the road isn't too uncomfortable.

Our press conference ended when the road manager, a former rodeo rider himself, came in to ask what "uniform" the band was going to be wearing. "Did we wear the wine last night…? Well, let's go with the blues." All the way to the bank.

Nighttime shows were held at the arena and lasted about four hours. As with the rodeos, they were orchestrated like a circus; any time the action slowed down, on came the clown with his dogs or Miss Something-or-Other, shooting down the track, flicking sharp salutes off the brim of her color-coordinated cowboy hat.

After the nightly chuckwagon races and some Roman riding by the sequin girls, on rolled the stage. This was a stage mounted on wheels and pulled onto the track like some musical ark. By this time, it was completely dark and the stage became a strange island of light and music out in the void. It was about my favorite part of the rodeo, sitting up in the press booth slowly sipping beer and looking out at those little figures below on the island while the spotlight beside me hissed and smoked…

But the music was bland and the audiences were, too. The real night spectacle wasn't out there. It was beyond the stage, in the distance, where Cheyenne pulsed and shimmered in the blackness, sucking you in. The real action was downtown.

It Goes Something Like This: "You know, I had just turned 14, and I was a rangy sort of kid to begin with, but nobody seemed to care. Frontier Days was really wide open then, with gambling and drinking right on the street. The only thing they cared about was whether you were able to reach that money up to the top of the bar. Well, I sauntered into the Mayflower and found a stool at the end of the bar. When the bartender finally got down to me, I was so nervous all I could think to order was gin and Seven-Up, which gives you an idea of how sophisticated I was. I didn't care, though, what I was drinking, never having done much of it before, and being so thrilled to be in there with all those real men, tossing them down just like the cowboys. Needless to say, I got a little carried away, and after about a dozen drinks, my critical judgment was somewhat impaired. You can imagine my surprise when I looked up and saw what appeared to be a band marching down the bar. Not only that, but it seemed to be a *very small* band.

"Now, there was this typical cowboy-type sitting next to me. We hadn't exchanged a word, but I figured I just had to find out about this decidedly midget band that was making its way toward me. I asked him if there was anything unusual about the band, and he turned around and took a look, turned back and said, 'Nope.' Getting a little panicky, I figured I better put it a little stronger: 'I mean, do they seem a little small to you?' He fixed me with a hard stare and *told* me, 'They seem perfectly normal to me.'

"For about a year after that I stayed away from gin. I figured midget bands marching down the bar was what happened when you got drunk."

That was the sort of experience I was looking for when I went to Cheyenne. I've always had a taste for the bizarre and what better place to look than in a town full of raving drunks, midgets and girls?

Curiously, the Mayflower was something of a letdown. It was big, for sure—three long rooms packed with sweating bodies, the beer lapping at my ankles—but it was also predictable. Like a big fraternity party where everybody gets loaded in a crowd and seethes around. The non-stop country music in one room was too loud, there were periodic fights, girls puking and guys passed out in the corners. OK...but all that does begin to pale a little when you get much beyond 25.

Also, things apparently were toned down this year because of a heavy rash of vandalism in previous years. The Mayflower crowd spilled out into the street and onto adjacent cars, but was pretty well beered into submission.

Sleaze, On The Other Hand, Was Certainly To Be Had. On my first night in town, I made a tentative circuit of the bars and was heading back to the motel when I passed a window that sported a sign imploring: "Girls Wanted." Rounding the corner and cutting through the lobby of the Palace Hotel, I came on a sight for sorely demented eyes.

Up on the bar of Sam's Place was a vision in pink chiffon and pale flesh—a vast expanse of pale flesh. A modified-bee-hived blonde was up there grinding away to "c and w," her head a few inches from the ceiling and her bare thighs a like distance from the brims of a line of straws. It wasn't a midget band, but it filled the bill.

I slid onto the first vacant bar stool and ordered up. The place was a study in contrasts: the bartender couldn't have been more than 20; she was dressed to the teeth and looked like nothing so much as a nervous high school queen—as opposed, say, to some very professionally-at-ease women at the other end of the bar. Aside from her lack of dancing ability, the dancer also had to contend with six pool tables in the other room. I remember thinking, "She's competing against high drama on the green felt and she's losing."

But not for long. She was dressed in a Barbie-Doll-type top (one wished it left more to the imagination) and a matching pair of panties. As old Playboy-type pin-ups were flashed on the opposite wall, she started to roll down the panties, letting a hint of pubic hair hang out, galvanizing every weathered eye in the place—and the battle was won.

Sam's Place also was the center of hustle in Cheyenne. There was some serious pool being played in the other room and a lot of obviously idle women were milling around. The place became my own "Outlaw Bar" in the town. I felt pretty comfortable there because it's the kind of outlawry I'm used to. While I was sitting there, two cowboys gave each other the soul/freak thumb-shake, a couple of black pimps paraded through in platform heels and pink denim pants-and-jacket outfits and a chorus of ladies in too much makeup successfully steered drunks in and out the door.

I Moved On to Another Kind of Nightlife. An evening far from the downtown section, at the Little America Motel and Country Club, convinced me that my stereotyped view of the cowboys and loners hovering on the fringes of respectability was, perhaps, bullshit. If anything, a large segment of the rodeo crowd is conservative and actively upward-mobile.

The two bars with the most authentically cowboy-rancher crowds were the Holiday Inn and Little America. We all know what the Holiday Inn is like; Little America is a giant place with hundreds of rooms all done in red brick, mock-Williamsburg Colonial. Inside the Manor House, looming in the middle of the compound, are the lobby, coffee shops, nightclub and a fancy restaurant, all reminiscent of a country club in an old Fred Astaire–Ginger Rogers movie.

One night my companions and I went to Little America for the Big Dinner. Aside from the obvious ego-boost of squiring two ladies around, I found my presence was apparently more acceptable in the company of such presentable women. We were ushered into a room literally saturated with red plush and crystal, those seemingly universal synonyms of taste that always give me a feeling of having stepped into a showroom for Castro Convertible furniture. It occurred to me that the red velvet booth we were sitting in probably folded down into a bed. But, hopefully, not during dinner.

The food was OK, just missing it all the way down the line. The only thing extravagant about it was the price. Still this was, we had been assured, the place for a fancy meal in Cheyenne, and the ranching crowd was out in force, particularly in the nightclub (the next room). Out in force, but not very lively. A group of the good burghers out for a good time. Hats and boots, but clean hats and shiny boots. The Marlboro man would have been appalled.

Wednesday Already? Time Begins to Fade and My Perceptions Blur. It's been too many afternoons of beer and foot-long hot dogs and being given the announcer's count of how many cars are in the parking lot from Nevada or Colorado or Maine. (He went through every state in the Union every day.) If I was going to take four more days of this, I needed some calculated diversion: the Carney.

I know all about carneys. They're all rip-offs. They always extract 50 cents from you and get you into some dumb tent to find—nothing. But I always go. I love it. Especially when you can say "screw it" and blow $10 going into every shuck show on the Midway, trying to guess just how you are going to be ripped off.

Even big Chloe. A very slick come-on, that lady. She had me on the ropes. Larger-than-life pictures depicting an Amazon telling her discoverers to "take me to your leader." A taped introduction explaining how she had been contracted for over $10,000 to appear with this fair, how she was "almost ten feet tall." How could a man of my diminutive stature (five feet, three inches) resist? The Ultimate Woman. I'd be able to gaze levelly, right into her navel.

Naturally, throughout the entire spiel, Chloe was always referred to in the present tense, a living, breathing monstrosity right in that tent, Rube. She was in there, all right. All ten feet of mummified bones. I came out whistling. Not a bad scam, I told the fat, bleary-faced ticket-seller. When he realized I wasn't going to demand my money back, a spark of recognition flashed into his one good eye. "Ah, so you can appreciate a good joke, my friend."

And so it went, right down the Midway. One folly after another. I saw the dog with skin like an elephant; the baby with two heads, four arms and three legs (mercifully, in a jar, since I don't think I could have handled that one alive); the chamber of tortures featuring "Slaves of Love." I saw everything, in fact, except one spook house. Nothing was going to get me in there. I had been in one of them at Excelsior Park in Deephaven, Minnesota—and I *knew* those things could scare you.

I really like carneys and the atmosphere around them. It's a speeded-up version of the old American Sell. All those people with angles trying to separate you from the long green. "Hey, Mountain Climber (a day pack over my shoulder), come over here. I want to talk to you." The Cosmic Guesser: "Just give me the first initial and I will guess the name of your girlfriend, your horse, your mother-in-law's birth sign…" That was pretty good, too. For $1 you could fool the Cosmic Guesser and have your choice of a rack of prizes, any of them easily worth 25 cents.

As The Week Drew to a Close, The Complexion of The Crowd Changed. There were just as many straight country folk as before, but there were more halter tops and hats made out of Budweiser cans. The number of dazed, vacant faces increased and drunks stumbled among the flow of humanity coming out of the stadium and heading for the parking lots, like salmon fighting upstream.

It was like a human demonstration of Brownian motion: in a given crowd, a certain number of people would collide with a predictable frequency.

After Big Chloe, I swore off carney women—until I saw Patti the drummer. Over in the beer hall, there were two bands alternating day and night. One was the weirdest country band I've ever seen: a giant black drummer who looked like he probably straightened auto bumpers with his bare hands when he wasn't drumming, but who sang in a soft, Charley Pride voice, backed by a stringy-haired freak on guitar and a raw-boned, slouch-hatted straight-out-of-the-bayou type on fiddle.

Patti played with the other group. The Speed Brothers, with little Patti up there pounding the skins. She was a vision behind those drums: a purple satin halter-top and matching mini-skirt, her straight black hair flowing down to meet her high, black vinyl boots. No matter that the two brothers were so wired that every time they announced a song, it came out unintelligibly through gritted teeth. Patti could play the drums, and she looked far out of place in the crowd of punk cowboys and beer-soddened college kids.

My palms began to itch about the fourth day. All that grease was beginning to seep out of my system through my hands. One too many corn dogs and more beer than I had consumed in the previous year. I don't normally drink beer, but when it's the only game in town...and the combination was beginning to get to me.

In fact I was reaching a very perverse point when I met the Crazy Whistle Man. He put everything into perspective for me. My first encounter with the Crazy Whistle Man was on Tuesday. After that, he became a daily ritual. I would hear a dissonant whistling and popping coming from the crowd below the stadium, and like a transfixed rat, followed the music until I could watch the act.

Usually surrounded by delighted kids and skeptical adults, a full-grown man was grotesquely distorting his face to create a whistling/singing sound with some kind of diaphragm in his mouth. He punctuated the whole thing by snapping his suspenders, popping his dunce cap up and down and grinning oddly. Certainly a man not of this world. His scam was selling those whistles at 25 cents a hit.

It's hard to approach your gods, but one morning, sitting in Your Father's Moustache, I did. After the crowds had cleared the tent and moved into the stands, the Crazy Whistle Man came in and sat down. He had a broad, Teutonic face and his neck had been ravaged by something like smallpox, but the overall image was of a warm person. For the last 17 years he had been the Crazy Whistle Man. It was his life.

"It offers me the freedom to do what I want. My family is here with me...my wife is over there in that hot dog stand and my son is selling ice cream."

With a sense of disappointment, he told me how he had just sent his other son back to school. "He doesn't like all this, he wants to be a surveyor." Hard for a man to figure, when the only life he has known since he was 14 is this one.

"I got tired of working for that." He gestured toward the Midway. "I wanted something of my own." The American Dream. But who else would want to be the Crazy Whistle Man except him?...and maybe me.

"It gets to you, let me tell you—after 17 years, the same song gets to you. But you come in, have a few brews and a little conversation, and go out and hit it again."

—*MOUNTAIN GAZETTE, #26*

FLETCHER MANLEY

Where the Trees Walk

BY
HARVEY MANNING

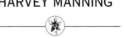

What my childhood would have been without Sax Rohmer there's no way of knowing. Had I read nothing but adventures of the Rover Boys and Tom Swift and Dave Porter at Yale, my experience of evil would have been limited to the sort of the naïve nastiness exemplified by the neighborhood bully. However, I strayed into pages of the *Saturday Evening Post* and got so sweaty about the unspeakable Dr. Fu Manchu that to this day I'm uneasy in a Chinese restaurant.

Rohmer's most devastating attack on my nerves, though, was a story centered on the Middle East. An outbreak of Arab unrest was suspected as being fomented by a sinister force, unknown. Simultaneously, British diplomats and generals and financiers were being murdered, one after another, under identical and baffling circumstances. In each case the victim, just before locking himself, alone, in a room whose windows were too small for a human to squeeze through, had been delivered a package a foot square on the end and three feet long. Hours or days later, when concerned family or staff battered down the door, they found the mutilated corpse, no sign of the killer, and the package—open and empty.

The hero-detective astutely discerned the packages were the key to the puzzle and patiently traced them to their common source—the Middle East. Instantly he realized his familiar antagonist, the fabled Old Man of the Mountains, was again plotting to conquer the world. He tracked the evil genius to his lair, or rather was lured there, whereupon the Old Man was delighted to explain how he had stolen babies and raised them on a diet of hashish, which stunted their growth and made them homicidal maniacs. He sent packages to leaders of the British Empire, with covering messages cryptically declaring the contents were to be revealed solely to the inner circle of the ruling class. Open the package and out leapt a drug-crazed dwarf waving a scimitar and snicker-snack and that was that. The assassin exited via the room's tiny window.

The bragging monster grew careless and the final hope of the Empire and the Western World saw a chance and cut out. The last scene, I remember vividly, is the hero running through the desert night pursued by a pack of little hashish fiends. He got away, of course, but then so did the Old Man of the Mountains.

One night while walking home along a country road I heard a noise behind me, looked back, and spotted a movement in roadside shadows. I kept walking—and heard footsteps. Another look back—a wiggling of bushes, as if someone had hastily jumped in. I broke into a run and now I had no leisure to glance over my shoulder, and no need. The hashish fiends were gaining and 100 yards from the house, I started yelling "MOTHER! MOTHER!" and she opened the door in the nick of time.

Country nights are dark, very dark, and a country boy cannot confine his travels to day. Often, far from safety when light failed, I ran for my life, the thud-thud-thud of little feet spurring me to great speeds. Only when I moved to the city to attend college did the fiends abandon the chase.

Comparable to Sax Rohmer as a molester of my childhood were preachers. Frequently I awoke screaming from a dream in which I was trapped in the basement of a house we lived in when I was six years old. If I could get to the steps leading upstairs, I was okay, but barring the way, crouched by the furnace that was his passage from Hell, was the grinning red Devil. I grew out of that dream and that terror, becoming convinced God was as good as everybody said and therefore He'd not tolerate a Devil in His world. Eventually, however, following the line of thought to the logical end, I realized elimination of the Devil left only one source of evil, and I was terrified by God. I decided Christianity was altogether too scary and gave it up for naturalism, which postulates a single order of reality, one not mechanically-materialistically simple and certainly not all good, but with everything out in the open, no spooks. The belief has since served me very well—the daylight and/or in a group.

Wilderness nights alone are something else.

I wish the hashish fiends were still with me on the dark trails. I'd welcome the thudding of their little feet. But they were scared off by the Others.

One spring I was the first hiker of the year up the East Fork Dosewallips River Trail. At Camp Marion, sitting by the fire, I heard voices in the forest and shone my flashlight down the trail, happy at the prospect of company. No hikers arrived. Who, then, was talking? Where? The "who" was unknowable but I discovered the "where." The voices were in the river. Listening intently, I caught occasional words, portions of phrases—and over and over again, my name. And hideous chuckles.

In the morning, plugging steps up snowfields from Dosemeadows to Lost Pass, I found the memory amusing. But not that night. I returned to Camp Marion, listening and watching through the long black hours. Nor the next night at Dose Forks, where I'd planned to camp before starting up the West Fork to Anderson Pass. With the first mention of my name I hoisted pack and fled to the road, pursued by the pounding Others, infinitely more dreadful than my old hashish fiends.

Not for some time did I again dare a wilderness night alone, not until I reflected that never, except beside loud rivers in black forests, had I been menaced. Thereafter, always camping in high meadows with a clear view in every direction, I enjoyed many a serene night.

Blithely, therefore, I left Cascade Pass late on a July afternoon and ascended Mixup Arm and the Cache Glacier, attaining shadowed Cache Col as sunset rays were streaming from behind huge Johannesburg, pinkening the narrow shelf holding Kool Aid Lake. In twilight I reached the lake, cooked supper on heat tabs, and—wood lacking for a campfire—crawled in the sack to read a while by flashlight.

So calm was the weather I hadn't rigged a tarp and lay in the open, vision unobstructed over the meadow bench. The night was eerily quiet. Downslope breezes were soft, not even stirring heather bells. The outlet creek, beyond a knoll, was a voiceless murmur, as was the far-below Middle Fork Cascade River. Starlight vaguely outlined moraines and the scattering of small alpine trees—wind-sculpted shrubs, really.

Peaceful, perfectly peaceful. Why wasn't I? Silence rang in ears. Eyes ran over print, fingers flipped pages, but no words penetrated brain. Scanning the surrounding tundra, I sensed some subtle change.

I noticed a shrub framed by the starlit glaciers of Formidable. Something was wrong with that tree. I focused on the book—and abruptly looked up. The tree was closer.

I checked the other direction and a tree there also had moved. I swept the circle continuously holding back the trees, which stayed put when I was watching, just as the hashish fiends used to dive off the road when I glanced over my shoulder.

Heartbeat by heartbeat I fought my lonely battle. Then, horrified, I saw Formidable's glaciers and cliffs were pressing steadily nearer from the south and Mixup and Johannesburg from the north. Hurryup and Magic, above the lake, were leaning outward, outward, into a virtual overhang. And the stars were lowering.

Mercifully, the eastern sky lightened, stars blinked out, peaks and trees retreated to daytime positions, and I slept, exhausted.

Never after did I feel secure alone in a mountain night. Boulders, sharp-edged and inert by day, became shifting, seething blobs. Creeklets giggled my name, ripples of tarns whispered threats. Bambi, soft-eyed in the sun, pawed turf under the moon, grinding his teeth.

My sanity may seem questionable, insisting as I do on repeatedly placing myself in mortal peril. It is, however, essential. I know, now, who the Others are, and understand that if I don't face Them in the wilderness, soon enough They'll crowd around my house, and some night when the family is away, will come right through the walls.

Last spring I deliberately walked into a trap, hiking from the Stehekin River up Bridge Creek. As night fell the nearest humans were miles distant and I was cut off from mankind in both directions by torrents difficult to ford by day, fatal to attempt in blind panic. Escape by running was impossible. I'd no choice but to stand my ground—

in the worse imaginable spot for a confrontation.

Dense forest and cloud-heavy sky merged into total blackness. The voices in snowmelt-thundering Bride Creek were the most savage I'd ever heard. In treetops moaned wind, and in bushes and duff rattled and splatted rain. The entire night was bellowing, groaning, babbling, cackling. And from the woods came a periodic thumping—when I shone my flashlight there I saw two balls of hellish green flame.

Yet I had not come to this test defenseless. My enormous fire thrust blackness back, back into forest, across the river, and above tree crowns toward swift-rolling clouds. And I'd heeded the teaching of Robert Burns:

"Wi' tippenny, we fear nae evil;
Wi' unsquabae, we'll face the Devil!"

Beside the roaring fire, rain splashing my face, I listened, tense, sipping lemonade. A power grew within me and I progressed from sipping to swilling. Bring on the hashish fiends! Bring on the damn Devil! I began answering the river shout for shout and by midnight was offering to rassle Them, one by one or in a bunch, I didn't give a shit. The later stages are hazy. I do remember that eventually we weren't so much yelling as laughing, and that in the end we agreed to continue the discussion another night.

Now, I've not the slightest desire to be the prophet of a new religion—or better say, of the most ancient of religions. Even a minimal organization, as implied by a "prophet," would ruin the whole thing. Each individual, if he chooses to go walking alone in wilderness—which is by no means compulsory—must find his own personally proper relationship with the Old Ones.

To none but the truly desperate do I suggest the Rites of Bridge Creek. Next morning the celebrant is sorry not to have been struck dead in the night. And over a period of time an excess of conversation with a river, whatever the benefits to the spirit, surely is injurious to the liver.

—MOUNTAIN GAZETTE, #19

MARC POKEMPNER

Breaking Free from the Human Potential Movement

BY
MIKE MOORE

I went the full week without getting touched, felt or laid.

So if you find that what I have to say about the Association of Humanistic Psychology and its 13th annual convention in Estes Park, Colorado, sounds at times a little grumpy...well, you already know the reason why.

I went there, let me admit right away, with a fistful of prejudgments, hellbent on writing a frolicsome satire that...who knows?...might even have found its way into Tom Wolfe's next anthology of "the new journalism." I went there thinking that this was just the right combination of mountains, people and madness for a *Mountain Gazette* article.

I also went there with a certain amount of fear in my heart. The prospect of spending five days...alone...with a gaggle of massage persons, transcendentalists, Rogerians (they don't even talk), Frommists, Zen Buddhists, nudists, sexual acrobats, Sufi dancers, acid heads, Jungians and existentialists, lesbians, pederasts and feminists, brainwave synchronizers, Rolfers, seers and magicians and god-knows-what new variations on your basic Elmer Gantry theme...well, I was afraid—afraid they might get me!

And maybe, in a way, they did.

WHO ARE THESE GUYS?

Somewhere, there has got to be a psychology that includes poetry, art and a movement toward social justice that will also help me understand myself—and nowhere does it exist, even beginningly—outside the Association of Humanistic Psychology.
—Rollo May

The first night of the convention I dropped into the Dunraven Inn, the only bar within easy range of our dry campus—the Y-Camp—for a bit of liquid inspiration before that night's activity, a pair of keynote addresses.

I sat down next to a grim-faced young man, who I later learned was a Vietnam veteran from western Kentucky. He had been living in Estes Park for the past six months, and he worked in a nearby meat-packing plant. He was drinking Wild Turkey, fast and straight. I've had a little barroom experience, enough to know that grim-faced young men who pop their Wild Turkey in that manner can easily become volatile, dangerous. So I sat silently, staring into my own drink, and tried to listen in on the conversation at the table behind me. It was an interesting one, about the Delancey Street Project in San Francisco.

I turned to find my stoolmate staring at the badge I had neglected to remove from my chest. "What the hell is the 'AP,' anyway?" he demanded. "At first I thought it was a bunch of A&P grocers, but Jesus, *these* people can't be grocers!" (Our $50 paper badges actually said "AHP," but there was a good reason to miss the "H." Some oh-so-clever graphic artist had reduced that letter to a single line running between the "A" and the "P." Just as with those silhouette drawings in which you see what you want to see—a vase or a pair of faces—so with the AHP logo. The "H" for *humanistic* could be there, or not there, depending upon how you looked at it. Several times during the week I was to ponder on the symbolic genius of that artist.)

"You're right," I told my new friend. "These people aren't grocers. They are humanistic psychologists."

"They're *what*? They look like a bunch of fairies to me," as he nodded in the direction of two middle-aged gentlemen, both attired in denim and soft, blousy flower shirts, one sporting a long, gray ponytail. "And all these women with their tits hanging out. You say they're *psychologists*?"

I tried to explain to him what was going on at the Y-Camp, who these people were, but I don't think I got the job done. My friend had his own idea of what a psychologist should look, act and dress like, and these people simply didn't fit his bill.

But it was a good enough question.

And it would have been an easy one to answer ten years ago when the movement—and that's what it is, a movement—first got under way. In the beginning, they were a band of rebels, led by Abraham Maslow, Carl Rogers, Rollo May and a few others. They set themselves off against the behaviorists and the mechanistic experimenters who then owned—and pretty much still do—the departments of psychology at virtually every major college and university in the country. They were existentialists, phenomenologists and humanists, opposed to the cold, rigid determinism of the stimulus-response, reward-punishment school of mind control. To a lesser extent, they were also in philosophical opposition to what they considered to be the overly structured methods of Freudian psychoanalysis.

The Esalen Institute in Big Sur, California, became the proving ground for experiments in the new therapy: encounter, Gestalt, mediation, the healing baths and the sensuous massage—all that we called "touchie-feelie" a few years ago.

Humanistic psychology—the human potential movement—is getting more, not less, complex. No longer firebrand rebels, the AHP has been accepted into the American Psychological Association as a "legitimate" discipline, and that has set them to the task of building their own edifice, with a theory, a methodology—no mean chore for a bunch of radicals who built their revolt upon open-endness, experiment, the "holistic" approach, a kind of joyful anarchy.

While the founding fathers are today talking about responsibility and epistemology, there is a new, more radical idea building within the human potential movement. It is a turn, not toward the left, but toward the East. In seeking a resolution to the old mind-body dichotomy, the new therapist is turning more and more toward spiritualism and transcendentalism. The mystics and the non-mystics were still talking to each other in Estes Park, but that split could finally prove to be the greatest dichotomy of them all.

SOME PEOPLE HAVE IT AND SOME DON'T

Gatsby believed in the green light, the orgiastic future that year by year recedes before us. It eluded us then, but that's no matter—tomorrow we will run faster, stretch out our arms farther...And one fine morning—
—F. Scott Fitzgerald

That I spent the entire week of the convention untouched, unfelt and unloved would appear to be my own fault. But I'd rather blame it on Rollo May.

I went up there prepared to take off my clothes, spread mayonnaise all over my body and climb trees if that was what was required to get inside the story. I went to Estes Park not knowing what to expect, and expecting just about anything.

Had I, on that first morning, passed by May's talk on American myth, and gone instead to the workshop of "Play, Games and Self-Awareness" or "Advances in Transactional Analysis: Regressive Techniques in Outpatient Group Psychotherapy," I would no doubt have written a much different report on the conference. I would also have missed my own "peak experience" of the week.

I suppose the idea of myth is largely Jung's legacy to humanistic psychology. I kept hearing the word "myth." It may be one of those 20 or so words—"space," "center" and "centered," "share," "stroke" are others—which at times that week seemed to make up the entire *language* of the human potential movement. In the hands of Rollo May, myth was something else again. He told us some nice stories: the story of Orestes; the story of Gatsby. (Great God, the Great *Gatsby*! The first morning of the touchie-feelie conference and I'm getting off on Gatsby!) (It is May's contention that Fitzgerald's *Gatsby*, and not Horatio Alger, is the true American mythic hero.)

Rollo May talked softly, unassumingly, but like an artist he made Jung and Spengler and even Kierkegaard come alive and dance. May's message was complex, and I'll probably destroy it in the translation. But I think he was telling us that we have no more myths. We still need them...need them badly. We are "crazy" (not May's word) without them. We make them up. ("Everything takes on a mythic quality when myth is denied," said May.) But these new myths are hollow, empty, useless, not the real stuff. May told us that we must become reconciled to living in this time of mythlessness. He also told us that we would have to go and meet our gods on the inside—inside ourselves.

I left the Rollo May sessions with a slightly spinning head. While I had come expecting "anything," I never quite expected *this*. I didn't recover from it all week. I stayed up there in the loft regions for five days instead of rolling around in the grass along with everybody else.

There was no way to actually cover this convention, to really report on it. There were as many as 20 workshops going on at any given time. You had to choose one. And I made some pretty bad choices.

I fell asleep during Ida Rolf's lecture on structural integration...Rolfing. She opened her talk with a history of physics—she was up to Einstein's relativity and Planck's quantum work when I nodded off. When I woke up, almost half an hour later, she was talking about gravity as something to be avoided. Sometime during the void I assume she built a "scientific" foundation on which to stand her Rolfing technique.

I did look around the room and observe that quite a few people had their eyes closed and their mouths open, and while some of them may have been in deep concentration on Ida's every word, a few others were audibly snoring.

I fell in love with Barry Stevens. The close associate of the late Fritz Perls, she is perhaps the *grande dame* of the human potential movement. A beautiful old Zen monk is what she is. Barry Stevens has it...the mysterious "it" that everybody in town that week seemed to be looking for.

Outfitted in her customary straw hat and smock, she stood in the center of a circle within a circle—the outer circle, a ring of impressive mountains; the inner circle several hundred friends, admirers and some people like me, there by accident, really, to experience this lady for the first time. Barry Stevens paced the center—calm, but full of energy, happy, funny, often throwing out enigmatic Zen conundrums, talking casually, conversationally, sometimes very intimately with us all. I remember thinking that this must be *real* Zen, or something close to it. It wasn't even necessary to *listen to her words.*

People kept asking her questions, about her life, her current thinking, wanting to know how she'd solved her own riddle, as she so obviously had. One fellow yelled out,

"Hey, Barry, I just like watching you." She laughed and threw her arms up over head as if to say, "Yeah, I know you do."

I also went to hear Werner Erhard talk about *est (Erhard Seminars Training)*, and I got the living shit scared out of me...but that's a story in itself. One I'll try to deal with a little later on.

NARCISSUS FOR PRESIDENT

Today our revolutions are against limitation of personal rights, the right of individuals at all social and economic levels to create, to chose, to achieve and fulfill, to realize their fullest humanity.
—AHP Program

A liberal is a Marxist with two kids.

The Estes Park convention was billed as a political event. But politics—just as you might expect with a group of more than 2,000 people seeking self-cure, self-mastery, enlightenment, personal growth, spiritual development—didn't come up very often. Indeed, political discussion seemed to be allotted to a single evening, the night of the keynote addresses. That night, John Vasconcellos, a state assemblyman from California, called the convention nothing less than "the most important political event of the year."

George Leonard followed him, and with Kennedy-like fervor, asked the audience: "Are you willing to get in touch with yourself and use our best and fullest powers to bring about political change?" He also suggested the price: "We must be willing to lay down our lives, our careers, our ease." (Ask not what the human potential movement can do for you. Ask rather what you can do for the human potential movement.)

There is a gaping paradox that lies at the bottom of a "politics of personal growth" and George Leonard went sailing over it like it was not there at all. I tried to ride with him as far as I could, but I finally sat scratching my head, wondering if I was hearing the politics of Freud or the psychology of Marx.

In the end I decided I was hearing neither, but instead a kind of grand, sweeping defense mechanism at work. While there may have been a few people in the room that night who were totally apolitical and self-involved, my own guess is that the large majority of them own a Volvo, vote Democratic, oppose foreign interventions and bleed over social injustice and world hunger. They are, almost exclusively, white, educated, middle-class and guilty. They are already politically active, involved.

Yet as a movement, not as individuals, they are accused of self-indulgence, of avoiding the large and critical political issues that face the species. These accusations hit home and they hurt—largely, I think, because of their accuracy. So we listened to a couple of speeches on politics, and never mind the leaps that had to be made in talking about a "politics" of "self." It made everybody feel better and quieted the noisy consciences. Therapy, they call it.

Or maybe George Leonard really does think self-actualization is going to save the world and that we should all set about organizing (however you do that) this universal self-awareness.

I heard Charles Hampden-Turner ask the day before, in a completely different context, "Are we to always wander in the wilderness...existential arrows, Lone Rangers of the Astroturf?" Well...why not? The best of the human potential is sifting down to the society and having a significant effect. Some are now beginning to argue, and with eloquence, that too much has sifted down. In any event, it seems contradictory, even absurd, to package something so private and personal into a political club and go running with it into the arena.

But politics was just an aside in Estes Park. The action was elsewhere, and I missed most of it.

NO NUDES IS BAD NUDES

And what is the point of revolution
Without general copulation.
—Weiss

Estes Park, a friend of mine said, is the Okies' Aspen. (I realize that is a rather snobbish way to describe the place, but we do have our symbols, and that one seems to work pretty well. Besides, the remark came from a local, Steve Komito, an enterprising mountaineer/businessman/bootmaker with ambitions of running for mayor of Estes Park. And if he can say it...)

The people attending the convention tended to look upon Estes Park as a cultural backwater, largely, I think, because of the state of shock the management at the Y-Camp (where the event was held) slipped into over the scattered incidents of nudity and random cavorting. (Komito quickly came to the defense of his town when he heard the "backwater" charge leveled by a pair of conferees in a bar one afternoon. "Listen," he said, "You're having your convention at the Y-Camp. That's Y-M-C-A. Those letters stand for Young Men's Christian Association. They have one of those in San Francisco, too. Go try screwing in the lobby there and see what happens.")

I saw a lot of *near* naked people almost everywhere I looked. But no dreaded penile member, not a single mound of Venus. I heard stories, though. Of maids stumbling over copulating couples in the fields while on their way to make beds in the scattered cabins; and maybe stumbling into some really bizarre California fetishes once inside the cabins. But, by and large, the conferees were much more...well, dignified than I expected or even wanted them to be. While the Y-Camp management quietly made the decision not to invite the AHP for another convention, I would speculate that their decision was based on the very general demeanor of the participants and not on any particular acts of outrageous behavior. I would also speculate that most of the conferees wouldn't return to the Y-Camp even if invited.

This was a convention, and in many ways no different than a gathering of, say, a bunch of hardware manufacturers or ski-equipment merchandisers. They came to talk shop, have fun, get laid.

The sexual opportunities were perhaps greater here than at most conventions. There was a balance of the sexes—with women conferees perhaps slightly outnumbering men—that you don't find at most conventions. These 2,000 people count themselves as among the most "liberated" in the country, you must remember. And, if those aren't enough reasons, consider that they look upon sex as *therapy*.

Most of the people I talked with, including a couple of AHP officials, were very open in telling me that, while the opportunity for sexual encounters wasn't the only reason they came to the gathering, it was certainly one of them. To have spent that entire week at the Y-Camp *without* a piece of some kind of action...well, that could well have brought on horrible traumas, deep feelings of inferiority and self-worthlessness. I managed to avoid the conflict—and the fun—by commuting to Estes Park...long, late hours on the lonely highway contemplating Rollo May's myth of mythlessness.

On the other hand, this group didn't seem to drink like conventioneers usually do. There was, no doubt, a preference for other drugs, but I also noted a specific disdain for alcohol. That's just as well, because the campus itself was bone-dry, and despite what thousands of beer-swilling college kids may tell you, Estes Park is no drinking man's town.

Indeed, the only half-interesting bar in town is The Wheel. But that one can get pretty interesting. I spent two hours in there one night drinking with a pair of the most decadent old men I've encountered in a long while. I walked out of the place thinking The Wheel must be where Tennessee Williams comes for his material. But that's getting off the subject. And despite what a humanistic psychologist might tell you, free association can be abusive.

And I've yet to talk about the single most disturbing event of the week.

THE HENRY FORD
OF HUMAN POTENTIAL

When Mrs. Pattycake comes to us to be
taught, turn that wandering doubt in her
eye into a fixed, dedicated glare and she'll win and
we'll all win. Humor her and we will all die a little.
—L. Ron Hubbard
(Founder, Scientology)

They wanted someone to set matters right again, to tell them what to do, and it did not matter how that was done, or who did it, or what it required them to believe.
—Peter Marin

One of the best-attended workshops at the conference was Werner Erhard's session on "est," Erhard Seminars Training.

Est was a point of major controversy at the AHP convention. The first day, at the opening workshop, I heard Rollo May take a public position against est, calling it just the sort of "anti-humanistic" approach to therapy that the organization should be standing against. He wondered aloud why Erhard had even been invited.

That night, George Leonard, ex-Esalen vice-president, ex-*Look* magazine editor, author of several books, including *Education and Ecstasy*, delivered the convention's keynote address. In his speech, he said that the AHP should be open to new concepts, dynamic ideas, like Werner Erhard's 'est.' (Leonard has taken the training.)

By now it was mid-week and I was a reporter without a story. I thought I might be able to make one up by putting Rollo May and George Leonard together at a table, throw est out in front of them, and just write down what happened. It was my feeling that such a mini-debate, however contrived, could stand well enough to symbolize the intellectual push and pull, the search for identity and purpose, that is now going on within the human potential movement.

I asked an AHP official if he could help me set up such a meeting. He raised his eyebrows and said he would try. In the next few hours I was approached by several people from the AHP office, warning me off the topic of est. The next day I was hosted to an impromptu roundtable of AHP directors, all trying to dissuade me from focusing on est in any article I might finally write.

This little gathering backfired, actually. Three of the four officials had taken the est course, and for a while, they spent as much time defending Erhard as they did in arguing that Erhard has "nothing, absolutely nothing" to do with the AHP. "Werner Erhard has been a good friend of AHP and that is all," I was told. "We certainly don't endorse est by having him here. There is a genuine and valid curiosity about his techniques among our membership."

There certainly is. Est has taken off like a missile, and at a time when Esalen and some of the more established institutions in the field are having trouble attracting needy psyches, Erhard appears to be turning them away by the hundreds. He says he is. Although for a vendor that is "temporarily sold out," he continues to do a remarkable job of rather sophisticated high-pressure marketing.

I was never able to arrange the date. May was willing and Leonard sounded like he might have been willing as well, but he had to leave the same day I approached him.

Werner Erhard scares me. I didn't like the idea of est before I listened to him that week. I'm even more uncomfortable about it, having heard him speak.

My familiarity with est comes from reading and from long conversations with several friends who have taken the training. I also attended what they call a guest seminar in preparation for the convention, but you learn next to nothing about est at one of those—except of course, that it is a mystery...a mystery that will cost you $250 to solve.

Erhard, and everybody who has taken the training, would also tell me that I have no right to write on the subject of est until I have been through it. It's not something that can be *explained*, we are told, but only *experienced*. We were treated to the analogy of the chocolate sundae at the guest seminar. How could we explain the chocolate sundae to someone who had never seen or tasted a chocolate sundae? How indeed? But then, the same problem would exist if we were talking about, say, a lobotomy. And I think then is the time to start trying to express the phenomenon in words and to hell with the experience.

I'm not equating est and lobotomy, understand. But on a scale between a chocolate sundae and lobotomy, est is closer to the latter than the former.

A fellow named Mark Brewer ate the sundae and didn't much like it. He wrote a very tough piece on est which was published in *Psychology Today*. (While Brewer was strikingly negative about the Erhard methods, the real force of the piece was in revealing, in some detail, what actually goes on at an est production. He lifted the veil on the mystic. And mystery has been an important factor in the success of est.)

I have a good friend who went through est the year it was started. My friend had his problems. He is a diagnosed psychotic. When this fact was "uncovered" during the training, he was quietly shown the door. (Est is not for the sick, the truly ill. They openly acknowledge that they screen out people with serious mental and emotional problems.) Anyway, my friend often speaks with an almost frightening clarity, and one day, in such a mood, he gave me his own three-sentence description of est, I liked the simplicity of it, and I pass it on for what it's worth: "They simply drag you down to the bottom of the hill, to the edge of the abyss, kicking you all the way. Just when you are feeling, or think you are feeling, worse than you've ever felt in your life, then zoooommmm, they send you rocketing back up. You come out of these dancing; you've never felt better; and you've completely forgotten that you didn't feel all that bad when you went there in the first place."

The est package is a complex stew, and a person could get heartburn trying to identify its many ingredients. I thought I recognized some borrowings from French existentialism (vague notions of "self-responsibility," the existential present, and "good faith"...but no mention, interestingly, of the idea of existential responsibility to the larger community). There is certainly some Buddhism (the idea of "no-self" and a jet-age ride to enlightenment). There may be some Wittgenstein (there are certainly a lot of tricks with the language). There's a great deal of Dale Carnegie ("you're perfect just the way you are.").

Also important is Erhard's own background in encyclopedia sales—he trained salesmen. That's a tough world, a world in which only the true geniuses of the flimflam survive. Erhard has been *training* people for a long time.

Erhard also has some experience in scientology. The scientologists, in fact, have lately been accusing him of having stolen their own system out from under them. Erhard laughs at this. In Estes Park he told us, "They'll call anything scientology if it works."

I knew next to nothing about scientology, save that they use tin-cans hooked up to simplified lie-detector arrangements (they call them "E-meters") to bring true confessions pouring from their subjects (they are called "preclears"). The system is the invention of a fellow named L. Ron Hubbard. With these tin cans he built a multi-million-dollar empire. Hubbard now lives at sea, on one of his yachts. He's been hounded with legal difficulties, and there are many shores on which his ships cannot land.

Coming back from Estes Park, I picked up a book called *Inside Scientology* by Robert Kaufman (Olympia Press). Another exposé, this time by a guy who'd "had it" for a while, but finally decided he'd been had.

In reading the book, I found what seemed to be a number of very evident parallels between est and scientology. In some ways est appears to be little more than scientology without the tin cans and all the sci-fi horseshit that Hubbard indulges in.

(Pondering all this at 3 a.m. one morning, I had a true paranoid's fantasy. I saw Werner Erhard as the creation of L. Ron Hubbard. Est was dreamed up, designed, on one of his yachts at sea. Erhard enters, Manchurian Candidate–like, to give the world the next, higher stage of Hubbard's forever evolving dianetics. That they appear at odds today is merely a clever cover to conceal the true origins of est. It didn't end there. That night I went on to dream that Werner Erhard was elected president of the United States.)

But enough of these dismal fantasies. The reality is dismal enough. Besides, I'm doing the same thing Werner Erhard does, which is to talk all around est, but never about it. Let me list a few of the known facts:

An est course costs $250. That's a fact. (Just for sake of comparison, that can be looked upon as a great savings over Hubbard's old ladder to enlightenment, which retailed for about $500 a rung. A committed seeker could lay out 15 grand getting to the top of that one.)

Erhard is working another side of the street—numbers, mass production. A son of Henry Ford. He may charge you only $250 to go straight to enlightenment, but you're not alone...There are as many as 250 other people making the trip with you. There are supposed to be—so far—about 40,000 est graduates at large in the world. Just run the math on that one.

But the money thing is pretty obvious, and Erhard is always ready for it. His technique, when challenged on money—or any other difficult question—is to disarm with candidness, openness. Sure, he says. You get help from me and you pay for it. With money. I'm not doing this to be a hero. The hero, he says, always gets killed.

Okay, you've made your investment. We can all agree that $250 is a small price to pay for enlightenment, after all. And maybe you're already well along on your way to a cure—because you aren't going to leave without getting your money's worth, are you? (But doctor, that placebo *worked!*)

The scene now becomes a large room. You are with your 250 fellow travelers. You will spend four days, about 15 hours a day over two consecutive weekends, in that room. The environment is very structured. You cannot, for example, get up and go to the john except at an appointed time. (There are two reasons for this, Erhard told us. It would break the group's concentration if people were constantly up and down. And, then, some people like to take off at just that critical moment when they are about to confront the truth about themselves: When the going gets tough, the clever duck out.)

But it is also part of the larger humiliation process. You are now officially an "asshole" (their word). You may think that $250 is a lot to pay to be called an "asshole." But, eventually, you'll "get it" (their phrase). You will "get it" by doing a lot of ridiculous things; by being ordered around, stared at and lectured to. Soon enough you will understand what you are supposed to be doing. You are supposed to be working on the dismantling of your value system, your belief structure, all of those ideals—notions of right and wrong, good and evil—that have been screwing up your life all these years. You haven't been aware of this, but you'll "get it," you'll "get it." Just throw away all that intellectual baggage, all those cumbersome values. Look past them. *Transcend* them.

Est employs several techniques to aid you in your breakthrough. There are a number of "processes" (a word, and a device, used in scientology) in which the "trainer" takes you through a series of standardized exercises, most of them a form of directed meditation.

Another technique is called "sharing." That is simply getting up before your fellow trainees and talking. Perhaps you'll want to share how you are benefiting from the training. Or maybe you'll want to confess—I mean share—what you see as your deepest personality flaws. ("Get rid of your shit, you asshole.") A reward, a "nice stroke," comes with it. When you share you get a round of applause from the room. It is not required that you share, "but just about everybody does, eventually..."

I was in an Aspen bar one night when an est program ended, just before midnight.

I noticed that the bar immediately cleared of all the happy alcoholics who had been laughing there all night. There was a real change in the mood of the place. Anyway, a woman sat down at the table next to me. She was very high, almost giddy. She was still talking about her "sharing" of the hour before. She had "shared her experiences in the womb." (For that, I thought, she must have been given a standing ovation).

Erhard doesn't want to wipe out your belief system so that he can replace it with another; he will assure you of that. He's not Hitler, out to set the world on fire. No. He will convince you he's the Buddha instead, out to light a small flame in your heart. But he's careful not to oversell this Eastern mystical thing; he doesn't like all those associations—those crazy kids in airport lobbies with their saffron robes and shaved heads selling "The Godhead." That could drive away customers. Est gives you the best of both worlds. You can have your karma and your station wagon, too.

An Erhard graduate, the fellow who led the guest seminar I attended, kept making these same vague nods toward the East that Erhard makes. Yet he would also make snide cracks about the poor devil who wanders his entire life with a begging bowl seeking the Truth. He kept saying it's right there for the picking, the Truth…"I've got it." I asked him if he really thought it was the same piece of truth that the Buddha was spending his life looking for. Was he even asking the same questions? He was a man on the go, he said (he was an airline pilot) and he didn't have time to screw around trying to find that out. Est got him there in a hurry and that, in itself, was one good reason to take the trip.

Erhard says some of the things the Buddhists say: "It is not what you know, but how you know it." But, then, I can say that, too. And *to say it is not to know it.*

Erhard is slick. He is very intelligent. He is very quick. He is probably quicker than thou. He is certainly quicker than me. He exploits language with an art. He can transform a question before you finish asking it…or make it disappear altogether.

In Estes Park, for the first two hours—the first half of his talk—he fell into some of the most inarticulate gibberish I've ever heard. He was facing a crowd of 800 people…intelligent, educated, *humanistic*…the great majority of them his natural enemies. He made a rule: In the first half of my lecture I'm going to give a talk, and if you have any questions, they must pertain directly to what I have to say or I will not answer them. Then, in the second half of the session you can ask me anything you want.

Then he let loose with this drivel, this jabberwocky. People would ask questions. He would refuse to answer them. People began to leave the room. They left laughing at Erhard. Their curiosity was satisfied. They had met the devil and he was a straw man. A hustler without a brain. No threat at all.

Hell…Erhard *designed* it that way. I hung around even though I wanted to get away from him, too. I'm glad I did or I would have missed the show. He came back a new man. There was a much smaller crowd. Maybe half of them were est graduates, on hand to pay homage and get their batteries recharged. And the new man was in high gear. Swinging. Still inscrutable, but now happily, aggressively, confidently inscrutable.

The new man. That's a lot of what est is about and a lot of what Werner Hans Erhard is about too. His name used to be Jack Rosenberg. He left his wife and four children in Philadelphia—ran away with a woman named Ellen—and changed his name. That piece of biography was in the *Psychology Today* article. And it came up again at Estes Park. A young psychotherapist, himself Jewish, was intrigued by the name change—not just the obvious implications of a Jew denying his Jewishness, but the heavier implications of a Jew going for a loaded title like Werner Hans Erhard.

So he asked him. And Erhard responded, like he did to every tough question, directly, smoothly, happily. I was never a Jew, Erhard said. I was not raised as a Jew and I had no Jewishness to deny. He said he pulled the name out of an *Esquire* flying across the country to his new life.

Neat. Rosenberg was never a Jew. But he had to have been *somebody*. He was denying something. His past, his personal history, his self. All the stuff that Erhard, the healer, now tells us is in our way.

And it is a fantasy we all have. The fresh start. The resurrection. The new man. That is a lot of what 'est' is about.

The feeling keeps creeping over me that all this polemic is...well, a hell of a lot of ado about very little. Werner Hans Erhard is hardly a national menace. That guy Kaufman, he thought Ron Hubbard was going to take over the world. I once had a friend who sat on the edge of paranoia his every waking moment, certain that Billy Graham was about to take the government by force. Mark Twain thought the Christian Scientists were spreading across the land and would rob us all of our minds, our culture, our humanity.

But then I keep thinking of the potential size of this thing—the implications if the grass fire spreads too far; the fervor of the est graduates—they're not proselytizers, they're zealots.

In Aspen, where they have one of the most politically liberal county governments in the country, all three of the county commissioners are est graduates. Earlier this year, the commissioners proposed that the county pay half the cost for any of its employees who would take the training. There were cries of "church and state" in Aspen, where est has almost divided a small town into warring factions. But Erhard has trained California schoolchildren under a Federal grant, and he is looking for business like that.

The est people see no conflicts in the training being used by political agencies. They say it's apolitical. And indeed it is. An est graduate can be a better Nazi, a better communist, a better candlestickmaker. An est graduate, a young woman, came up to me at the guest seminar and, with all the style and grace of a Los Angeles used-car mover, tried to get my name on a contract. She was as inarticulate as all est graduates finally are in trying to state her gains from the training. She ended by saying: "It's like this.

I run a lodge in Aspen. And before I took the training, I couldn't look my customers in the eye. When I showed them a room, I was always looking down at the floor. Now I can look them square in the eye!" When I suggested that, if her prices were comparable to the prices of other Aspen lodges, she was quite right not to be able to look people in the eyes, she moved on without so much as a "thank you" or a smile.

There is common sense good in almost all psychologies, and est is no exception. But I keep feeling that, with this one, there comes more harm than good. I'm no card-carrying humanistic psychologist, but I see Erhard's est coming down on the wrong side of the line that divides freedom from determinism.

I just thought I would *share* that with you.

HUMANISTS CAN BE HUMAN, TOO

If we would have pure knowledge of
Anything, we must be quit of the body.
—Plato

Apart from the body, life is an illusion.
—Alexander Lowen

I want to warn you Indians that honesty
Is the last weapon of a desperate man.
—William Eastlake

There exists in all of us a mind-body dichotomy. As a society, we are generally out of touch with our bodies. This I believe. Most of the time I cannot find my own, I will freely admit.

But still...

I had expected, even anticipated, the communal love that was all around me at the Y-Camp that week. People didn't just say hello to one another, they embraced, fondled, kissed, touched, felt. But even expecting this, I found myself to be a little troubled by these displays. I kept wondering, if this is how they say "hello," what do they have left when they want to express something a little deeper, to make a gesture of *real* intimacy? There was, finally, something contrived, forced, phony about it all. And something passing, ephemeral, about it, too...like the "love generation" of a decade ago.

This was illustrated in a way by the relatively high incidence of small-scale emotional "breakdowns" I witnessed walking the grounds. I recall one young lady falling quite apart—weeping and moaning—when she discovered she had lost her program. She was immediately swarmed by people, all strangers I'm sure, who smothered her with embraces, comfort and kindness. That seemed to help. Oddly, nobody thought of simply getting her a new program. Similar incidents of coming apart seemed to be happening all around me, I suppose because there was so much *help* everywhere you turned. (I was reminded of a day when my own youngest daughter cracked up on her bike. First, she looked around to see if there had been witnesses. Seeing none, she shook off the hurt and remounted her machine. She had only pedaled a few turns when she spotted me and, of course, immediately broke into tears.)

This aid-and-comfort approach was even institutionalized in the form of an ongoing "nourishing Touch Program—for people who want caring." You could just drop by and get a hug whenever you needed one.

At a workshop on creative anger—a workshop that was apparently panned by many of the professionals present, but one which I thoroughly enjoyed—there were perhaps 200 of us, gathered outdoors. We were using techniques designed to bring out suppressed anger—a most useful therapy for depression. With the first technique, we were to go back in time, to again become three or four years old, and to throw a fine, bloody tantrum. One young man got deeply into his own three-year-oldness, or seemed to, and put on a magnificent display of screaming and shouting. He kept getting louder and louder. The people around him stopped their own kicking and ranting to watch him in wonder. Soon, all 200 of us were craning to see into the center of the circle where the young man had by now lost all control and was in a full rage. Jesus. We were worried, concerned, that some deep place had been touched in this young man, some horrible wound opened up. One of the therapist team—a big, strong, gentle guy—went to the now weeping child and held him tightly, speaking to him softly. He was calmed. We were all moved and rather impressed by the experience.

But with the very next technique—we became dogs and cats, a hundred of each, and made aggressive and menacing gestures toward one another—the same kid did it again. He was no cat. He became a roaring, mad lion. And at that point he also began to lose his audience. Muted word passed among us: "Oh, he's just a primal."

He proved to be a very bad actor. He'd lost his audience but would not stop his performance. By the end of the session, he was still carrying on in a highly exaggerated fashion, but he was ignored by us all. In fact, he had won our hostility. He was an embarrassment. And we were resentful about having been taken in the first place.

It's just a feeling I have (after spending a week with the human potential folks, there is in me an irrepressible urge to express my feelings), but I sensed an emptiness while walking among all those joyful, happy, feeling people. In reaching so far to reaffirm their bodies, their spirits, their human playfulness, many of them seemed to have over-reached. They've come to deny their equally human minds, their sadness, their aloneness. And perhaps they are now feeling the loss of these things as deeply as, before, they felt the distance from their bodies. Or maybe they are simply trying too hard; trying so hard because they sense the therapy isn't really working.

In all their talk of "peak experiences" the human potential folks seemed to pass by the notion that in order to have a peak you must also have a valley. In order to reach a great height of joy, or anything else, you must necessarily come out of the darker places down below. Not depression or despair. But just plain old human hard times, sadness, the blues. The people I talked with in Estes Park—many of them, certainly not all of them—seem to be trying to make life on the summit. They want to stay up there—high up—in a state of near-continual bliss. My own feeling is that it's a nice place to visit but...

AND THAT'S NOT THE HALF OF IT

> *Consciousness is a congenital hallucination.*
> —Blaise Cendrars

As I said, I missed a lot of what was going on that week in Estes Park...maybe most of it. ("Or all of it," I can hear Werner Erhard saying.)

I missed the magic. I've since read in *Rolling Stone* that, while the conference was going on there, the First World Congress of Sorcery was being held in Bogotá, Colombia. Stanley Krippner and some parapsychologists in Estes Park tried to put themselves in telepathic contact with colleagues in Bogotá. Rolling Thunder, a Shoshone medicine man, was somewhere in Nevada trying to plug into the Colorado/Colombia circuit. I missed that one completely.

I also missed The Women. Or, rather, I saw them only out of the corner of my eye. I felt their presence, but I didn't give them my full attention. I should have. The Women were probably the most important thing going on at the AHP conference. They seemed to be a step away from power there; or maybe they already have it. I had a murky view of strong, attractive and self-possessed women and meek, powerless men. That's both a vague and a mighty generalization, of course. And here's another one: it seemed in Estes Park as though the sexes had not met together in that valley of hoped-for harmony; but had simply passed by one another on the trail...the women on their way up the mountain, of course. I should have watched that one.

I also missed the star attraction of the week—Jonas Salk. Or rather, I went to hear him talk, but missed much of what he was saying. Jonas Salk was not in town with a vaccine for loneliness, but something almost as far removed from "basic medicine." Salk has worked out a large and ambitious theory of biological evolution that would appear to explain the transition the world seems to be undergoing at present. It's a complex and interesting theory, and I don't dare go further into it without reading his book on the subject, which will be published soon.

I also missed *Hearts and Minds*, a film that was exhibited several times during the conference week. But I'd already seen it. It's a powerful piece of film, and I talked to a number of people who were thoroughly shaken by it. That was the true "political event" of the week.

Finally, it was those two men who were the focus of "my" conference. There was Rollo May—who might be called the dinosaur of the human potential movement, because so much of the movement seems to be away from what he stands for—telling us that we must know ourselves through our own histories, through the histories of the civilizations that came before us; telling us that we are all imperfect, flawed—but to be otherwise is to be against ourselves, to be inhuman.

And there was Werner Erhard—call him the serpent of the human potential movement—telling us that we must deny our histories, both cultural and personal, telling us that we are perfect just the way we are.

I had already written this story when the October issue of *Harper's* appeared. In that issue is an article titled "The New Narcissism" by Peter Marin. I mention it here because I think it is a very important piece of writing, and I recommend it. While I think Marin goes too far in his criticism of the human potential movement—he leaves no room for May, for Stevens, for the obvious good that has come with self-discovery—he talks clearly, eloquently about the inherent dangers in the movement, "the ways in which selfishness and moral blindness now assert themselves in the larger culture as enlightenment and health." He talks about "the warm winds of forced simplicity blowing away the tag ends of conscience and shame." He writes: "We proclaim our grief-stricken narcissism to be a form of liberation; we define as enlightenment our broken faith with the world. Already forgetful of what it means to be fully human, we sip still again from Lethe, the river of forgetfulness, hoping to erase even the memory of pain. Lethe, lethal, lethargy—all of those words suggest a kind of death, one that in religious usage is sometimes called accidie. It is a condition one can find in many places and in many ages, but only in America, and only recently, have we begun to confuse it with a state of grace."

I think he's "got it" there.

—MOUNTAIN GAZETTE, #38

BARB BOMIER

Wild Red Dharma Pickup Truck

BY
LACEY STORY

Prologue

This is for the mystery mechanic in loving memory of my last truck, a '91 Toyota 4WD 4Runner, which he repaired and shored up through its dying days. She thankfully retired on top of Baldy Mountain at 12,000 feet in the company of a Tibetan Buddhist monk while looking for retreats way back. My mechanic just might, maybe, appreciate the contagious lure of the Wild and the religious ecstasy of backcountry fervor in any season. He should know, even though unbelievable, I am the shyest of the very shyest New Hampshire small town farm girls underneath the false bravado of title, occasional sparkle, and survival personality. I am pathologically shy, even, and don't take to parties and never, ever, ask any man out, even for skiing, with only one exception in my life before Xmas this year. This would take too much of a shift and the fear of rejection is far too strong. I travel alone, study alone, work alone and get right into the wilderness at every possible moment. So I related to his dog, Sara, instantly, with her shy girl wildflower way, and when I looked at her looking away quickly with her head down I looked at myself. I thought that if Sara could find love maybe I could, too, if I ever dared to consider the possibility.

However, I was not born a dog in Summit County. Wild can be introverted although that's not the usual association. The mystery mechanic should not be intimidated by any wordsmanship here knowing that a quivering aspen leaf of a girl who spooks easily in social settings wrote this. If he ever looked at me straight in the eye, if I could ever look back, he'd see things that perhaps could write another vignette, a tale of far off places from another time, or a tale that only he knows deeply in his heart. Somewhere there is a connection. Hey, listen, I'm trying here.

It's not important to know what it is, perhaps, but maybe it is to him, so I warmly welcome his response. He will do with this as he wishes. What follows is a disclaimer. He should not, under any circumstances take this personally or deduce this means anything, because, after all things are examined it really doesn't matter in light of the Greater View. Be forewarned. Inspiration is the head of literary free rein. Artistic license gives one enough rope to choke oneself with and that would be missing the point entirely. This writing is neither true nor not true. It walks the Middle Way on its own, on the razor's edge. Have courage. Think Red. Enjoy the journey. It never ends.

My longest relationships are with pickup trucks. Four wheel drives usually last the longest. A vintage 1985 red Toyota, extra cab with a topper is the current flame. Red enough to create passion off road even with a stick shift and a thick pile of topo maps between you and a likely hard core companion (preferably a mechanic with tools, long hair and a few days growth to tickle tender, soft flesh). Red enough to go all the way. Up the mountains, across the divide, higher and higher to highest. Red enough to handle a scattering of spark plug boxes, emptied Chinese deer antler and Chang Bai Mountain ant extract bottles on the floorboards interspersed with Tibetan language tapes, empty Mountain Sun organic raspberry juice containers and a used up Big Daddy Red lipstick case. Red enough to know that pink satin thongs, a kiss me relative of red, are the correct attire under jeans for steep boulder-littered climbs to the secret cabin in the woods. The only better moment for wearing them is with a tan, a thin white wet T-shirt under a waterfall on a summer day near a rock big enough for two. Add blue columbines, a liter of rhodiola tea, a fresh watercress salad, venison sandwiches, and voila, you have an afternoon. This is never far from red: action. Enough to make Padmasambhava beam. Even beyond Red, just beyond to the Land of No Return, to the farthest shore. Let's make it deliriously excitable and a true out-of-body experience.

It is winter. These are the days to add snowmobiles to the list of other ways into the backcountry just to spice things up in an already perfect and committed relationship with one's truck. Envision reclining back on the extended black leather seat with the trees as a backdrop to the perfect entry of hard deep unzipped bliss, the snowflakes gently released from high spruce boughs falling onto his long hair tied back with a black and white bandana. If you think about it too much at all, the opportunity vanishes. The rule is non-attachment. The method is volition. The path to heaven plays a catchy tune sung as a mantra that chants the words "I give up" in the repeating stanza. Surrender completely and the world is your oyster. Loosen into the genital groove of all that is or ever will be. A physical consort is better than none at all. The liquid light of unconditional love unravels any preconceptions or obstacles that you might be carrying around like lodestones. Fruition comes only through letting go. Lighten up.

Alas, men do not come in four wheel drive anymore. Let alone with split shift. A manual model is becoming obsolete. No computer chips for me. No automatics. No preprogrammed undesirability. Something you can touch, feel and are partial to lovingly caress all Sunday afternoon is a requirement for total satisfaction. Years of devotion build the greatest of loves. I prefer four, preferably five, on the floor with a good heater and a sturdy body that withstands a few dents and branch scrapes, definitely capable of enduring an equal amount of days without a shower or an oil change.

Things are better in the woods. Water is sweeter. Men are wilder. And it's pickup trucks that get us there. To the back bowls, to the Divide, to that nameless peak along the gulch road. Edward Abbey says, "Wilderness is the only thing left worth saving." That includes me and you, wildness embodied in human form. Read and you discover that the Wild is strong in the equation of biological imperative. It's the primitive, dark horse deep river, the pheromal unpredictable lucky star, that wins the evolutionary race to transcendence and the perpetuation of thankful genetic shifts nearing the Ultimate Sphere. Everything I've ever loved has happened in the wilderness, in remote cabin shangri-las or in pickup trucks. Some fortunate beings are even conceived in pickups, the most auspicious vehicle for the continuation of our species. In New Mexico, if you're really onto lightning bolt luck, you find a box canyon or a ghost town with a perfect ledge to line up tin cans and plink at them with a .357 and a case of .38s for the second best part of an afternoon. The reverberation in your cerebrum induces ecstasy if taken with scent of after-the-rain sagebrush. The echo lasts forever. The flash of rapture ricochets in your cavernous mind like a pinball wizard's eternal game. The tailgate is down. The passenger door is open to the Indian blanketed full front seat. All things holy happen here. The music of the celestial spheres seduces you with enchantment and beckons you in with a warm embrace. View the Sangre de Cristos from a remote access dead end off Buckman's Road to the Rio Grande through the thick willows. Don't even think you've ever heard of clothes on the sandbars a few miles in towards Jaconita. Here, eagles pronounce your name in an accent you recognize from past lives together. In a pinch, Moab, specifically White Rim, will do. The deserted cabins in Futurity, Colorado, are a close third. A yearning curiosity arises for the inner sanctum lakes of the Gore Range and Holy Cross is looking better and better these days. Awaiting spring thaw seems to be an eternity. I simply cannot confide the true number-one locations. Find your own, sweetheart.

Rename some mountains and some creeks. Make them yours. The geographic change will do you a lot of good. This simple act provides for a vacation on your own turf. An understanding of impermanence is pivotal for the necessary changes we need moment to moment. Baldy is Grizzly now. Copper is still Copper. Wise is still Wise. Santa Fe can still be Santa Fe, if only for the vestigial memory that Canyon Road was once hard packed red dirt without the sort of folks that are there now. Illinois is now Sugar just because we all need a Sugar Mountain. Deer Creek can stay the same, as can North, South and Middle Forks. Every mountain and creek has the unique privilege to have secret names, like old Navajo clans or Tibetan Deities. Mountain of the Lotus Born, Flicker Breath Canyon, Buddha Fingers Creek, Guru Rinpoche Gulch, Yeshe's Pool of Transcendence, a bottomless well as black as jaguars. Never say their secret names aloud. Keep the noble silence. Be breathless at the mere thought of their existence in the same life as yours. When nightfall comes, rename the stars. Your secret name is written in the sky in case you never noticed. Create a new mythology, a legendary saga for every year you've journeyed around the sun.

    I graduated Most Unpredictable with high honors. Translation: Most Likely to Be in the Woods or somewhere maybe on the Tibetan Plateau in a Mongolian ger drinking fermented yak milk and listening to Russian tunes on the accordian. Somewhere in Montezuma or Sts. John out of cell-phone range contentedly in a twin-zip sleeping bag with no one anyone would imagine me with, a regular reformed badboy quiet homeboy with the spirit of a deep river current. Deer Creek is as close as I can come these days. As a young girl I was happiest in a meadow clearing, lying on my back, surrounded by princess pine or in the white birch grove amongst intoxicating Mayflowers by the hidden spring. I was so quiet, so small, so blended in energetically that the deer would browse within a few feet of a barely breathing body. Tread lightly with Indian moccasins as supple as baby white oak leaves in the spring. Later on was the discovery of friction climbing Pawtuckaway Ledges for a view of the hawk migration and thousands of orange-bellied salamanders, Franconia Brooks Falls and haunts where you can still find picture agate on Hurricane Mountain. My graduate thesis was in Comparative Geographic Hideouts. Today I specialize in Rocky Mountain ghost mining towns. Always the wilderness, the Shambala. Always the driving force to keep the wild still wild. Still. Wild. Silent and Wild.

    The best sex is in trucks. A five-star rating often includes a Vietnam Vet point man from the Texas panhandle or an off-the-map Oklahoma ranch near Antlers. His name is sometimes initialized, like R.J. or J.R. An evening of sunset teasing in a split window '49 flatbed Ford is a good beginning to the discovery of how good it can get. No need for dual 36, cherry glass pack mufflers. Add 10,000 points for every 1,000 feet over 8,000, up to 14,000, not that you need to attach a number. Adorn the rear-view mirror with a red lace bra and the door handle with unmatching flutter panties. Consider the perfect red polish, chipped to just enough near imperfection on your toenails, an old jean jacket with a Nirvana Now pin, a pair of worn elkskin boots with a riding heel and a delightfully sheer voile dress on and off again. Honey dust and royal jelly are always welcome. A bonus gift includes the scent of pine needles and spruce anointed way up, between the legs. Mingle it with manly musk.

    The best relationships require a perfect balance of body, mind and spirit. You know it in your brain, taste it always as the nectar of immortality deep back in your mouth (a personal favorite place) and feel it in your heart as one with the infinite expanse. The recurring pattern of continual quivering occurs when the mixture takes on alchemical consummation. Hear the bells. Take a red pickup truck and balance it with trees, mountains, sun, sky and water, and you have perfection. All five elements in harmony. Metal, Wood, Fire (red), Water and Earth. Take spruce trees, tall ancient sentinels of wisdom and combine with compassion, the emptiness of spacious clear blue sky. The Wilderness Drive-In movie is playing *The Six* plus *Four Perfections*. Generosity in sparing no cost in truck maintenance. Joy in truck ownership. Energy enough to change a flat on uneven ground in the dark. Wisdom in carrying emergency fan belts and flashlights with working batteries. Meditation upon envisioning the truck upright and in good form. Discipline with carburetor adjustment and the right amount of air in tires. Right means in choosing the path that gets you to the most perfect of perfect places. Vows to maintain the truck in peak condition and honor the pristine environment. The manifestation of the ten powers in getting you out into the backcountry with a snap of the fingers, and finally, the knowledge of true truck dharmas. *The Six* plus *Four* shows every day at the perpetually spontaneous matinee in your mind. Instant replay insures achievement of enlightened truckness. Correct view is the paramount star. Always, in all ways, seek the Wild. Add a creek, hot springs or river bed and fuse them with your awareness. Open sesame. Abracadabra. This is it. That Such-ness, That Which is That Dharmakaya Being with Light. This is it. Right Now. Love the moment. Paint it in the rainbow body shop a pure red. Watch it glow in the dark and light up the sky on full moon nights as you ride, ride, ride to the highest heights. The vehicle of your dreams takes you There, Right Here, to the Now Tao of 4WD Truck Delight, the Abode of Pure Visions. Truck body, truck mind, truck spirit.

These days I am truckless. Each night before I go to sleep I pray for truckfullness in physical form. Perhaps tomorrow, if I just let go of today, it will appear outside my cabin window in grateful union with a big red bow, a single red rose and a small card addressed "For Lacey." On its trailer is another guiding star, a Polaris snowmobile, just for diversity, just to keep things spicy. If you're feeling plush, throw in a CD player with "Days of Future Passed" and Lucy Laplansky's "Secret Journey." Remember the bumpersticker "No rules over 10,000 feet." It is not an option. Maybe even add an invisible "If you're not the lead dog, the view never changes," in memoriam to a past truck love. Stick it on, then quickly take it off. A bell, prism, hawk feather and Tibetan turquoise on rawhide adds a nice touch to the rearview mirror. The keys are on the seat. The keyring has your number on it. It reads a scrawled, "Call me. Now." I decode intentions encrypted in the handwriting analysis. Be reality in a dream and a dream in reality. This is my prayer. I beseech the Powers That Be to deliver the red 4WD pickup truck, to make manifest the vehicle of bliss, heaven sent, maybe, even, from you. Screw love. Give me a truck. Truck Love Lasts. I call it Supreme Joy. I ride the Red Road. Come and join me. My refuge name is Unity. My secret name is Lucky. My truly secret of secrets name, unspoken, is Wild. Going my way?

With thanks to R.J. in 1967, a psychedelic Oklahoma rancher, who taught me to ride tough in the saddle, roll my own while four wheeling in the driver's seat of a dilapidated excuse of a truck, read the Tibetan Book of the Dead & Kerouac to me on LSD-25, never complain when your hair's full of hayseed and your hands full of grease and who first planted the gentle hint that what I wanted to be when I grew up was enlightened. I am happilyeverafter.

—*MOUNTAIN GAZETTE, #81*

Obituary
The Deceased: Lucette K. Car
Born: 1985
Died: 2000
Cause of Death: Internal Combustion

BY
B. FRANK

It's time for "closure," as they say in the latest self-help psychobabble books. Last time I saw her, she was waiting for the operation, a transplant of sorts that would put her back on the road again, with slightly newer guts and a new clutch thrown in. But let's get back to the beginning of the parting, when whispering rattle turned to fearsome shuddering, the grinding of internal disarray; then complete failure. That mournful morning of the diagnosis, the specialist had come out shaking his head wiping Lucette's fluids from his hands. His prognosis? Terminal.

"Cost more than she's worth," to quote him apocryphally (at my age, verbatim quotes are mythology).

But then I met Vince, who said he could get her going again in no time, in trade for another old nag he desired. This one was just sitting there, too; \my partner's old steed, now a rat's nest of useless parts and memories. So we made the trade, and I towed Lucette over to Vince's shack. I left her in the yard, pending the operation. Next time I saw her, the hood was up, her tranny was on the ground, and Vince was seeing if he could scam my partner out of the other clunker, without delivering Lucette back to us. We declined his kind logic, and waited for the next move. It came, in the middle of the night. Vince (and Lucette) were gone. More about Vince later, for I come to honor Lucette, not this scoundrel.

Lucette was a common sort, the Pinto of the '80s, a Plymouth K-car (Iacocca's David). I bought her off a car lot in Santa Fe, scamming the dealer out of $1,000 of sticker mark-up robbery with a $30 compression tester as a weapon. She had 50,000 miles, a burned clutch, and cigarette burn marks on the upholstery, courtesy of the nervous bomb scientist who had used her as commuter chariot at Los Alamos. I fondly imagine his flipped stray cigarette butt starting last year's forest conflagration, and visualize him slipping Wen Ho Lee's computer files into Lucette's trunk, for transport to the Dumpster. Lucette as accomplice to international intrigue and red-faced official federal accusations, admissions and retractions. She was that kind of car.

Low-slung and front wheel drive, she went places where most SUVs fear to tread (lest anyone get me wrong, she disdained making inroads on the land, preferring to wait patiently for my return at the end of the tracks) and held a survivalist's wet dream of gear, while getting 40 miles to the gallon on the open road. A double advantage over recent trends in auto making, as one could afford to get to the wild places AND not have to hike past oil wells drilled to feed the monster once there. Sheltered me and dog from an exploding marriage, and more. She provided safe haven in the dens of post–Mardi Gras New Orleans and the soft white beaches of a southern spring. A literal doghouse on wheels, and smelled like it. Later, she did quite a few river shuttles, waiting at river's end for her load of raft gear, stinking river-rat and dog. When I found a partner, Lucette somehow made room, if we tied a little more gear on the roof and trunk lid.

    One thing I hate about reading obituaries. The departed always comes off as Mother Theresa, or Martin Luther King without the imperfections. Lucette was no such animal. Her kind was slapped together on the cheap, built to fall to pieces and be parted out long before she actually met that fate. If she did. Her carburetor was always on the fritz, making her vapor lock at the damnedest times; her heater kept cutting out on the coldest drives; and Chrysler must've thought vacuum tubes would never start to leak, so they ran them all over the engine compartment to replace electrical switches. Bad idea. I never did get a new clutch and have the flywheel resurfaced, so nobody else could drive her without chattering the gears and killing the engine in the middle of a busy intersection. She was cheap and tattered, but she was mine. Mine alone.

    I put the dent in her gas tank on an ill-starred exploration of the Grand Canyon's South Rim fire trails. Somewhere above Lizard Head Pass and Trout Lake, another rock banged the floor, impressed my son and made a handy spot to perch your accelerator foot on a long downhill. That one just barely missed taking out her shift linkage. I should have seen the pattern of risk-taking behavior she had concerning her trans axle then. There came a few steep hills too, as she got older, that defeated her, leaving me to hoof it to the top for a view over the other side. Her bronze paint faded to a nondescript tan, the ceiling fabric peeled away in strips. She took some dents, and avoided killing countless panicked deer, a few incautious drivers, and one death-tempting pedestrian. Her odometer, which only worked part-time, ground past a quarter million miles. Inside, the gears were grinding toward her end; but the dog and I were showing gray too, so maybe that's why the crisis came as a shock.

    I heard it as we rolled home to the San Juans from a river trip, an increase in the muted grinding that had disturbed traveling reveries for several months now. Caution would have dictated immediate diagnosis. Instead, we started south next day, headed through the Rez of the People (Dinetah) with faltering steed and overweening confidence. We made it to Kayenta, replaced a drive axle and limped into Flag. There her transmission ground itself into a pulp of scrap. We towed her south to our winter hideout in the Sonoran Desert and used her for storage until Vince's offer changed her fate…

We know Vince disappeared, with my Lucette in tow. I have given up hope of hearing from him the end of this story; but with the aid of a little sleuthing, write this apocrypha. According to a source (reliable for his lack of interest in the matter), Vince is a coyote. This not only describes his dealings with my partner and me, it is his profession. He is, like me, a guide-for-hire, except that my clients usually have their proper papers and more disposable income. Vince is one of the entrepreneurs you only read about when the van's front tire blows with a Border Patrol SUV in pursuit, killing several of the 20 or so occupants. Our government prosecutes the coyote for homicide, deports the surviving passengers, and thus deprives yet another view-padlocking trophy-home owner of cheap labor, clean toilets and white (oh-so-white) sheets to be bedded upon. Until the survivors return with another coyote, to resume their "service-industry" careers, sans benefits and a living wage, I'd add if I were a flaming radical.

Yes, Vince is a coyote, and Lucette joined him in his profession. My Lucette is not Lucette anymore, has been re-christened, perhaps, with one of those revolutionary nicknames so loved by the Latinate peoples of this hemisphere. La Comandantina? She now rattles along the high and low roads of El Norte. With a new coat of paint and tinted windows. She clambers past roadblocks along the militarized border, on midnight highways that become sandy washes by day. The "underground railway" for a new generation of economic slaves. She slips onto a two-lane blacktop, then stretches out on I-10, I-25, or one of the other national defense highways of the Soviet-fearing Eisenhower administration. She delivers her cargo of hardworking laborers, demons of the demographic nightmare of any God-fearing xenophobe, then turns south again.

"I may stay south a while this time," Vince thinks, "maybe be a taxi for the gringos along the coast, around Puerto Peñasco." But they'll come back, until our INS (the oddly named Immigration and Naturalization Service) agents catch Vince again, toss him back in prison, and pretend they want to stop the illegal influx that justifies the departmental budget. Vaya con Dios, whatever names you travel under.

I much prefer this reality for her over the ignominious crush of anonymous recycling into an overvalued Sport Utility Vehicular-homicidemobile. I drive an old Suby now, blending as well as possible into the changing culture of these mountains, my home. She has real four-wheel drive, and cost $300. Her odometer will pass 250,000 on the next trip south, somewhere past Dinetah, just before we drop down into the heat of the desert. She's just broke in, I figure.

A few thoughts on Lucette's apocryphal end, and the word apocrypha, if I may. Now and again I see an old two-door Plymouth K-car with a cargo of brown-skinned workers on Dallas Divide, commuting to work in the trophy homes of Telluride; and wonder. I pass by the roadside roundups of these latest immigrants, feeling helpless to change the story being played out. One of them will be labeled coyote, prosecuted for fulfilling the labor desires of this society, the others deported without trial. During the

16th century Christian Reformation, 14 books of the Septuagint (an early version of the scriptures of the Hebrews) were excluded from the Protestant Bible, says my dictionary. These are the Apocrypha. If some stories are excluded, can a full history of the people be remembered?

Our ancestors chose to excise the Apocrypha from their Bible, made the word synonymous with untruth. Others of them ignored the codified knowledge of this continent's indigenous peoples, also our ancestors, and called them pagan. Skin color determined humanity. Many turned away as zealots took their neighbors to internment camps. Eventually, newly branded aliens' eyes were avoided as death camp trains passed through Europe. A world away, and in the past? I wonder. Now many in these mountains are silent as brown people are chased through the deserts and valleys, herded into deportation buses in our towns, vilified as aliens and denied a right to tell their story.

What is the crime, who is the perpetrator, who will turn away? I check my own family history and find that even the least desirable of these people and their stories are my story, too. May Lucette be swift and sure on her journey with coyote. May we all journey with our own gods and goddesses, in grace.

—MOUNTAIN GAZETTE, #81

M. JOHN FAYHEE

Crossings

BY
M. JOHN FAYHEE

"*Eventually, all things merge into one, and a river runs through it. The river was cut by the world's great flood and runs over rocks from the basement of time. On some of the rocks are timeless raindrops. Under the rocks are the words, and some of the words are theirs.*
"*I am haunted by waters.*"

—Norman MacLean
"A River Runs Through It"

"*I do not know which to prefer,*
the beauty of inflection
or the beauty of innuendo,
the blackbird whistling
or just after."

—Wallace Stevens,
"Thirteen Ways of Looking at a Blackbird"

The Thames

Tomorrow, just before the pubs open (this was scheduled intentionally by my fore-thoughtful relatives), we will spread my grandmother's ashes in the River Thames, and, no, I never did remember to ask anyone why this particular waterway is named backwardly, as though it flowed through Mexico rather than the UK. My grandmother, who had just turned 80, actually died almost a month ago, but there's been a horrendous flu outbreak in Southern England this winter, and it has taken a heavy toll on the aged and infirmed. This is a country not only of socialized medicine, but, it seems, of socialized cremations and funerals. There has been a backlog of undisposed dead old people, and, tomorrow, it's finally Grandma's turn. I don't know where she's been in the interim—stacked up somewhere like so much socialized cordwood, I suspect (hell, since she's gonna be toasted, how do we even know we've got the right corpse?)—but finally, she's on deck.

We will meet at a little chapel close to where one of my aunts lives, close to where, when you get right down to it, all the Martins—my deceased mother's clan—live, have always lived, and, unless there's a major shifting of the tectonic plates, will always live. After the service, we will all stroll through the winter half-light mist to the banks of the Thames, and my mother's mother, who I only ever met a couple times, will enter into the fetid waters next to which I now sit to join her husband and several of her previously departed offspring.

I'm parked on a bench that bears the name—by way of a small memorial plaque—of one of those offspring: my Uncle Harry, a wonderful man from all accounts who had trouble handling this world. The cousins all understand and forgive, and shake our heads at what it must have been like for our grandparents, parents, aunts and uncles to have grown up right here in Kingston-upon-Thames with the Nazi bombs falling and falling and falling. Just sitting on the curb near the Thames, and hearing the roar of the planes and the deafening explosions and seeing the fires. Day after day, lifetime after lifetime.

It is the wish of a great many of my aunts, uncles, cousins, sub-cousins, sorta cousins, nieces and nephews that, when they pass from this mortal life, they, too, be incinerated and then made one with the Thames. The Martin clan is a big one; should everyone who has indicated an interest in spending eternity in this tired, diminutive river come to that end, England's most famous waterway will become little more than Martin sludge, more wet ashes than H_2O. (I wonder how many dead people have found their way in one manner or another into this river. I mean, hell, all the way from druid and King Arthur days. Beheaded people, speared, sworded and jousted people, drawn-and-quartered people, bubonic-plagued people, bombed-and-burned people. Man oh man, ancient stuff makes me sweat; death to all things old!)

This river is in my family's blood, and, thus, I guess, by default, it is in mine. Even after my mother crossed the Atlantic, with scarcely a glance back toward England, she talked often about the Thames, like it was the river mentioned in Genesis 2:10. Hardly ever talked about family, but talked often about how much she missed the Thames.

My grandfather, a lifelong member of the British Merchant Marine, a man who had traveled the high seas from Malaysia to Cape Horn, spent many years building a cabin cruiser that he intended to navigate up and down the Thames, as far as I can tell from family recollections related simultaneously and boisterously by 50 people all bellied up to a bar, forever. When not cruising, he planned to moor his craft and live along the banks of his beloved Thames, in front of Hampton Court Park, as so many people do in a wide array of whimsical vessels that look like they are the spawn of a drunken, though cordial, mating of the fictions of J.R.R. Tolkien and Lewis Carroll. On acid.

On its maiden voyage, with much fanfare, my grandfather launched his craft and almost immediately piled into Kingston Bridge, which I can see from where I sit on my departed uncle's memorial bench, irreparably damaging his boat, which promptly sank, necessitating a much publicized rescue that traumatized many Martins as the event played out for some days in the local press. When my mother heard about the sinking, she simultaneously cried and laughed.

She, too, talked about having her ashes dispersed into the Thames, and, since she knew well in advance that her time was coming via slow-moving cancer, my sister and brother and I tended to take notes when such pronouncements were made. In the end, though, she stuck to the lot she cast more than 40 years prior when she left England; she opted to be planted in the New World, in a high-and-dry cemetery filled with tall trees. I have only visited her grave twice, but both times I wondered if we did not err, if we should have taken her remains home to the Thames to be with her people and the river she used to swim in as a child with her siblings.

My grandfather was crestfallen when his newly christened cabin cruiser went down. His river, which he had dreamed about all those years out there on the lonely ocean, had betrayed him. He never recovered and passed away soon after. Yet, he still wanted his ashes to be cast upon the muddy waters of the Thames. I don't get it. This river looks spent and used up the way most urban rivers do. The Chicago, the Cuyahoga, the lower reaches of the Hudson. All the same, all city rivers, all near dead. The water of the Thames is cold and mucky. This is where the bombs fell around his children. Why not someplace more exotic, warmer, cleaner, with less bad memories?

The answer is simple, I guess: When it comes to rivers, one man's Cuyahoga, one man's Styx, is another man's Ganges.

The Saranac

The last two years of my life in the North Woods, when I was 11 and 12, we lived in a small town named Morrisonville, New York, just outside Plattsburgh, between Lake Champlain and the Adirondacks. Through the middle of town flowed the Saranac, a winter river, mighty, swift and frigid, a river capable of floods and drownings and whirlpools and mysteries.

For most of a year, my mother simultaneously dated two men who did not live in the area: the staid English professor and the dashing fighter pilot. They alternated weekends, taking turns staying at our house with the little brook in the backyard, while the other sat at home and, I guess, wondered. They both wanted to marry my mom, and though neither was an outdoorsman, they both pretty much had to play that part in order to suck up to my mom via her run-amok-wild-through-the-woods eldest child.

One surprise weekend they both showed up at the same time. That was an amusingly tense two days. The professor, the fighter pilot and I ended up fishing on the Saranac during spring flood. The plan was to cross to the other side of the river to a promising-looking eddy. We were all wearing chest waders. Two weeks before, the husband of a close friend of my mom's drowned in the Saranac while fishing by himself, while trying to cross the river. He walked into a hole and his waders filled and he was sucked down and it took 'em three days to find the body. When they finally did, it was bloated so badly they had to use dental records.

When the English professor, the fighter pilot and I got to the raging middle, the water was deep enough that it began to fill the rubber death suit I was wearing. As the current began to pull me down, I remember asking the sky: "Why did we come here?" I was the only one of the three who knew the Saranac, who knew the ways of the North Woods. I would not have tried to cross the river at this point, at this time, the time of whirlpools and fearsome tales. But somewhere deep down I suspected one of these two men would become my next dad, and I needed to act tough, I needed to keep my big mouth shut, to not speak unless spoken to, to make sure I got taken out fishing again when these two men were finally winnowed down by my mom to one. The last man standing, as it were.

The professor grabbed one of my hands and, possessed by a nightmare fast in the making, started pulling me roughly to the opposite shore. The fighter pilot grabbed the other hand and started pulling me back toward the shore from which we had come. A many-leveled tug-of-war was transpiring there in the middle of the Saranac. Both men had the same look: How will we tell her what happened? And: How will she learn who saved the day?

The professor won, and we made it to the other side, gasping, dazed. Did I hold onto one hand more tightly and let go of the other? I don't remember. The fighter pilot soon crossed too, and an argument ensued. I thought maybe there would be a fight. It had been so long since I had a dad that I wanted two just to make up for lost time, two willing to fight over me. But they weren't fighting over me. I was the player to be named later. As their yelling reached a very disconcerting anticlimactic shoulder-shrugging anticrescendo, I looked out into the Saranac and saw trout hitting the surface. Still shaking, I got up to cast my Mepps Spinner into the flow. It was only then that I realized my pole was gone, a casualty of the mid-river terror.

After that weekend we never saw the professor again. He had won the tug-of-war but lost the battle. Perhaps the fighter pilot recounted the story more gloriously, perhaps, in a case of occupational irony, he painted the timid professor as too wild and irresponsible, the way he pulled me toward the other shore. Perhaps in that frantic few moments in the middle of the raging Saranac, the childless professor, never before married, learned what it would truly mean to suddenly become a family man. Either way, that crossing of the Saranac River taught me that, in this life, you can rarely have it both ways, though sometimes you want to so badly.

The Kentucky

Every day at five o-clock, the bourbon distilleries that lined the shores of the Kentucky River in and around Frankfort released whatever it was they released into the slow-moving current. I believe it was called mash, but it might have been called something else. During summer evenings, the kids of Frankfort would descend upon the murky green river and gulp as much of the mash (or whatever)-filled water as their bellies could hold, believing on some sub-level of intellect that can only be mustered in the craniums of young people playing freeze tag with hormonal onslaughts that the distilleries were releasing into the river some manner of bourbon, and that, if we drank enough, we'd get not only a buzz, but a free buzz. The fact that, if there was one thing the distilleries would likely not release into a river it would be bourbon, did not enter our heads. They were distilleries, by gollies, the thing they made was bourbon, and that's what they must be pouring into the river for our enjoyment. Sure, maybe it was defective bourbon, but our standards weren't that high.

The Kentucky is a lazy, meandering river eroded several hundred feet into some of the most verdant hills this side of central Belize. Near the state capital, pretty much in the middle of town, there was a place called Melody. Up on the hillside there was the crumbling foundation of an old building, and, just below it grew a monstrous tree. Rumor had it that, years ago, there was a flood that filled the entire valley, and, while the water was as high as anyone could ever remember it being, someone bolted a full length of telephone pole guy wire to one of the highest branches of that skyward-reaching tree. To that guy wire, 40 feet of heavy chain was doubled up and bolted, and to that chain was tied a 20-foot section of the kind of rope we used to have to climb 20 times in gym class when we were caught peeking into the girls' locker room. It took two people to get that long section of thick rope, attached to the chain, attached to the guy wire, attached to the tree all the way up to the crumbling foundation. It was well known in Frankfort that you'd never be considered a man (read: you'd never, ever have sex) until you grabbed hold of that rope, jumped and soared high above the Kentucky River and let go. This was our Sioux Sun Ceremony, our rite of passage. Even grown-ups estimated the apex of the swing at more than 50 feet, a height that, when combined with the speed of the swinging action, resulted in belly and butt flops of the worst kind imaginable, as well as acts of almost unbelievable physical acumen and beauty—flips, twists, air ballets.

The best thing about Melody, once you were confident enough in your ability to take the ride any time, day or night, drunk or sober, was the fact that, very often, when they surfaced, many females had mysteriously become separated from all or part of their attire. We always assumed it was on account of the forceful impact. The seeming impossibility of losing a bikini bottom while hitting the water feet first caused me to consider taking up the study of applied physics. Some years later, I realized I could more quickly arrive at the truth-of-the-lost-bathing-suit-parts-matter by studying poetry and literature. No matter the cause, sometimes those parts sank, sometimes they floated slowly to the surface.

I lost my virginity one languid summer afternoon right at Melody, right in the warm water of the Kentucky River. Not next to it, or on a boat or dock on it. RIGHT IN IT. There were only six of us down there, three of each gender. Four were partnered up, which left me and the other girl, who was not, by all accounts, a virgin, and had not been for quite some time. We were both 16 and had known each other since the fifth grade. She was lovely and had really dark eyes that looked like alligator eyes when they're moving in for the kill. Seductively predatory eyes. It took longer than it should have for her to surface after she hit the water after the long floating endless fall from the rope. The impact had apparently separated her from her bikini bottoms. I had prayed to the river for this turn of events, and the river had answered me with a hearty thumbs up. She did not seem overly flustered by the fact that, try though we might, we could not find the lower half of her swimwear. The river had itself another souvenir. By the time she breast-stroked her way to where the water was only breast deep, her alligator eyes, just millimeters above the water line, torpedo-locked on mine. The other two couples were already face-to-face, undulating their way toward riverine bliss, murmuring in a way I had never heard in real life. Two sets of rhythmic ripples were lapping on shore.

It was with some degree of chagrin that I learned later there had been a whole herd of onlookers sitting up on that crumbling foundation, taking it all in, making all manner of interesting bets. (Though, later, I realized that they were not voyeuristic spectators but, rather, witnesses to the tale I planned to recount, and did recount, at least 12 million times.)

Shortly after the girls left, the five o-clock whistles blew at the distilleries, and the mash was released into the Kentucky. Don't know why, but I never drank from that river again. Something had changed. It was now holy water.

The Ware

It was not the first time my then-closest-buddy Norb and I had decided that the best education we could gain lay not in the stuffy classrooms of Tidewater, Virginia, but, rather, along and in the muddy estuaries of the Chesapeake Bay. The Ware River, more than a mile wide where it emptied into the Bay, and prone to all of the current-based vagaries that can be mixed into one swirling liquid mass, was generally our hooky venue of choice. Norb, whose parents thumped the bible with regularity and aplomb, were not exactly privy to the fact that we often spent school days educating ourselves in the way of waves and tides. My mother, now married to and in the process of divorcing the fighter pilot, figured that, if I was out and about in the great outdoors, then at least I wasn't back in jail. You find good where you can, I guess.

 The Ware did not stick out in any remarkable geographical way from the other estuary rivers that cut deep into Virginia's Middle Peninsula: the Severn, the North and the East. It did, however, stick out insofar as its shores were home to the most affluent class of people in the vicinity. Old-money mansions loomed from source to mouth along the Ware, and in many of those increasingly decaying mansions dwelled the bored and unproductive progeny of the idle rich, some of whom were our age, female and, like Norb and I, predisposed to spending as much time out of school as in.

We had spent the early hours of this particular non-school day poaching tennis courts at the august Ware River Yacht Club (where we were not exactly dues-paying members) and decided to swim across the river to investigate a rumor that a particularly tasty feminine dish was pining away for company over in a place named Zanoni. (It makes me dizzy to think there was a time in my increasingly flaccid life when swimming a mile-wide river on a preordained fruitless nookie quest was something I did without hesitation.) We were as brown as the mud that lined the Bay's estuaries. Halfway across, Norb, twice my size, cramped up hideously, and I had to perform one of those arm-across-the-chest sidestroke moves we learned in Boy Scouts. It took a long time to get to the other side with one arm and two legs propelling two people, and, by that time, the tide had turned and was moving swiftly out toward open water.

We fetched land on a windswept point, with no shade, with the sun frying us, our many stinging nettle burns painfully swelling our skin, with the deer flies and saltwater mosquitoes descending on us in droves. Because the tide had moved us so far downriver, we were nowhere near the dwelling of the young lady we were seeking, and seeking to impress. We were nowhere near anything, save sun, mud, bugs and tepid water.

"I guess you're going to tell everyone we know about this," Norb said, smiling, lying on his back and massaging his knotted calves.

"I promise I won't if we only smoke your dope for the entire next month."

"Deal."

Only problem was, the dope, as well as most of our belongings and our car, was on the other side of the river.

"You ready?" Norb asked, his cramps finally dissipated.

"Yeah."

Without another word, we waded out into the Ware and began our long and leisurely swim back to the yacht club. We were seniors, and, in less than a month, we would stand facing the unknown, which is both great and not great. We had played football together, we had been arrested together, defended each other in bar fights and camped and hiked together. But we had no plans save crossing to the other side of the Ware.

Norb ended up in the Pacific Northwest, while I ended up in New Mexico and then Colorado. Though we have visited 15 countries together since and had some mighty adventures on and around other rivers—the Yuna in the Dominican Republic, the Yangtze in China, the Urique in Mexico—those experiences took place among and between old friends, rather than friends. When the adventures were over, he went back to his life, which I have never visited, and I went back to mine, which he has never visited.

The meetings became less frequent, until, now, we haven't been in contact for more than five years—the total amount of time we knew each other before we swam across the Ware River that hot and muggy Tidewater day.

There's no doubt, though, that the lesson we gained over the years from the Ware is with us both still to this day, because, once this particular lesson is learned, it's with you always: Playing hooky is not an elective; it is a required course, a prerequisite in the curriculum of sane living.

The Gila

Two fantasies met, merged and mated at my feet, though I still don't know which one was on top and which one was on bottom. Throughout my life clear up until now, I have always slotted people into two inviolate categories: those who had the nuts to run away from home and those who didn't. When I interview prospective employees, I always ask if they ran away from home, and those that did get extra credit, even if they just finished a 15-year murder sentence. I can always tell if a person ran away, especially when they're lying about it. I grew up under the assumption that, though they might say different, all kids wanted more than anything to run away, even if only for a few hours, even if the move was purely symbolic, even if it was the proverbial cry for help and/or attention that adults mostly believe running away always is. To this day, I have never been able to, I don't know if "forgive" is the correct word—maybe "understand" would do better—myself for never having had the stones to up and run away.

Back in Frankfort, Kentucky, my main miserable kid cohort, Hank Senn, and I talked about running away incessantly. We even had a plan: We were going to save up to buy a Folbot Super kayak (we had a catalogue), assemble it, float down the Kentucky River to the Ohio, then to the Mississippi, then out into the Gulf of Mexico, through the Panama Canal and all the way across the great Pacific to some islands we picked on a map because they were about as on the other side of the world as you could get from our lives in Frankfort. (If you're gonna plan a real runaway, plan big, that's what I say.) We saved our lunch money for an entire school year. Thirty-five cents a day each, $3.50 a week between us, something like $140 by the end of the school year. About half of what we would need to buy the Folbot, rig it with all the necessary ocean-crossing gear and leave forever.

We never even got to the point of ordering the Folbot, and I have always considered myself a lesser human being because we never progressed past the dreaming stage. I still get impatient with people who only dream and never do.

A decade later, David Wright and I were unloading a very well-used Folbot Super kayak—the first one I had ever actually touched—from the top of Dave's 1972 tan VW van. I really wished Hank Senn was there to go with us, but I don't know whatever happened to him, the same way I don't know what ever happened to me. The Folbot, bright yellow, belonged to our buddy Jay Scott, who had traded some of his artwork for it. The boat, 17 feet long, had sat unused and dry-rotting in Jay's backyard for numerous years, till David and I asked to use it to float a four-mile section of the Gila River in Southwest New Mexico.

The last words we heard before we entered the complicated runoff-swollen little river was "those stupid moron idiots are dead meat." We, of course, were the stupid moron idiots in question, and the man uttering those words was a professional river guide down from Taos who simply could not believe his eyes when he checked us out. He and his cohorts had already been on the river for three days, having paddled all the way down to the Middle Box, our put-in, from the Gila Cliff Dwellings National Monument. They had whitewater kayaks and a little raft and a look of knowing and understanding, though they were no older than us. We had a 17-foot sea kayak, no life jackets, one two-bladed paddle between us (Dave had his half, I had mine) and a six-week-old puppy I'd just adopted from the pound a few days earlier. To the guide's stupid moron idiot comment, we retorted with the retort of the lame and doomed: "Don't worry, we live here," like our mailing address would somehow protect us from the vexations of water moving from high snow to low desert with reckless cheer and enthusiasm.

The guide had used terms that we pretended to comprehend: sweepers, strainers, suck holes of certain doom. We nodded sagely and tried to muster wry grins that let this man know we were men of the Gila Country, faux Aldo Leopolds, reincarnated Ben Lillys, Geronimos in well-worn Army-surplus attire.

The only good thing that happened during the Gila trip was actually a non-horrendous thing: We swamped for the first time (there would be many, many more) just around the very first bend in a place where the guide and his entourage could not see us, though I'm certain they knew what was happening. Were they formulating a rescue? By the end of the journey, we had trashed Jay's boat, rescued my puppy, who was being swept downstream, literally by the tip of his tail, lost two shoes (one each) and a camera. That we did not die was flat-out a miracle.

At the time, I was writing for the *El Paso Times* and had blown off some dumb-ass self-serving press conference–type assignment to go on our ill-fated Gila River foray. My editor, an old-school, whiskey-drinking, foul-mouthed, good-ol'-boy from Abilene, who later became a monk, believed that every young reporter's middle name was "goddammit!" He was displeased when he learned that I had dunked the press conference assignment he had personally laid on me. I told him it was cancelled because the mayor had died tragically from a killer bee attack or some such nonsense and that, being the ever-vigilant reporter I was at least pretending to be, I had endeavored to seek out and destroy an even better story than the one I was assigned and, were it not for the fact that my camera was now floating its way toward Phoenix, I would have had the story of our Gila high-water misadventure to him ready to run tomorrow.

To my eternal amazement, the editor was intrigued. He sent a courier up from El Paso the next day to bring me a new camera and told me to get my posterior back on the Gila ASAP and don't come back till I was either dead or had good photos. A few days later, I was back at the put-in with that irreparably damaged Folbot, camera in hand, Jay Scott, whose boat we had demolished, acting as my model. The river had gone down several feet; there was not enough water to even float the boat, which couldn't float anyhow, on account of the fact that its entire port side was crushed. So, I sat the Folbot on the bottom of the foot-deep river, made it look like Jay was paddling, took the photo, sent it off with a story that was half exaggeration, half outright lie and half completely inaccurate misrepresentation, and saw what probably was the main fantasy of my life since I gave up on the notion of running away all those years ago reach back-assed-fruition: My first first-person outdoor story saw ink, and that lead to a fairly fruitful and ultimately frustrating career as a bullshitter of the backcountry, a bona-fide rent-paying outdoor writer. That one river trip, simultaneously ill-advised and well-advised, crossed me from the realm of "this is what I'm going to do" to "this is what I do." And, no matter when or where that happens in your life, it catches you by the throat and makes a big point.

And I realized two profound things when I saw that story in print (a copy of which I still have). First, I realized that the main sin is a poorly told lie—especially if that lie has any ties whatsoever to a river. And I realized that I had, indeed, run away after all. It just took a while.

The South Platte

The first freelance writing assignment I ever got in Colorado was for a little Denver alternative-alternative weekly called *Up The Creek*. I had spent my first few weeks in the state a fish out of water, a small-town person suddenly living in a big city, getting to know a new place in a way I believe is endemic to economic refugees. When you move somewhere because you're destitute and have no choice, you sniff the air trying to pick up the scent of salvation, trying to avoid the stench of further humiliation. I had been unceremoniously laid off my job at the *El Paso Times* and had been offered a couch in Denver by an old friend from Virginia.

I walked up and down the paved bikepath that follows Cherry Creek all the way downtown to the South Platte, waiting for something magically to appear. Under every bridge along Cherry Creek dwelled the unshaved and unwashed, and they ended up being my salvation. I decided to live among Denver's street people for a few weeks, to tell their story to all 12 of *Up The Creek*'s loyal readers. I hooked up with a group of particularly odious ex-cons, child molesters, falling-down drunks and left-field psychotics that hung out around Confluence Park, where Cherry Creek joins the South Platte. There was much cheap alcohol, and the threat of sudden violence hung in the air, verily permeated the air, like the stench of the Mile High City's infamous brown cloud. I did not tell these street people I was not really one of them; that they could not tell both made me proud and made me recoil. Somehow, I had become a bottom-feeder chameleon, able to lie myself into the underbelly without even lying. I had no money, was living off the largess of an old friend who clearly, and understandably, regretted offering to put me up. The cheap wine was starting to taste not only good, but normal. The research phase of the story was taking much longer than anticipated. I missed several deadlines.

These were people I had no need to be embarrassed in front of; they did not care that I had lost my cushy job and had to leave a place I loved to move north, penniless, like someone in the dust bowl days. I did not need a valid driver's license to open a bank account with these people. I worked menial day labor, pawned a camera and a typewriter—the tools of my old trade—and sold plasma several times. Found out it was mostly used to make the base for shampoo. (Unwashed as I then was, I thought that was funny. Consider it less funny now.) Good money, no work. Just lie there and get your body fluids sucked away through a dirty tube in a roomful of glaze-eyed winos.

My gaze started to drift more and more to the South Platte, the closest thing to a touch of nature then in my life, a river, though sad looking and beaten down by its frequent run-ins with dams, irrigation ditches and foul urban discharges, that still retained a semblance of the kind of dignity that things born and raised in the high mountains always retain. The fouled South Platte through the heart of Denver in those days would have had every reason to throw in the towel, to just give up and proclaim, "I am no longer a river, I am now a ditch, a sewer, a prisoner!" But it didn't, at least not to me.

It is our tendency to mentally follow the flow of a river down toward the point where it is no more. I have always preferred looking upriver, to the place of the river's beginnings, to the place where you can step across the mightiest of waterways without getting your feet too wet. I am a backpacker by nature and disposition, and, like most backpackers, I approach and perceive rivers from the sidelines, coming at them in a perpendicular fashion rather than parallel, more quick to dry off than to get wet. I camp next to rivers and look at them as sources of hydration, or obstacles that need crossing—sometimes dangerously—via wading or rock-hopping or flimsy and slick log bridges. Or I look at rivers as ambiance-enhancement units that are nice to listen to while I'm eating ham sandwiches on the side of the trail. I am more at home in the places where rivers come from, rather than where they go and how they get there.

My homeless pals spent a lot of time in the Denver Public Library, where it's always warm, and so did I. One day, I looked in an atlas and found the source of the South Platte. Up near the Continental Divide, where it's always cold. Closed the book. Wrote the story for *Up The Creek*. Moved up to a town just west of the Divide, the spine that tells even the mightiest of rivers where to go. The South Platte taught me that, if circumstances can't beat down a river, they can't beat down a man.

—MOUNTAIN GAZETTE, #80

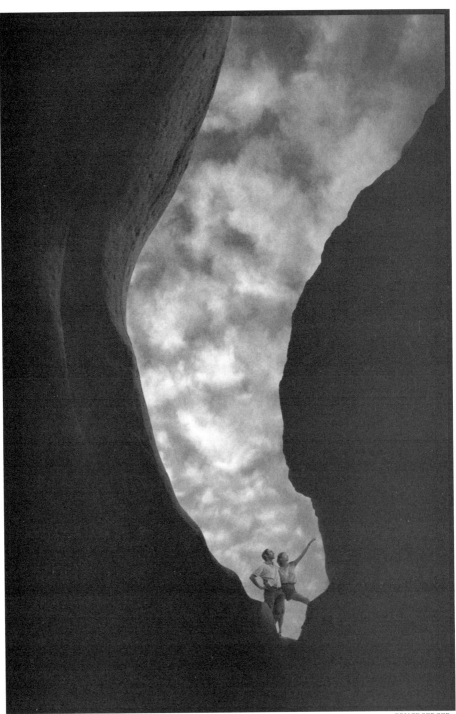

BRUCE BERGER

There Was a River

BY
BRUCE BERGER

The following records a two-week trip through Glen Canyon in October, 1962, months before a dedication ceremony for a dam served simultaneously as a canyon's obituary. The journal also records, less explicitly, a discovery of the English language which, like many things we accept from birth, we only find again with a kind of surprise. My rivermates took snide notice of my scribbling while grease hardened on the dishes it was my function to wash, and even made nasty asides while parching for pothole coffee water as I tried to snag metaphors before they disappeared.

Typing it fresh took my mind off a week of diarrhea contracted the last day from settled river water, when potholes ran out, and I revised once, a year later. But I could think of no interested publisher, and my stupor continued into my second year's subscription of *Mountain Gazette*, over a decade later, when a piece on the desert brought home the obvious: that they weren't hemmed in by the mountains. I have since tried to render some baby fat, but have had to stop short of rewriting a subject which is gone forever. So the narrative should be seen from its occurrence in 1962.

October 6

Beginnings are a bore. If a trip really begins from the first blow to the brain, the memory begins with the push-off: The preliminaries make dull reading even if one is lashing balsa logs on the slopes of western Peru. As it was, we made most of our preparations in a supermarket in Grand Junction, Colorado. Perhaps, as the world is leveled for their benefit, it is in the supermarket that our future adventures lie. Suffice to say that it carried everything from bonito flakes to a 60-cent copy of *The Celtic Twilight*, that I bought the latter, and that we wound up with $80 worth of green stamps, a sales receipt over six feet long on which I was able to write a very newsy letter and an acute case of museum feet.

The idea of the trip was simple—four of us were to float approximately 120 miles down the Colorado River, entirely within the confines of Southwestern Utah. The distance in a straight line would be considerably less, since rivers never aim where they're headed, but our own distance would be more, following curiosity as well as gravity.

Calm water, uninhabited land—all of our passage was scheduled for inundation by Glen Canyon Dam at the close of a year that was already nearly over. I was prepared for beauty and desolation, and felt privileged to see it before it was destroyed. What I didn't know was that the trip was to bring so much into focus, to tinge its pleasure so strongly with the bittersweet.

Three quarters of our party drove from Grand Junction to Blanding, Utah, to pick up the supplies not destined to be eaten. It was an improbable trio for that dusty Chevrolet. From my own driver's license: male, 24 years old, blond hair, blue eyes not permitted to drive without corrective eye lenses.

The leader of the expedition was Miss Katie Lee, a professional folk singer who had already traveled the river 14 times and acted as if she owned it. She has explored many of the Glen's side canyons, and given 25 of them the names that appear on the map. She got her start folksinging by learning the guitar as she waited to perform bit parts on Hollywood sets; when it became possible to make a living in the real world, she evacuated. She crested a vogue of sick songs—a brief period when people enjoyed Freud as much in the night club as on the couch—but came into her own singing songs of the West, improper ballads, and Spanish songs that allow her to show that, unlike most folk singers, she can play the guitar. Her temperament is volatile, by turns gentle, amused, bitter and impossible. I recall a period when she tried to water down her vocabulary and wound up saying things like "Oh shit! I mean hell." She was unquestionably in charge, and the rest of us, more quiet by nature, were able to absorb her fire and be warmed by it. One of my favorite people.

Natalie Gignoux (the name is Huguenot and pronounced jih-NOO, like French for the knees) was in the process of selling the hugely successful taxi business she started in Aspen, against the wisdom of those who told her that, a) the town didn't need a taxi business, and, b) it couldn't be run by a woman if it did. She took tourists on Jeep trips into the high country, managed a staff of unpredictable drivers who would have preferred to find their salaries floating in bottles down fishing streams, and specialized in civilizing the impossible. She is as strong-willed as Katie, but is more likely to have her way by just sauntering off and having it. Also one of my favorites.

Since Leo was described to me as a river rat, I looked unwittingly for the rodent, and found it. He is tall with a face that accentuates the vertical, but his close-cropped hair and roundness of feature—including a small mouth and two prominent front teeth—add a certain woodchuck inquisitiveness. He looks like Humphrey Bogart.

Leo works as a garage mechanic in Riley, Kansas, but his job is merely a springboard for the next leap. Last year, Leo hunted jade in Wyoming and discovered the second largest piece ever found in the United States—2,200 pounds. He propels his own life as if it will never shake the world, neither will his world fall to pieces. Next year perhaps Baja, California. Lotta country out there a fella might look into…We picked up the raft, the cook box and other equipment from a good friend of Katie's who runs a weather station in Blanding between trips through the Glen and Grand canyons. We passed up his invitation to stay the night, for a neutral and deserted motel. We wished to say goodbye slowly to a tile bathroom...

October 7

    We meandered the road from Blanding to White Canyon (pop. 2), approximately 90 miles of sandstone and sagebrush in endlessly refocusing patterns. The pavement gave out after ten miles. We passed the turn-off to Natural Bridges National Monument, and the road continued like a bad scar. We bounced up and down in the seat and mentally I rerouted the road in straighter lines, wondering vaguely why it went where it did, knowing there was some not very interesting reason. Roads like that set you adrift.

    We rounded another crest, and there was the river. It was about what I expected—a width of muddy brown water flowing through one more depression in sandstone, this one a bit deeper, lined with a vibrant and shocking green. Save for the vegetation it was neither inspiring nor breathtaking; it was merely there.

October 9

    Two days late, we have finally blown up the raft, labeled all the cans with magic marker in case the paper soaks off (four successive cans of peaches were labeled Peaches, More Peaches, Still More Peaches, Peaches Goddammit), piled box upon box on the raft and—best of all—shoved off. The raft, the apparent property of the Museum of Northern Arizona, is a 16-foot monster of billowing black rubber, shapeless as a dead whale. An interior skeleton of wood has been improvised to hold the motor, so that the bulbous mass remains rigid, fends off water and doesn't buckle in the middle. There is a constant pull and give between the rubber and boards, as if a fat woman had swallowed her corset. It is designed as a 12-man raft, but with the heaping gear there is only space to seat two abreast on each side between the supplies and the 35 horsepower outboard. Otherwise one must move about the rim or jump in the river.

October 10

    The mellowness of the night before was rudely shattered by a trio of blunders that gave us a sickening first-morning-out-in-an-awfully-long-parade sort of feeling. The campfire flames leapt up to the bacon grease and made the grill look like the Chicago fire as Nat sweated blood for every charred hotcake. As we pushed off the propeller fell off the motor, and for half an hour we waded in icy water scouring the mud with our feet, performing a sort of bump-and-grind through the ooze. After we covered the same area three times Leo put on our only spare, and we would have to accept whatever its fate. Off for the second time, Nat's hat blew into the water, and we barely pulled the raft around in time to retrieve it. We hoped it was true that mishaps stop after three.

We spent most of the day drifting on the river I have put off describing because it is so difficult. And so simple: a flow of brown between sandstone cliffs, with strips of willow and tamarisk along the banks. That is all—and nothing. First light brands the rimrock with cayenne, then bleaches—depending on time and the locale—through saffron, peach, salmon, sienna and rust. Light slants deeper into the bas-relief, trapping shadows between illusions of free-standing figures. The sun climbs, light flattens. Glossy streaks of dark, shining blue in the sun, plunge from the rim and sweep like murals of Spanish moss. Called desert varnish as a phenomenon, tapestry walls as an effect, it forms as water seeps from the rim, leaving streamers of lichen and deposited minerals. And below all this moves the endless serpent of water, remixing each eddy and backwash, running coffee, chestnut and sable into caught blues, receiving the sky and tossing it back mottled in beige.

Between water and stone lies a brief transition, a talus crumbled from the heights and projecting like toes of balance for the sheer walls. The oldest of this debris has broken into loam for the willow and tamarisk that line the banks. This green, so shocking in contrast, sheers water from rock the way the horizon sheers rock from sky. All landscapes interlock, but here each piece is also honed to a unit.

From the shore Glen Canyon is spectacle, but floating in passage it invades the five senses. A 35 horsepower motor sounds aggressive, but in deep current it provided little speed. Its function was rather direction, to avoid the thrashing of canoe paddles while guiding us to campsites and side canyons. Still, the scream of gears deadened the river's voice, along with our own. If one had to communicate one began with the hands and concluded with a voice pitched for hog-calling.

But if the river by-passed our ears, beneath we could always feel the swell and slide, the meanders and eddies where power insists. It assaults even the nose with a penetration of slurry and mud. One is forced into the landscape until it evicts the self and becomes, as Eliot said of strong sensation,

> *Music heard so deeply*
> *That it is not heard at all, but you are the music*
> *While the music lasts.*

The river has doubtless been sharply defined by minds down it many times. Side canyons become known, formations stand out as landmarks, all is anticipated. One knows to look up when old granaries merge with the rock, when tributaries loom. But I will know it once, and am glad now to be worn out. My senses are exhausted, grateful for simple flames, hot rum and a swelling moon...

October 12

We spent the day exploring areas that promised Moqui ruins. If the destruction is weeks away, this is the last chance. We tried a bluff that seemed to retreat along a shelf, hiding the wall behind it, thinking it a spot likely to have been missed. What we found was very much alive: a two-point deer. Mixed with my thrill was a burst of ego, since Katie had never spotted one during her 14 ferret-eyed trips on the river. But up ahead Leo gestured to us frantically, and two more bounded gracefully from beneath his perch. We watched them leap and freeze, leap and freeze till they disappeared. Several minutes later we glanced back and caught their heads arching silently over the current as they swam the river.

We climbed to Leo's perch and found the deer had been startled from a beautiful emerald pool. Katie announced it was movie time, so we waited while she returned for the cameras. After they were dutifully arranged she descended to swim the icy waters, while we worked the machines from above, ringed on the upper tier like med students at an operation. (As it turned out, the operation was successful.)

We named the pool Two Deer Spring, though it seemed ludicrous to label a spot anonymous for millennia and fated to succumb within the year. We rounded a bend and came upon another sight that had nothing to do with ruins. We suddenly found ourselves dwarfed by an enormous hollowed-out dome with a round window on top, a sandstone Pantheon, plunging to crazy formations like lava frozen into basins and whirlpools. Around the dome was a faint seepage of water from which fell a delicate band of maidenhair fern, repeating its motif like decorations for a Pompeiian dining room. It was cool and damp as a wine cellar and secret from the river. I have decided that to really see it one must walk up and down both banks, which would be physically impossible even if there were time. No wonder we are about to lose this—how can you tell people when you can't even find it yourself?

October 15

For once we spent the day entirely on our feet, exploring the opposite bank. Our first destination was a place called Music Temple, a cavernous room that opens dramatically from a brief and narrow canyon. The floor is a pool that mirrors a hemisphere of black rock, and the only exit is a cleft that begins 75 feet above the floor and reaches the sky through hundreds of feet of twisting precipice that must assemble in the mind from one glimpse below. But its greatest qualities are not visual.

Katie stationed herself against one wall and began to strum her guitar. The whole canyon suddenly filled with the sound of vibrating strings, as if the single guitar had gained the resonance of an orchestra. Her voice entered and bloomed with its own resonance, yet never blurred with echoes and overtones. The music, like the light, came from nowhere, but filled the whole canyon with the body of its sound, as if Katie's song swelled from the canyon's own throat. (It must be the distemper of the times that Utah, with Music Temple and the Mormon Tabernacle securing its primacy in acoustics, glorifies what it built and shrugs off what it was given.)

On the way back, Nat and Katie climbed toward the back of Music Temple to see if they could look into the chamber where Katie sang this morning, but they found another chamber in the way, hanging in mid-distance. They returned to Leo and myself, and we began the descent. All of us suddenly realized how tired we were, and Nat was so abandoned she started sliding on the sandstone, missing an occasional Moqui step, and caused the rest of us to miss an occasional heartbeat. When we got back we were so worn out we nearly had seconds on the rum (I could down five any night, but the supply doesn't permit loose behavior). As final proof of weariness Katie, for the first night, did not sing.

Yet even in stupor it is amazing how much less time it takes to do things than at first. Campsites become home in a hurry. After the non-humanity of so much stone, we wallow in the domesticity of a well-screened grate and selected bedsites. The firelight itself throws up walls one must brave, in opposite directions, to get or pass water. Tonight, everyone works like cogs of a greased machine. Crass orders are still sometimes the rule, but no one in the party would dream of bristling under crassness. Part of it, of course, is knowing when to stay out of the way. One does not impede Nat while she is cooking unless she wants something, in which case you better jump like now. But out here in this physical and mental health I'm getting so that I relish the wood-gathering and dishwashing, the way others enjoy cooking, mechanics and bartending. It works—tonight we are in bed by 7:30.

> *Mend my holey pocket*
> *And I'll caulk your leaky bucket;*
> *Life's a mutual fixit*
> *Or a universal fucket.*

October 17

After nearly ten hours of untroubled sleep, I woke to find it still dark and a soft rain falling, the sort of rain sometimes referred to as Scottish mist. Sometimes even in clear weather we waken late if there is a high wall to the east, but with no clarity to lure, you could spend the whole day curled in your bag, avoiding the chill where the rain has seeped in, wallowing in your imagination.

We all felt slightly sodden at breakfast, and lingered over coffee making snide remarks. (What did you use for a coffee pot, Paracutin? It must have a hardness of seven. And that old standby, it's like the Missouri River—too thick to drink, too thin to plow.) Katie looked up at a raincloud and said, "I don't think this is a 24-hour rain. I think this is a 48-hour rain."

Leo looked at her blankly and said, "In thet case, ah think ah'll draw the string to my poncho."

When we finished stalling for dry weather, we crossed the river to Twilight Canyon and spread our sleeping bags out to dry under a broad dome. From there we hiked the canyon turn after turn, unable to stop, because one must always see around the next turn, even if it's only another turn. At last our feet, exhausted from clawing their way over boulders, informed us it was the end. When we got back to the dome we ran excitedly to our sleeping bags and found them still soggy and rank.

After lunch, the old routine—look for one thing, find another. We climbed to a notch reputed to hold a scaffold house built by the Moquis, and caught no trace of it. Gazing casually across the river our eyes were stopped instead by an enormous rock perched on the wall's edge, greyish green, nearly round with an elongation toward the vertical.

Elephant Ass Rock became the subject of much speculation. The binoculars revealed a surface of strange cracks almost like canals that laced it like a roadmap. Perhaps a relic from some earlier age, like so many of the smaller ones caught in pockets of all strata, it looked more like a space ship about to take off. Katie swore she was going to get up there the next day to investigate, but on reconsidering our schedule, our devotion to science waned. You can't expect to understand everything.

We intended to push on as far as Forbidding Canyon so that Nat and I could hike up to Rainbow Bridge the next day. It is the only well-known spot on the river, being a National Monument, and we had it in our minds that we really ought to see it. We already knew what it looked like from many pictures, and it hadn't the glamor in our minds of the little-known side canyons. Katie, by small remarks, let it be clearly understood that she wanted to sail on past it, but we pretended not to notice and dutifully pulled up at the campground. There was mansign, the first real evidence we'd seen since we started. We could make out two outhouses (his and hers? guests and guides?) and a tent where tourists could stay for the night. Yes, real tourists; they are transported by commercial launch from downstream at Kane Creek so they can hike to the bridge, then repose in comfort—a soft touch from our point of view, rugged for modern tourism. We beached timidly downstream, and Katie went up to arrange for one of the guides to give assurance of our approach to the people who were to meet us at Kane Creek next Sunday. Meanwhile the rest of us began to have some doubts about wanting to stay. The outhouses glimmered ominously. Katie returned.

"What news?" asked Nat.

"There's two fat bitches, a drunk and a cretin and another party coming in at nine tomorrow."

"Let's get outta here!" It came out like a chorus.

"Ha-hooooo!" shrieked Katie. She leapt, scissored, fell back into the raft and we shoved off.

We camped at the edge of a broad soggy beach, farther than we wanted to from our rubber beast, the weather still bleak. After dinner, the stars came out and once more we thought the storm was over. Katie sang her first concert in three nights. By bedtime the sky was black again, but we were grateful to have sailed on from our glimpse of civilization. It left a bad taste in Katie's mouth, and we newly valued our own small tribe. I have wondered since whether the people were as nasty as they looked to Katie, and whether the peace and stability of the river wouldn't have shown any visitors in an unfavorable light. The trouble with mental health is that it finally makes the entire world look sick. Perhaps the river will make misanthropes of us all…

October 20

Day dawns in a dozen shades of deepening blue—last night's blood and thunder was the storm's farewell. As the trip neared an end, our thoughts turned to a suitable finale for Nat's movies. The scenario: a shot of Nat cooking breakfast, the camera zeroes in on the skillet, the spatula turns over a pancake on which is written THE END, Nat flings it over her shoulder into the river, the camera pans it as it floats downstream, the muddy water washes over it and slowly engulfs the writing as it vanishes. We fried a pancake till it was brittle and wrote THE END on it in magic marker. The camera began to roll. A close-up of Nat turning the pancake over. It says THE END. With a look of utter nonchalance she flips it over her shoulder and out of range. The camera swings wildly, fishing for it, finds it floating face down in the river. It sinks. We push off for our last complete day on the river forever.

We spent most of it making our way up the right fork of Dangling Rope Canyon, another canyon Katie named and would be saying farewell to. By this point our methods of getting up canyons defy English prose—though perhaps Oriental characters could suggest some of the possibilities. Foot-feeling, ass-wedging, straddling, jimmying, spread-eagling, hand-stirruping, knee-boosting, shoulder-bracing; scaling rocks, trees and each other: after two weeks we move like a team of polished contortionists.

On April 11, 1956, with a grin, President Eisenhower triggered the first blast, and on January 21, 1963, the dedication was celebrated by the Bureau of Reclamation's Floyd Dominy, with the Sierra Club's chagrined David Brower in attendance (those interested in that moment are referred to a delicious description in "Encounters with the Archdruid," a three-part *New Yorker* profile on Brower by John McPhee). But even completion of the dam did not end the bickering. As it filled, the Lower Basin states complained they weren't getting their share, Udall opened the floodgates, the Upper Basin threatened to sue if they didn't get their water back, the Lower Basin threatened to countersue, and so on.

It is said that if the world were reduced to the size of an orange it would be still smoother. But to us, the specks that crawl its surface, it is full of staggering peaks and chasms, towering bluffs and shoreless oceans. These are the places we love and return to, that we will walk, drive or climb miles to see, and carry them home inside us. With their sense of the past stretching beyond us and transforming under our eyes, we watch in them the same process that flows in our blood. When we destroy them it is part of ourselves we freeze. It was years before I could bring myself to go back, and see firsthand the slow encroachment of Lake Powell. But as the world's new comfort is enforced, we become desperate for the wild remains. So I have been back to hike what's left, up from the lake or toward it from the rim, unable to avoid the slow erasing of Glen Canyon.

Its veins bleed out beneath
A winding sheet of water. Here and there
Cottonwoods grapple for air
In side canyons, their leafless tops
More like upturned roots, and a few trapped deer
Come to the end and stare
As the cool impartial waters rise
And include their terror-stricken eyes.
The rest is rippled stone
As the new creatures water-ski
Over the mask of death—
Though doubtless some catfish still ask where
Such rapturous mud has gone,
And sandstone, if it remembers, still
Dreams a return of the sea.

—MOUNTAIN GAZETTE, #31

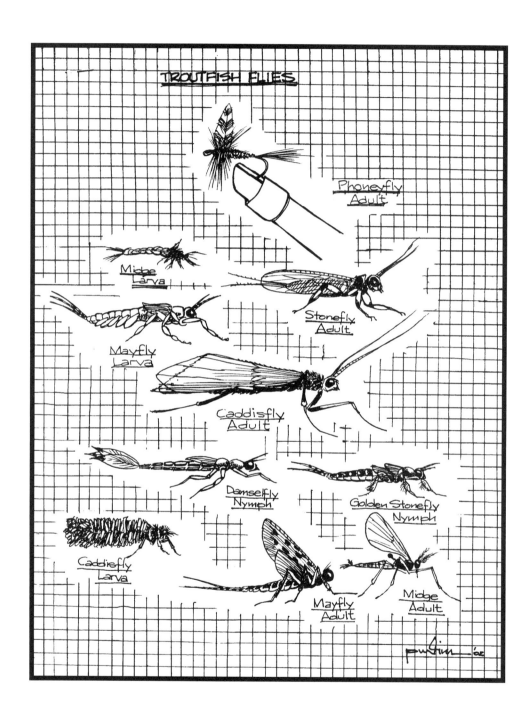

For the Sport of It?

BY
GAYLORD GUENIN

My daily official duties are such that, after completing them, any more writing is a painful labor; but yet, in view of the fact that I take a deep interest in all legitimate matters connected with fly-fishing; that I have been a lifelong lover, and for very many years a keen follower, of the gentle sport; and further, that I am indebted to you personally for several nice suggestions or hints in your writings, I feel that it would be simply churlish of me not to give you a few words in reply to your courteous letter of inquiry. However, I shall ask you to give full credit to my statements as to lack of time, and consequently to make due allowance for haste, condition, and brevity.

—From *Favorite Flies and Their Histories*, Mary Orvis Marbury, original printing 1892. The above is a portion of a letter from Gen. O.D. Green, Fort Leavenworth, Kansas, to Charles F. Orvis, and it also expresses the sentiment of this writer.

Like a ginseng root, the old man, so weathered and bent. He was a Montanan, perhaps the first Montanan. At least the second. In the early 1900s, maybe before, he filled huge cans with fingerling trout fish and mule-packed them into the high mountain lakes above his valley. He loved the mountains, the lakes, the trouts, and the sport of it. Before the ginseng root, there had been no fish in those lakes. Now there were, and the Finlanders and squareheads, when they were not digging coal and dying, tossed dynamite into those lakes and blew the trout fish out and mule-packed them back down to the valley to be salted and dried and eaten when it was time. It was their way, to blast the fish, when times were hard, when times were good, whenever it was time. But the lakes did not die, nor did all of the fish, and the old man who looked like a root always returned for the sport of it. Even then in the early 1900s, or maybe before.

And he showed others, and he showed me, and it was for the sport of it.

THIRTY-ONE: *A week later, Lobo Neves was appointed governor of a province. I clung to the hope that the decree would be dated the 13th. However, it was dated the 31st, and the digits, by this simple transposition, lost their influence. How deep are the forces that control our lives!*

—Chapter 110 (In its entirety)
Epitaph of A Small Winner
Machado de Assis

But first, for me, was my grandfather, a hunting and fishing guide and a dedicated fly-fisherman, and my father, not a guide but no less dedicated to fly-fishing.

So count three reasons for my lack of success—three fly-fishermen, one who looked like a root, set me on this trail. While other five-year-olds were being schooled in the fine art of war profiteering and influence peddling, I was receiving instruction in the business of fly-fishing, for the non-profit love of the sport of it.

The reader is burdened with that background information only to demonstrate that I am not a Johnny-come-lately to trout fish killing. It should not, however, be concluded that a Johnny-come-early to a specific endeavor is necessarily an expert. What I know about this is what I know. Undoubtedly, others may know far more, and they can kindly keep that fact to themselves. In truth, this article will offer little instruction in the business of trout fish killing. Anyone who can master a pegboard can capture trouts. I have waded Montana and Wyoming streams for more than 30 years, which is, I agree, an excessive and unnecessary amount of sloshing about. But it is done and during that time, I acquired an hysterical fear of water, a festering suspicion that in the next world anglers will have to account for all of the trout fish killed in the name of sport, and an uneasy fondness for the same bugs that trouts love to eat, and those bugs are more or less what this article is all about, more or less.

Caddisflies, craneflies, dragonflies, damselflies, Mayflies, stoneflies, no-see-ums, mosquitoes, springtails, gnats, midges, moths, back swimmers, water scorpions, and god only knows what else, make up the list of these mountain bugs. Most of them, the most beloved, are aquatic insects, and in truth they are not singularly mountain dwellers, but it should do no great harm if we think of them as mountain bugs, for a while at least, for the sport of it. We hunt the trout fish because it allows us to go to the mountain, that is what I remember. There is no redemption if the trout is hunted away from the mountains.

A place to begin?

With the salmon fly! That should be obvious.

Western anglers call them salmon flies, willow flies and, on occasion, troutflies. The smart asses in the lab say they are *Pteronarcys californica*, and they are members of the stonefly family.

Yis, Henry, that's it; and queer enough it seems to a man of the woods. Lord! I guided a man a year or two ago that knowed everything that books could tell a mortal. He was a full of figgers and facts as a hedge-

hog is of quills, and if ye poked him up a leetle with a question or two, he'd shed 'em faster than ye could pick 'em up. But when ye got him right down to it, he didn't know nothin', Henry.
—Adirondack Tales
W.H.H. Murray

Baby salmon flies (nymphs) are called hellgrammites, which is actually the nymph of the dobsonfly, but who really cares? The nymphs are ugly as sin. If you were unfamiliar with them and found one in your bed, your heart would turn cold and pump wet fear. The nymphs, however, are not dangerous, they will not attack, or bark, or bite, or form a subversive political group, or even cause a rash.

Their fame, and they are famous in angling circles, is not due to anything they have ever accomplished. Caddisfly larva are more deserving of fame. They are extremely clever architects, building tiny stream-bottom homes out of sand, bits of gravel, fragments of wood, and pieces of root. But the salmon fly has our attention because of its size—it may grow to two inches or longer—and because it appears in such huge numbers.

Adult salmon flies are called adults, the first practical name the smart asses in the lab have ever come up with. The adults have wings, the nymphs have none, which is true of most aquatic insects.

The nymphs live in clean, well oxygenated streams, mostly under rocks, for about two or three years before hatching. When the time comes, they crawl from the stream, shed their nymph uniforms and become adults. As a bonus, they get their wings. The whole affair, which normally occurs in June, improves their looks.

Outdoor and fishing editors are inclined to write things such as this about salmon flies:

Then he said I should pick some of the big natural willow flies off the bushes along the stream, tear the wings off, bait them on a hook and throw the affair out so that it would drift back into the deep holes but near the bank of the river.

Tear the wings off! Jesus Christ!

After the bugs become adults they hunker around the stream, sit on bushes and twigs, fly around a bit, crawl into your ears and down your back, mess up windshields, mate, drop their eggs into the stream, and die.

Not much of a life but no less futile than that of the trout fish killers who thunder from stream to stream in hopes of becoming a part of it.

So when you are going after the large trout when the willow flies are hatching, which is the same thing as your salmon fly, you will find that the trout much prefer the female flies as bait.
—More outdoor editor talk

Female salmon flies are likely to be loaded with tiny eggs that resemble caviar but taste like something an insect would lay.

But the finest characteristic of the salmon fly is that it does not become something else if you happen to be on drugs, or some of Austin Nichol's excellent Turkey, or are troubled by visions.

A friend of mine in Montana used to get into LSD now and then before he went trout fish killing (honestly), and I once found him curled up in a hollow stump babbling about things that were after his salmon fly. Sure as hell, he had a salmon fly cupped under his tongue, offering it protection, I assume.

One once saved my life, but I can't tell you about that.

During a major hatch, there may be millions of the bugs in the air. The Minneapolis-vacationing-musician-trout-fish-killer came stumbling up a path. He wasn't running but it was obvious he was escaping. He had been fishing the Yellowstone near Big Timber when the salmon flies began to move and now he was tumbling and swatting and trying to understand it all.

First memories—my dad had walked upstream and I was fishing out of sight below him when the flies began to move. I didn't know what in the hell they were and most certainly did not want them crawling all over me, which is precisely what they wanted to do. I did not know it was for the sport of it.

The escapee-to-be wanted to know what the bugs were. I wanted to say they were flying Rocky Mountain spotted fever ticks so his heart would stop and I could have the Orvis rod he was carrying, but I didn't. We told him we didn't know what they were because we wanted him to leave. We were off to a trout fish kill of our own. Some of the Chinese who lived in a nearby railroad town had the habit of placing illegal set lines in the river, heavy lines that you weight and bait and leave overnight. We, a friend and myself, were off to see if we might pilfer a few trouts from those sets.

The root would have approved. My father and grandfather would not have approved of the set lines, or our cleaning the trouts off of them. The Chinese anglers would have been appalled at such a crime.

But that was long ago.

Mrs. Little Kid's Neighbor Lady Who Was Nice always said, "Insects are our friends." She always shook her finger in your face when she said that and she always said that after she had caught us with a jar full of bees, which we were shaking as hard as we could, or when we were jamming a water hose into an ant pile, or putting a spider into the same jar as a lady bug. She used to let us eat rhubarb out of her garden. She was the only lady in the neighborhood who would. But we still never swallowed the stuff about insects being our friends. Until later.

Maybe it was an accumulation of Turkey in my blood and brain, maybe it was the ginseng root's attitude towards things natural, maybe it was my father's belief that even trout fish and things such as salmon flies, like buffalo and grizzly bears, would not last forever, not even for the sport of it, but whatever it was, something brought about a change.

Salmon flies cannot be domesticated. They do not live long enough. But no matter.

They won't hurt you. None of the under-the-rock-in-the-stream-bugs will. Fish feed on them. It is the natural way. Trout fish killers use some of the bugs as live bait and would probably use more if they realized their potential...

Next spring, I intend to send you an insect we call here the salmon fly. It is mostly killing bait, and it ought to make a very successful fly.

—Written about 1889
By a Montana Angler
Favorite Flies and Their Histories
Mary Orvis Marbury

...but most of them don't. Anyway, the majority of the nymphs are too small to be shoved on a hook and they are difficult to catch in quantity without a partner.

Hey! You really don't have to catch them and stick them on hooks. If trout fish killing does enhance the soul as poets and presidents and philosophers have claimed, then it can be fully accomplished without ripping wings off bugs and shoving hooks through their bodies.

Oh the brave fisher's life,
It is the best of any,
'T is full of pleasure, void of strife,
And 't is beloved of many:
    Other joyes
    Are but toyes,
    Only this
    Lawful is,
    For our skil
    Breeds no ill,
But content and pleasure

—Izaak Walton
1593–1683

Bullshit!

For the sport of it. The ginseng root saw no contradictions there. Perhaps there are none in the sport of it. It just seems there should be, somewhere.

It was, for them, socially important. The trout fish brought home, the size, the number, were important. They put on their man clothes for the weekend and Winnebago-packed their way to the high mountains, taking their *busy* with them, the first thing that should have been left behind. Fight for a parking place, rush to the stream, hate the angler there before you, curse the bugs that swarm around your head, damn the wind blowing down the canyon, the damp that seeps into your sleeping bag, the tick sucking on your thigh, the skunk that got your dog, the leader that broke, the rain that brought lightning, and goddamn and goddamn and goddamn those trouts you cannot kill. Gather up your *busy* and Winnebago-pack it back to town.

    Man pants off, patio groovies on. Trout fish killed and counted are now to be bragged about over cocktails, to be discussed for the sport of it. No recall of graceful Mayflies, of the smell of the rain that brought the lightning, of weird bugs drifting by in the stream, of tiny mushrooms beside a dead tree, of the things the root always remembered.

    But he never took his *busy* into the mountains. He could go trout fish killing and never bother with it, spending time just sitting and watching whatever it was he saw, and feeling whatever it was he felt, and he always came back with more than anyone.

    The Winnebago-packers returned with nothing but numbers—inches and pounds. They attacked it for the sport of it like a golf foursome or a bowling team, keeping score but nothing else.

Lord, keep my memory green.

    The tourists in camp (and me, too) were always in a rush to get on with the fish killing each morning. Grandfather and his partner were easy. It took a long, long, long time before I understood their ease, their relaxed, studied approach to the day. There had to be time to see the day before you jumped into it so that you could remember what it was when the time came to jump out of it. The sport of it was not in the killing of the trout alone, as it is in bowling a 300 or shooting sub-par rounds, but I didn't understand that then.

    Rock Creek near Missoula was overrun with vehicles and anglers. The word was out that the salmon flies were hatching.

Big browns and rainbows were rolling to the surface everywhere but they would not take the adult salmon flies we were using as bait. We gathered some large nymphs from under the overhang near the stream and placed two or three on our hooks, adding a couple of small splitshot. It worked like mad.
    —More outdoor editor talk

Big trout fish killers were rolling up and down the road in their cars but they would not take our bait. We gathered some shotguns and placed them in the hands of a crazed ecologist we found under the overhang near the stream. It worked like mad.
    —Indoor editor fantasy

    The damselfly nymph is more attractive than the salmon fly nymph. It is about the only nymph that is attractive. You might call it almost sexy, but you would never be able to convince a doctor as to why you almost called it that. Damselflies are found in quiet water, such as ponds and lakes. They are long and slender. Unfortunately, some varieties are said to be cannibals, a thought that tends to make them somewhat less attractive, but they remain a delight to watch.

The contradiction exists. Trout fish killing, for the sport of it, becomes more enjoyable as you reduce your participation, as you spend more time watching those things the root watched, and hearing what he heard, and being easy in the mountains.

You don't have to keep score anymore, just for the sport of it.

The legal limit was ten trout. Being a good sportsman, NRA member and all that, plus being a sniveling coward, the local trout fish killer would not have thought of being caught with more than his limit. So he played a little game: when he had nine fish in his creel, he would throw the smallest fish he had caught into the bushes if number 10 were larger. It was an endless scene—keeping number 10, if it were large enough, and reducing his total catch to nine by chucking the smallest dead trout fish into the weeds. He was an obscene, 20th-century Johnny Appleseed, skipping along the banks of Montana streams and scattering dead trout everywhere he passed.

He also was a greedy bastard, but he considered himself to be a good sportsman. This sport breeds greed. Really! You have to fight it. Always!

Spirit of Pitsford Mills, Royal Wulff, Queen of the Waters, Rat Faced McDougal, Beaverkill, Wichams Fancy, Greenwell's Glory, Whirling Dun, Badger Spider, Royal Coachman, Iron Blue, Cahill, Gray Fox Variant, Breadcrust, Zug Bug, March Brown, Strawman, Isonychia, Martinez Black, Quill Gordon, Carrot, Jock Scott, Popham, Black Doctor, Red Ibis, Green Weaver, Gracle, Hill Fly, Mooselucmaguntic, Parmacheene Belle, Oquossoc, Silver Doctor, Seth Green, Saranac, The Time, Sheenan, Ben Bent, Bluebottle, Bissett, Cinnamon, Cow Dung, Equinox Gnat, Esmeralda, Furnace, Ethel Man, Gosling, Golden-eyed Gauze Wing, Great Dun, Golden Monkey, General Hooker, Hare's Ear Imbrie, Jungle Cock, Jenny Lind Blue Wing, Lady Sue, Neversink, Prime Gnat, Cisco, Puffer, Blue Professor, Quack Doctor, Silver Horns, Widow, Welshman's Button, Hammond's Adopted, Silver Sedge, Academy, Cracker, Beaufort Moth, Dark Flaggon, De Gem, Golden Dustman, Polka, Max Von Dem Borne, Toddle-Bug, Premier, Maid of the Mill.

Just names. All names of flies used for trout and bass fish. Nothing very important. Call it a diversion, something to get my mind off greedy good sportsmen.

This article lacks substance, so maybe it is time for some standard, outdoor-writer-type-article-information. Here it is: the most dangerous animal you are apt to meet while trout fishing in North America (excluding North Americans) is the cow. Honest to god, it is. Moose, bears, wolverines, rattlesnakes, chipmunks, turtles, etc., etc., are really overrated. For one thing, if you are in the woods, you should be aware that things live there that may not welcome you, but who in the hell is ever going to suspect a cow? I have seen guys knocked down by cows, chased out of fields and shoved into ditches by cows, and I mean cows, not bulls. Nothing could be more nerve wracking than having three or four stupid cows following you—they don't growl, or bare their teeth, or hiss, or rear up on their hind legs, or do anything else to give you fair warning they are about to attack. They simply give you the standard, dumb, contented cow look, and then, before you know it, one of them may be standing on your foot or attempting to shove you down a bank. Oh, they do give one warning. If a cow sticks its tail straight up in the air, you know it is going to take a dump, so don't stand behind it.

Beware of cows! Consider that a tip from the wild woods.

Reduce your participation by increasing your participation at another level. Less casting and catching for the sport of it, and increase your joy of it.

The contradiction lives.

K-Mart stores furnish the trout fish killing kits and Ford urges you to the open road with pleas such as this: "Travel in your own motel-on-wheels! The Econoline makes just the camper you want—equipped just the way you want it."

Who supplies the ethics? Woolco? Sears? K-Mart? Ford? GMC? Nixon? CBS? McDonalds Hamburgers? Ron Ziegler? Head Komics?

They are all scorekeepers.

Don't keep score.

Not for the sport of it.

Trout fish killing needs more losers.

Little trout fish, particularly little brook trouts, are excellent eating. Most trouts, however, have flesh that tastes like attic must. A little lemon, salt, wine, Wild Turkey, pepper, mushrooms, onions, will correct the taste. Trouts that maintain a diet of freshwater shrimp, which are not shrimp, by the way, but scuds, according to the smart asses in the lab, usually have an excellent taste and bright, pink meat. Hatchery trout taste like mud.

The girls, college age, were skinny dipping in a large pool on Lolo Creek. Trout fish killing was immediately forgotten. Mad rushes of evil acts overwhelmed both of our imaginations.

This was trout fish killing at its finest. Hiding in the trees, lusting. God, it was open-mouth breathing time. But what to do? Action as well as caution was required. Should we strip and charge out of the woods in full rape? Should we casually stroll upon the scene, act surprised and embarrassed and attempt to apologize our way into a relationship? Should we sneak closer and just keep up our voyeurism? (At this point in my life, I was a sophomore in college and had spent three years in the Marine Corps, my sexual experiences and carnal knowledge amounted to little more than a few giggles.) Should we...it didn't matter now. Our hesitation was our downfall. While we pondered it all, the girls had put on their jeans and were walking away from the stream. My fishing companion, also a sophomore in college and life, was outraged at the girls. He erupted, suddenly screaming vile accusations at the girls, who stopped and turned to see a fly-rod-waving-chest-waders-and-fishing-vest-wearing-out-of-control maniac accusing them of being whores and teases and bitches and sluts. I wonder if those girls ever understood why they were the targets of such abuse.

Nevertheless, it was a memorable day. Can you hear me, Izaak Walton? It was one hell of a fine day.

> Oh, the brave fisher's life
> It is the best of any,
> 'T is full of pleasure, void of strife,
> and 't is beloved of many...

And we didn't even bruise a trout.

This-is-my-most-favorite-sport-of-all historians seek the superlative, assuming that all is well with their sport because all has always been well, they think:

This is one of the oldest and perhaps the most artistic forms of sport fishing in the world.
—From the introduction to the chapter on "Fly Fishing" in McClane's *Standard Fishing Encyclopedia and International Angling Guide.*

Those same historians can tell you about men who wrote of fishing with the bait fallacious before the birth of Christ.

So, we know trout fish killing pre-dates rock. But who supplies the ethic?

The ginseng root is gone, and he was one of few who understood that the sport of trout fish killing requires very little killing for the sport of it.

In 1969, 29,855,000 fishing licenses were sold in the United States. The cost to the anglers was $87,501,000.
—Pocket Data Book USA 1971
A United States Department of Commerce Publication

Who supplies the ethic? Trout Unlimited? The Justice Department? Mother Bell? The NFL? The Upper Clark Fork and Vagabond Inn Seal Hunters Society? The salmon fly?

Maybe the salmon fly, or the Mayfly, or the caddisfly, or the damselfly. Maybe they will, maybe they won't.

The root seldom said anything, but I remember once that he said of his mountains, and lakes, and trouts, "It's a nice day." We were sitting in his tent, above timberline, camped on the Hellroaring Plateau, shivering. Outside it was cold, raining and blowing.

No trout fish killing this day, not even the squareheads would have been out, if any of them still lived.

He was right, it was a nice day. A gentle day for the trout fishes, a learning day for the trout fish killers.

Don't keep score!

My daily official duties are such that, after completing them, any more writing is a painful labor; but yet, in view of the fact that I take a deep interest in all legitimate matters connected with fly-fishing; that I have been a lifelong lover, and for very many years a keen follower of the gentle sport...

...I shall ask you give full credit to my statements as to lack of time, and consequently to make due allowance for haste, condition, and...

—MOUNTAIN GAZETTE, #11

Confessions of a Butterfly Chaser

BY
ROB PUDIM

I should have gone to work. I should be doing a lot of things. But here I am, Monday morning, chugging up the mountains, not for anything dramatic such as rock climbing or wrestling a grizzly bear. I am chugging up the mountains in front of a dust cloud to chase butterflies. I should be working.

Being a collector of anything is strange unless you do it for money.

A garbage collector collects garbage because people pay him to do it. People are happy to pay him to collect trash. In cities like New York, not only is he paid well to collect garbage, but during hot stretches in July, citizens gratefully make special donations of filthy lucre in plain, brown wrappers to their garbage collectors. There is so much money to be made collecting garbage that the Mafia has muscled in on it.

People collect paintings, coins, stamps and certain butterflies because they can turn them into cold cash. There are dimes worth $30,000. Ratty purple stamps, poorly printed, can bring a half million. Who would not collect them? From this perspective, there are good financial reasons for collecting campaign buttons, comic books, jewelry or antiques.

Recognition, speculation, and greed are pure, simple and understandable. The problem is how to understand those who are interested only in the collecting, not in the money. Try, for instance, to explain how somebody comes to collect dinosaur shit—excuse me, coprolites. The essential point is that most collectors of this sort do not choose to be collectors. You do not choose to be a butterfly collector and willingly trot around in public with a net in hot pursuit of an essentially worthless, inedible bug. Who, given all the rational alternatives, would choose to look for petrified dinosaur road apples? Nobody chooses to be a collector. Perhaps the collecting urge is caused by cosmic rays.

Somewhere in the human brain, among our ten billion neurons, the grey-matter givers of consciousness, is some DNA. When the DNA is cosmic-ray zapped, it rearranges itself slightly. The altered DNA in the brain cell begins ordering its 20 million or so messenger RNA slaves into producing new proteins. These proteins, in turn, effect transmitter molecules in the synaptic cleft between neurons. New links and pathways are created among the 60,000 to 300,000 nearby neurons and something happens.

The zapped DNA produces strange proteins which slightly alter normal human brain holograms. This newly created hologram is identical to those found in animals that compulsively collect things—animals like magpies, octopuses and pack rats. It causes heretofore normal people to compulsively assemble, classify and store all sorts of godawful stuff. They get zapped, and their brain begins to change. They become like freshly hatched ducklings wandering about waiting to be imprinted. A duckling will accept the first thing that moves as being its mother. A recently zapped human will collect the first thing it sees.

Zap! A butterfly collector.

Zap! A dinosaur turd collector.

Deep down other people know collectors have a medical problem and know collectors cannot help themselves. They know that at any moment they, too, can get zapped. They laugh at butterfly collectors, ball-of-string collectors, empty-beer-can and tinfoil collectors. But they know that there, but for the grace of God or chance or a random cosmic ray, go they.

The bog I have driven to lays between two mountains, just below the tree line. The mid-summer winds coming over the bare peaks have a wintry smell. It never vanishes. You have to haul ass into the high bogs to get there in time. Too early and there is a glaze of ice around the grass hummocks. Too late and the clouds build up to the west and the sun disappears. In between there is a brief time when the sun warms the bog and small brown butterflies and flashy blues and sulfurs emerge.

I put on my Mekong Delta swamp boots, one of the few good things to come out of Vietnam. The Vibram soles are good for walking and the holes in the side of them allow water to run in and out. The nylon-net sides permit your feet and socks to dry once you are out of the bog. They are also cheap.

I put glassine envelopes in a shirt pocket, drape a forceps on a chain around my neck and assemble my net. I look carefully around, not for butterflies but for people. I light a cigar. I tell myself that it keeps the mosquitoes away, but I know it is to show I am a macho sonuvabitch should anyone happen to see me. The full beard adds to the effect.

There is nothing dumber and sillier than a butterfly collector. Hundreds of cartoons testify to the character, armed with a net, traipsing around a meadow in foolish pursuit of a Painted Lady. Any real man would be pursuing real painted ladies in downtown New York or Las Vegas. There would be dignity traipsing around that same meadow, armed with a rifle, in hot pursuit of a defenseless, inedible crow instead of a defenseless, inedible bug. Hundreds of movies testify that butterfly collectors are batty in a pleasant sort of English way and, without exception, they are males who speak with a high voice and gesticulate with a slightly limp wrist. The outfit is usually a safari jacket, shorts, a Boy Scout hat and thick glasses.

Every butterfly collector is embarrassed about being a butterfly collector. People are embarrassed for you when they discover you do it.

Collecting butterflies is something like masturbating. Each time you do it you feel relieved and somewhat ill at ease. You try to do it in private, away from roads, from people, from casual on-lookers. You would never do it at a party or in a stadium except in the most extraordinary circumstances. The rationalization is that you want to collect butterflies in untouched areas, areas unaffected by man's pollution and presence. The truth is you would rather not be seen collecting butterflies by other people. And the worst part of all is that after you are finished collecting, like masturbating, you know you are going to do it again.

This may be why butterfly collectors like to call themselves lepidopterists. They think they can hide behind the Greek roots *lepid* or *lepis*—"scale"—and *pteron*—"wing"—and be something dignified. Deep down lepidopterists know that a scaly-winger is a butterfly collector. The rose is not going to smell any sweeter because it is given Greek roots.

Butterfly catchers never talk about the pleasure they derive from butterflies and the joy and satisfaction they get from the pursuit of them. They talk a lot about embarrassing situations—about the time a moth flew in their ear and they had to go to an emergency room to have it removed, about the time a rancher caught them sneaking up on some fresh cow patties and no butterfly was in sight. They talk about where they have been and the dangers of rain forests in New Guinea, high mountain passes in Alaska, or jungles of the Amazon. They talk about everything but pleasure, joy and satisfaction.

This is not exactly true. There is an exception.

"This is ecstasy," Vladimir Nabokov once wrote about standing in green woods among rare butterflies. "Behind the ecstasy is something else which is hard to explain. It is like the momentary vacuum into which rushes all that I love. A sense of oneness with sun and stone. A thrill of gratitude to whom it may concern—the contrapuntal genius of human fate or to the tender ghosts humoring the lucky mortal."

I never felt this when collecting butterflies—even in the presence of a rare one. I am not sure I know what the hell he is saying. I have felt something connected with a sense of oneness with sun and stone. But that has occurred a lot of times under a number of different circumstances.

I remember the first time I caught a certain black swallowtail in my net. It was not rare but I had tried to get one a number of times. Holding a swallowtail between your thumb and forefinger, there is a tremor like a low-grade electric current. It buzzes silently at the end of your fingers as the muscles beneath the hard exterior skeleton try to drive the wings to flight.

The sunlight as it strikes the compound eyes reveals deep inside a honeycomb pattern, magnified and moiréd as the insect is turned. The eye is alive and completely alien, like a wife's eyes during a divorce proceeding.

Held between your fingers, there is a thick and pungent smell of freshly crushed parsley. It is as much a smell of summer to me as creosote from July railroad ties or an asphalt pavement in August is to others.

But like a good collector, I am prowling the swamp catching small brown butterflies. I am swearing a lot. To catch the little bastards, I have to watch them, not taking my eyes from them as I trot after them to where they land. This means tripping over hummocks a lot and, once and a while, stumbling into a bog hole or creek up to my ass. Tree line water is cold. Elephant hunting would be a helluva lot easier.

When you are up to your ass in a swamp, it is hard to remember the object of the exercise is to empty the bog of butterflies. Some conservationists and the Department of Interior do have a Genghis Khan fantasy about unfortunate, zapped collectors like myself. The fantasy starts with equating butterflies with Bengal tigers or African elephants.

****Butterflies First: Butterflies are the first insects to join the ranks of the U.S. Endangered Species List. Of the 700 kinds of butterflies in the U.S., the Department of the Interior has put 41 on its list of threatened and endangered species. This marks the first step toward protecting the butterflies from interstate shipment, commercial sale, and mass collecting.*

—NWF Conservation News: 5-15-76

Until recently we worried about such vertebrates as whooping cranes, whales and bald eagles. Now, in addition to the more than 100 kinds of vertebrates on the Endangered Species List, there are some invertebrates on it, and things will never be the same. Invertebrate preservation poses a unique set of problems to the conservationist. It is a set of problems which the current set of vertebrate answers do not satisfy.

It is difficult to make an insect extinct. This is not to say they have not become extinct. The fossil record proves they have. There are no longer dragonflies around with 12-inch wingspreads. But there are only two known cases—one in America and one in Europe—of insects becoming extinct because of man.

No insect has become extinct or even been threatened by extinction because it has become part of man's menu or wardrobe. It has not been threatened by over-collection, natural predators, artificial insecticides or insect diseases, or because it has been stomped on in enormous numbers. As a matter of fact, the insecticides, such as DDT, have done more harm to animals and birds higher on the food chain than they have to the roaches, mosquitoes or beetles against which the poisons are directed.

Given high birth rates and short life spans, invertebrates like butterflies are adapted to life in short-term environments such as the brief sub-alpine summer or the growing period of a specific plant. Their population explodes to take advantage of brief but favorable conditions. This also allows them to evolve defenses rapidly to counter the punches man throws at them. As long as they are in some sort of equilibrium with their food and predators, they can adapt and survive.

Remove the predators, and insects increase in numbers and devour their food plants at such a rate that starvation follows. Some survive, the plants recover, the predators catch up and the equilibrium returns. This cycle occurs regularly in the Painted Lady butterfly (*Vanessa cardui*), and the mountains are treated to enormous northern migrations as the adults seek their food plants elsewhere.

Add predators—this includes humans—with all their ingenious devices for attacking bugs—and insect numbers are kept to manageable numbers and an acceptable level of food plant destruction. The predator-prey balance is healthy for the survival of insects. After all, the butterfly is both prey and predator simultaneously—the former in all stages of its development and the latter in the voracious caterpillar stage.

Fluctuations in numbers of insects—enormous fluctuations by vertebrate standards—are normal and are usually due to environmental changes. Some of these changes are due to natural events and others are due to human impact.

It has been said by a number of biologists that butterflies are barometers of industrial civilization and man's burgeoning numbers. They are, for example, indicators of atmospheric pollution. There is a phenomenon called "industrial melanism" that describes the darkening of wings in response to dirty air. It has been observed in places as separate as Manchester, England, and Pittsburgh, Pennsylvania. One Pennsylvania moth, for example, had gray wings to match the tree trunks upon which it landed. These wings darkened, becoming almost black, to match the soft coal begrimed trees and buildings in the Pittsburgh region. Now, after the antipollution program has cleaned things a little, a lighter form of the moth is becoming characteristic of the area.

The Xerces blue did not fare as well. A decade ago developers bulldozed land in the San Francisco Bay Area for housing sites and tore up a wild lotus native to the place, which was the food plant of the Bay Area Xerces (*Glaucopsyche xerces*), a small butterfly first described in the early 1800s by Jean-Baptiste Boisduval. The wild lotus disappeared. The blue disappeared.

Commercial sales did not do it. Over-collecting and interstate shipment did not do it. Something as ordinary as a housing development, a shopping center and a parking lot made the Xerces blue the first insect in North America to become extinct as a direct result of human impact.

In England, another butterfly, a large copper (*Lycaena dispar*) became extinct in the latter part of the 19th Century. Again, the culprit was not over-collection. The destroyer was a human need for peat, a need that required draining the bogs in which the butterfly lived.

The historical message is clear. If the habitat of a species is destroyed, that species will disappear, no matter what steps are taken to preserve individuals of that species.

The general truth invertebrates teach us, according to Lee D. Miller, is this: Habitat preservation without restrictions assures the survival of the species, but prohibitions without habitat maintenance assures extinction.

Miller, as a lepidopterist, is aware of a paradoxical thing about butterflies. Most butterflies are rare in spite of their adaptability and prodigious birth rate. A butterfly's occurrence is local and restricted. Most butterflies are delicately adjusted to their immediate environment and bound to a single food plant. A single female can produce several hundred offspring. A few thousand females in a few summer weeks can create a population of millions of adult offspring—but all will be found along a single mountain chain, or on a single island in the Gulf of Mexico or in one bay on the California coast. The unusual butterflies—such as the Cabbage white or the Painted Lady—are those with a wide range and numerous kinds of food plants.

The Department of the Interior does not need an Office of Endangered Species, it needs an Office of Endangered Habitats. Butterflies are telling us that habitats are being destroyed, but that they and other animals can adapt to an altered habitat if the changes are gradual.

Butterflies are also telling us about the shortcomings of having an Endangered Species List and its protections and not an Endangered Habitats List. This can be further illustrated in the recent attempt of the Bureau of Land Management to save a butterfly, the *Nokomis fritillary*.

To quote from the Spring 1976 issue of *Our Public Lands:* "The species is dependent on a particular violet (*Viola nephrapaylla*) that grows in steep meadows on National Resource Lands in BLM's Grand Junction District in Colorado. The female lays her eggs on the leaves of the violet, and the emerging larva feeds on the leaves while waiting for metamorphosis."

Habitat restrictions make Nokomis extremely vulnerable to certain kinds of development of National Resource Lands. Springs that feed the seep meadows are sometimes tapped to provide water for man or livestock. When this is done, the meadows dry up and the violet cannot grow in the drier soil. In some cases, whole colonies of Nokomis have been wiped out.

Studies are now being made to determine if Nokomis should be included on the list of endangered species. If it is placed on that list, definite parameters will be set on any management decision affecting the butterfly or its habitat.

Tom Owens, a BLM district manager, is quoted later as saying, "I never thought the day would come when butterflies would be one of my problems." Butterflies such as *Speyeria nokomis* are not Owens' problem. Owens' problem is, and always was, an endangered habitat.

Reading the article, one would never realize that a butterfly is a basic link in a food chain, not at the end but in the middle. The butterfly's prodigious reproductive capacity is not meant to produce caterpillars to eat violets, but to provide other insects, birds, frogs, toads, lizards, spiders and even small mammals a food supply. The *Nokomis fritillary* may be important because it is food to another invertebrate with an even more restricted range.

Moreover, there is little reason to believe Nokomis could be included on the list since there are flourishing colonies of it elsewhere. The whitewater area in question may not have many *Nokomis fritillaries*, but this is by no means the limit of its range. Thus, no parameters can be set on this particular habitat to preserve it.

This example also illustrates that even saving the endangered habitats of a single species may be too optimistic an endeavor. Emphasis on rare and endangered species has resulted in several land reserves for the protection of individual animals. This emphasis—for instance on the Whooping Crane—ignores the more common plants and animals. This is an expensive program. A bog in Gilpin County, Colorado, some prairie land in Weld County, the southeast corner of Middle Park—it would run into millions of dollars to save only three modest butterflies in one state, Colorado. It is one thing to persuade people to preserve at astronomical cost a large and easily recognized bird like a Whooper or a plant like a Joshua tree. Whether the same people will be willing to underwrite ambitious programs to preserve brown dingy butterflies is another thing.

What the butterflies are telling us is that we need to preserve communities—an assemblage of a population of plants, animals, bacteria and fungi that live in an environment and interact with one another, forming together a distinct living system with its own composition, structure, environmental relations, development and function.

Evolution does not normally carry a species toward domination of its environment. If this were true the world would be overrun by a number of successful animals like roaches, sharks or butterflies. The goal of evolution is not the survival of the fittest species in a dog-eat-dog world. The goal of evolution is the persistence of a species in an ecosystem. Evolution does not produce butterflies or horses; it produces communities of living things in equilibrium with the inorganic world around them.

The way to preserve communities is through the concept of the "megazoo"—a concept advanced two years ago in an article in *Science* by A.K. Sullivan and M.L. Shaffer.

Sullivan and Shaffer argue that we need to look at a system of primary wildland reserves to ensure a diversity of plant and animal life in the future. Existing reserves are presently inadequate in size and number and are clumped in one geographical region. The megazoo approach would provide a planned network with several levels of reserves of varying size, starting with first and second order watersheds large enough to support stable populations of large carnivores. The megazoo reserves can effectively cut the per-species cost of preservation. The price will still be considerable, and it will not be easy to sell to a large segment of the populace. The megazoo concept offers the best and only chance for the continuation of many species that otherwise will become extinct in our country.

The presently conceived regulations to preserve butterflies and habitats are unlikely to succeed. Similar regulations do not work well for the vertebrates, for whom they were formulated. The Xerces blue experience is warning us that the way to save the timber wolf is not to save the wolf but the timber and all the animals and plants in the food chain for which the wolf is the endpoint.

There is a passage in W.J. Holland's *The Moth Book*, published in 1903, which expresses a notion we all have in the backs of our minds: "When the moon shall have faded out from the sky, and the sun shall shine at noonday a cherry-red, and the seas shall be frozen over, and the ice cap shall have crept downward to the equator from either pole, and no keels shall cut the waters, nor wheels turn in mills, when all cities shall have long been dead and crumbled into dust, and all life shall be on the very last verge of extinction on this globe; then, on a bit of lichen, growing on the bald rocks beside the eternal snows of Panama, shall be seated a tiny insect, preening its antennae in the glow of the worn-out sun, representing the sole survival of animal life on this our earth—a melancholy 'bug.'"

This is turn-of-the-century romantic nonsense and ignorance we must purge from our minds. When man goes, he is going to take everything with him. And being the gentleman to the end that he is, it will be the birds, mammals and butterflies first.

As you look across a meadow, seeing a butterfly is both a commonplace and remarkable event.

Butterflies are showy animals. They are common animals found all over the world. There are a lot of different kinds of them. Next to the beetles, they are the second largest order in the animal kingdom.

Butterflies have no aggressive characteristics. They do not bite, sting, puncture, pinch or tear. Their defense is flight or protective camouflage, coupled with complex behavior patterns or, occasionally, distasteful body juices.

That butterflies have survived so long so well with so little is remarkable. They were there before the first fish stuck his head into the poisonous atmosphere and decided it was better than the terrors of the deep. They watched the disappearance of the dinosaurs with the same dispassionate stare that watched the arrival of humans.

But in truth, that is a romantic notion. The insects are no more aware of our presence than they were of the stegosaurus' disappearance. In the shifting kaleidoscope world of their compound eyes, humans, the dinosaur and the Kenworth tractor are no more than a mosaic patter of ultraviolet light. Humans do not see the same wavelengths as butterflies, they do not hear the same sound frequencies, they do not taste the same chemicals or feel the same touch.

Humans forget that reality, the real world, is the creation of a nervous system, be it a few fiber-connected neurons as in insects, or an elaborate tangle perched on the neck of a man. Reality, this creation of our nervous system, is a paradigm, a pattern of a possible world. Reality is an invention that allows a nervous system to handle the enormous amount of information it receives and processes. According to K.J.W. Craik, an English psychologist, this is the nature of explanation. The "true" or "real" world is specific to the species sensing it and dependent on how its brain and neurons work.

Robert Pirsig, in *Zen and the Art of Motorcycle Maintenance*, talks a lot about the classical and Apollonian in contrast to the romantic Dionysian concepts of reality in humans. This is a penny-ante distinction compared to the perceptual worlds of butterflies and men. The German biologist, Jakob Johann von Uexkull, argued that the work of the brain was to create a model of a possible world and transmit to the mind a world that is metaphysically true. Different worlds are constructed by different species and they are all true worlds.

Butterflies do not have much of a brain, but the butterfly's world is no less real than man's. It is not more true or more false when compared with "reality." It is devoid of human beings and the contrived order we have discovered in the randomness around us. Someone once pointed out that all ordered relationships are merely a subset of total randomness or, saying it another way, order is a special kind of chaos. The disturbing thing about butterflies for us is that, despite all our ordering, thinking, planning and calculating, we are a haphazard event in a butterfly's life. The gridwork cities, the ordered rows of crops and the systematic use of pesticides surrounding the butterfly do not exist in that butterfly's reality.

From this, an observation about life on Mars or on planets of other stars can be made. It has been said that we may not be able to recognize life elsewhere because it may be different. It might be silicon based instead of carbon based; it might use ammonia instead of water; it might be energetic instead of material; it might be asymmetrical rather than symmetrical. But behind all this is the assumption that even if we do not recognize the beastie, we will recognize his communication.

We listen to the voices of the stars at the hydrogen wavelength for a signal from them, confident we will recognize it as a signal because it will be repetitive, intentional and meaningful in some way.

The irony is that an extraterrestrial's choice of pattern may be indistinguishable to our nervous system from the chaos around us—just as dit...dit dit ...dit dit dit...is patternless to a butterfly. It may very well be that other intelligences are transmitting messages but all we perceive is a hiss from space, a white noise from multicolored stars.

There is one more observation to be made. The more advanced and sophisticated a society is, the more random its order becomes. To the Chinese all Western music sounds like a thump-thump march. The regular beat in Western music is seldom found in the complicated beat patterns of Indian ragas. They also have more tones, so that Indian and Chinese music is closer to modern electronic music with its bleeps, wheezes and burps than it is to Stravinsky or Beethoven.

Language has evolved from sounds to a written code of symbols that have little or no relationship to the sounds. "A" has a number of sounds associated with it, some of which are identical to "U" or "O" symbol sounds. These symbols, in turn, have been converted to pulses on a wire or to electromagnetic frequencies, modulations, or amplitude variations. Bursts of laser light or the warble of satellite radio telemetry is random sound or a flickering light to a primitive man.

Thus the music of the spheres the ancients talked about and the hiss of space the astrophysicists hear may be the same thing. While the cricket hears it through his leg and saws his wings in time, a man merely listens, shakes his head and feels alone.

—MOUNTAIN GAZETTE, #54

The South Side of the New England Soul

BY
JOHN SKOW

There is no sense to maple-syrup making, no sense at all. I am not sure, but that may be its attraction. What I know is that if you want gentle exercise in the open air, stump-pulling is less strenuous, and if profit is your goal, the shell-and-pea game is a far sounder investment. If all you are after is syrup to put on your pancakes, the only sane course is to humble yourself, stop at a roadside stand in New Hampshire or Vermont, and buy a quart. The cost will be absurdly high, but it is a near-certainty that the callused philosopher who sells it to you will be losing money at that price.

If the philosopher has made the stuff himself, however, he will have gained something more valuable than money, and if I think for a moment I will remember what it is. Self respect? No, a man who has satisfied himself for the 20th or even the 40th year in a row that he is a damn fool (syrup-makers are long-lived and stubborn) does not speak willingly of self-respect. Self-knowledge is another matter. The syrup-maker had more of this commodity than he needs, and in March and April he sometimes shares the excess, at the top of his voice, with his dog, his sap buckets, his snowshoes, his sap-sloshed boots and pantlegs, and the trees themselves. A good deal of winter-weight swearing gets done in spring in a sugarbush.

All very true. Yet I have never heard of a New Englander who, having made syrup once, did not go on to make it, muttering to himself, year after year. Until very recently, a man in my New Hampshire town led his oxen out on the lake each January, and, using a big crosscut saw, cut himself a year's supply of ice. This rite was much respected by the rest of us (as it was intended to be), and although the man owned a perfectly good electric icebox and freezer, no one thought him strange. Ice-making was what you did in January, never mind the sense of it. And syrup-making is what you do in the spring, when a combination of cold nights and hot days causes a watery, almost tasteless fluid to run in astonishing quantities through the white sapwood of the rock maple tree.

This is not good city logic, but by last winter I had enough of cities, having strayed from New Hampshire, by mistake, to live in one of them. I told some friends back home that I would help work their sugarbush. On March 9, I left the city to its own burned-out devices, and the next morning at 9 a.m. I reported for work. I was the first soul at the sugar camp. New Hampshire, a state of bluff, honest motel owners, no longer keeps farmer's hours.

A farmer once owned this sugarbush, and doubtless had his sugarhouse where the present one is, sheltered below a steep hill of maples. A sugarhouse, where the boiling is done, is always below, and a clever man does not need more than five minutes of lugging sap buckets to see why.

Sugarhouses burn down with some regularity, because of the great wood fires that roar beneath the evaporating pans, and because carbon cakes up inside the sheet-iron chimneys and burns, and because boiling syrup itself can catch fire just when you don't want it to. One that stood on this spot burned down—or sideways, since the shed end wasn't touched—and after that a big oil-fired evaporator was installed. It hasn't paid for itself yet, in terms of gallons of syrup produced, but it hasn't set the sugarhouse on fire either. This is as far ahead of the game as the syrup business lets you get.

The sugarhouse that burned was almost certainly not the original, or even a first cousin of the original, but in any case the farmer who once owned the sugarbush was not around to care. His dwelling house, not below the hill but triumphantly at the top of it, burned at the beginning of the century, and what remains is a cellar hole, grown with saplings, and a well, still good. Not thinking about it, working out a series of switchbacks to let my new showshoes take me to the top of the hill, I arrive near the place where I found the cellar and the well one afternoon three years ago. I think I have found it again, but there is three feet of snow now, and it is hard to be sure.

A yell from below; my friend John is here with the rest of the free labor. John is a prep-school teacher who has volunteered for extra duty; he is teaching a term-end "mini-course" in the economics of syrup-making. Clever students will discover—although not, John hopes, before some work is gotten out of them—that the granite underpinning of syrup-making in New Hampshire is free labor.

One hundred years ago on this spot it was the free labor of the farmer and his family, who had little else to do in March except the chores.

Now the owners sneak off from work to lug buckets—John's brother Bill also teaches, and their father is a surgeon—and recommend the sport of syrup-making to their friends, who show up dependably on sunny Sunday afternoons, but not in any quantity on rainy Wednesday mornings. As a hobby, sugarbush owning is not more expensive than yacht-racing, and it has this advantage: when it is finished, everyone gets some syrup. (It is said that in Vermont and New York, large syrup combines make a profit and pay their labor, but in New Hampshire these rumors are widely unbelieved.)

Now in a syrup class there is much to be learned. A boy whose home is in Florida learns to his surprise that he can move in an upward direction with a stack of 20 sap buckets on his back. Then he learns that on the third or fourth step the snow crust breaks, one leg drops out of sight, and the buckets roll down the hill. Back for snowshoes. This time he reaches what he started out for, a clump of maples where we have started tapping.

I appoint myself chief tapper and learn that a 20-pound gasoline drill gains so much weight in an hour that a man is reluctant to pick it up. There are other subtleties. The south side of a tree flows first. Tap under a limb and about two inches deep, the old-timers say. Don't tap at a comfortable chest height, because when the snow you are standing on melts, the sap bucket will dangle over your head, and people will laugh when you walk into the drug store to buy your paper. Watch what you tap; your maple might be an oak or a beech. I pass the gas drill to a student who looks as if he needs instruction.

The day has been fairly cold, and my first tap showed only a promising wetness gathering at the edge of the wood. Now, toward noon, it has warmed, and when the drill binds in a hole and stalls, there is a small new sound in the air: this slow *plink* and answering *plunk-pink* of sap beginning to drop into empty 16-quart buckets. Then a yell brings everyone skidding down the hill to the sugarhouse; John's wife Sue has turned up with a grill, charcoal and hamburgers.

During lunch a problem that has been skulking about the edges of the sugarbush all morning comes into plain view and leers at us. The fact is that no one knows exactly where the pipeline is. Snow is not a total surprise to New Englanders, and there are stakes to mark the pipeline that is supposed to carry sap to the sugarhouse, but this year's snow has been unusually deep and persistent, and most of the stakes are buried. A search party is sent out to take soundings. There is much to be done. The fittings where the big 36-quart funnels will be screwed in must be dug free, and the pipes must be flushed with hot water and checked for leaks. There is no telling how much time we have. The sap running now could run all night, or stop in an hour. But already some of the buckets have stopped plunk-plunking. Sap is falling quietly into sap.

Up to this point the syrup-making class has been a success. The boy from Florida, for instance, has responded well to what educationists call the learning situation. He has worked hard, watched closely, asked questions, and now he feels that he understands how to tap a sugarbush. Good, his quick mind tells him, on to the next lesson. And sure enough, the sugarbush has not run out of things to teach.

The new lesson, however, is that the old lesson will continue. The boy from Florida finds this idea difficult to grasp. Yes, he is told, there is a fine grove through the juniper pasture over the top of the hill, and then when we finish the home orchard, we follow the road for about three-quarters of a mile tapping both sides, and then there is a sugarbush on the backside of the property that you get to through a swamp. We have hung something like 200 buckets, with perhaps twice that many to go.

To improve his spirits, I let him work the gasoline drill, which has not gotten any lighter. A short while afterward a cheer arises. The Florida boy, in his agitation, has tapped a beech. But the hooting dies quickly. We have all snowshoed up and down the hill too many times to laugh much. When we quit, a few minutes after four, talk is brief.

Tiredness: the distance runner floats loose above his body, the tennis player feels pleasantly done in, and the golfer's unused muscles are not so much tired as bored to distraction. But the tiredness of syrup making is, after that of mountain-climbing, the most agreeable I know. This is because it is rotten with self-satisfaction. The syrup-maker's carcass, normally city-bound, has put in a day's work. It feels proud of itself.

By morning some of the self-satisfaction has been replaced by stiffness, but the sun is hot and there is a great rush of life in the sugarbush. The snow is still deep, but beneath it there is the quickening sound of water pouring down the rocky hill. The sap, which stopped during the night, is running again. It is shirtsleeve weather, if a man keeps moving. Stiffness does not matter.

For reasons no better than sun and sap and the fact that the earth is turning—for no reason at all—there is excitement in the air. The pipeline still is buried, and there are still 200 or 300 holes to be drilled, galvanized iron spiles, or spouts, to be knocked into the holes, hooks to be hung on the spiles, and buckets to be hung on the hooks. Yesterday afternoon our mood was edging toward mutiny. Now we are all loony with cheerfulness.

By mid-afternoon, most of the buckets are hung, and one branch of the pipeline has been excavated. The other branch is found, and hot water poured through it. The water does not reach the holding tank at the bottom. We dig to find the break, down through last week's eight inches of snow, now melted to four, through layers that mark the monstrous storms of February and January, down to the fossil snow of the November storm that started it all. A 20-foot section of galvanized pipe has split lengthwise. We wrench it out and replace it with plastic tubing. I am glad, with my city man's stern sense of New Hampshire tradition, that there is no more plastic in this sugarbush. Some nervously up-to-date syrup-makers have used plastic to eliminate buckets and sap-carrying altogether. They run the transparent tubes from their pipelines all the way to the drill holes in the maples. Deer and snowmobile delinquents sometimes knock down the tubes, but except for repairs, the up-to-date operator had little need to round up free labor. And has not much fun either, would be my guess. Such a man runs a danger of making a profit; he might as well be running a frozen orange juice factory.

By late afternoon, although there is no need for it, since no bucket is more than half full, we gather our first sap. It is clear and tasteless. A thirsty man can drink it. It gurgles down the pipelines to the holding tanks with a sound like the rush of water beneath the snow.

We talk of boiling. But the next day no sap runs, and the day afterward almost none. It is hard to say why; the motives of maple trees are as murky as those of the syrup-makers. A reference work written 50 years ago offers the information that what makes sap rise during some periods of cold nights and warm days, but not during others, and indeed what makes it rise at all, is not fully understood. More recent reports assert confidently that some trees are more productive than others, but add that why this is so is not yet known.

This murkiness is one of the most pleasant aspects of syrup-making, because it means that one man's opinion is as good as another's. Both opinions may be worthless, but no one can say for sure, and the citified newcomer who at the beginning of March desperately seeks advice will be lecturing confidently to tourists by the end of the month.

For two days the sap refuses to run. Then it begins, slowly. Now, five days after we started work, the holding tanks are full, the buckets are filling, and my friends John and Bill are ready to fire up the big evaporator. And I must leave New Hampshire to keep a journalistic appointment. That is the trouble with free labor.

When I am able to return, three weeks later, I still have not tasted a drop of syrup, but everyone else has had more than enough of it. The chart tacked on the sugarhouse wall records the production of 80 gallons. This is far less than in other years. For some reason, the sap flow has been thin. Still, since about 40 gallons of sap must be boiled down to make one gallon of syrup, my friends have poured some 3,200 gallons of sap, weighing about 15 tons, down the pipelines.

The sugarbush is deserted. Spring vacation is over for my two teacher friends, and although this is Saturday, each has morning classes. Their father, Dr. Bill, has been busy patching up end-of-the-season skiers, and no one has collected sap for several days. Half of the buckets are full and spilling over. I set out with the two 24-quart collecting pails, a funnel, and a piece of cheese cloth to filter out the moths, meadow mice and bark chips which—so say the old-timers—give maple syrup its special flavor.

Sap-collecting is hard and humbling. With a full collecting pail dangling from each hand, a man lugs some 60 pounds. The old-timers eased the strain with carrying yokes fitted across their shoulders, and two of these yokes hang on the wall of the sugarhouse. But they are dried and split, and none of us has taken the trouble to repair them. So our shoulders ache, which is the beginning of humility. The middle and end of humility is that there is no dignified way to move around a sugarbush. Snowshoes are awkward on steep ground, do not work on rocks, and cannot be maneuvered with grace over a barbed-wire fence.

Now they cannot be used at all, because large patches of snow have melted away. What remains, however, may be three or four feet deep. The sap collector may walk daintily across this snow when his buckets are empty, but on the return trip, when they are full, he sinks to his wishbone at every third step. When this happens there is a moment when the collector thinks he can prevent his sap from sloshing, and a moment immediately afterward when he realizes that he cannot. In a very short time he is soaked with sap. It all works downward, and you can tell how long a collector has been working by measuring the level of sap in his boots.

Syrup-making, however, is not all humility. Short of piloting a steamboat down the Mississippi, I know of nothing so well calculated to improve a man's opinion of himself as taking the controls of a big maple-syrup evaporator. The thing is chest high and broad as a pool table, fired by two roaring oil burners. Cold sap is piped into the boiling pans at the operator's left. It circulates, boiling, through a maze of pans, over which a great haze of steam rises toward the open ports of the cupola roof. As it loses water, the sap gains color and density. Maple syrup must weigh exactly 11 pounds a gallon and it must boil, at sea level, at 219 degrees—seven degrees higher than water. When sap reaches this temperature it is syrup, and a sensor opens an automatic valve that squirts it into the top of a filtering tank. When the syrup soaks through a filter of paper and another of felt, it is ready to be canned.

It takes about two hours of violent boiling, at a guess, to propel a given molecule of sugar to the filtering tank. When everything works right, the journey is automatic. But the currents are treacherous. If the sap runs shallow, the thin stainless-steel pans will burn through in the time it takes an agile man to open his lunchbox. The operator must be ready to ladle sap into these shallows and also to subdue boilovers.

It is an out-of-control boiling pan that gives the gallant evaporator pilot his most splendid moment. It is a fine Sunday afternoon, and tourists have driven all the way from Massachusetts to give their children a glimpse of country life. The pilot, assuming a yankee twang and saying "a-yup" several times, obliges them with a short lecture on sap-making. One of the pans commences to boil over. The pilot pretends not to notice. The tourists are terrified. A heaving mass of hot foam stands six inches above the pan. Just as it is about to slop over everything, the pilot takes a stick about as long as a six-iron and touches the splattering sap lightly with the end of it. Instantly and humbly the boiling subsides. The pilot goes on with his story. "Salt pork," he explains, when his awed listeners ask for an explanation, and it is true; a piece of pork fat is tied to his stick with a shoelace. A small amount of milk or fat knocks boiling sap down magically. Once, when a raccoon had stolen our last bit of salt pork, I used flecks of butter from a sandwich. It worked, although I suppose chain saw oil would have done just as well.

A day comes, toward the middle of April, when a batch of syrup comes from the filtering tank too dark to be called Fancy or Grade A. It is Grade B and it has a stronger, better maple flavor than the other kinds. But tourists don't think that anything labeled Grade B can be better than Grade A, so we finish off the last batch for our own use, and at 117 gallons call it a year. As we wash buckets and holding tanks, we boil the last bit of Grade B over a wood fire and let our children eat sugar-on-snow until they can't eat any more.

I leave, obliged to go back to work, before all 600 buckets have been washed. If I can, I will make syrup again this year; it is what you do in March. My friends will be there, of course. I think they enjoy each year being the butts of the same primordial joke. Sap is a crop that requires no planting, fertilizing or cultivating. It is there for the taking, free. My friends are New Hampshiremen, and something-for-nothing fascinates them. But they do not really believe in it, and I believe they are reassured when 5,000 gallons of free sap and six weeks of donkey-work produce red ink in their account books.

Or perhaps it is simpler. It may be that in that dark place, the New England soul, there is a sense of obligation. If the maple is fool enough to produce sap, the New Hampshireman must match the foolishness: He will collect it.

—*MOUNTAIN GAZETTE, #7*

Mountain Gazette 45

75 CENTS

I. HERBERT GORDON

Lobster Fishing in America

BY
GEOFFREY CHILDS

Three figures appear dimly in a field at the end of the path. They are stooped over at the waist, searching for something in the wet grass with their hands. They seem to shuffle, taking a half-step in one direction and then turning the opposite way, brushing past one another without ever touching. They are gray, sexless and ageless, through the gauze.

"He's an artist," my friend tells me a bit too breathlessly. Her mouth is very close, and her hair is backlit by the houses behind us. Moisture beads at the end of a recent Afro already turning gray at 23. "He rents the house every summer," she whispers as we move closer. "Nobody's ever gotten to know him very well. I think he's Jewish. He's from New York." Her emphasis on "New York" seems to explain everything: How he can afford to rent the house, why he is an artist, why no one has ever gotten to know him, how he came to be a Jew...

The scene comes clear as we approach—a man, a woman, a child. She is wearing a dark sweater draped loosely over her shoulders, a long skirt, a baggy peasant blouse, and she is carrying a wicker basket full of mushrooms in the crook of one arm. The boy, whose age is difficult to tell, is wearing an Allman Brothers T-shirt. He sees us first and smiles. His mother stands and nods her head in a short, European gesture. There is no condescension in the act, only a *noblesse* that people were once born into. "Good evening," she says quietly with a very thick French accent. "Good evening," we both reply. Her husband is still searching for mushrooms and berries—they are what he has come here looking for, not neighbors. We grin uncomfortable goodbyes and separate. A short distance later my friend and I pass his house, a small, plain Cape with a square green lawn and painstakingly cared-for garden dividing their property from the road by a long line of azaleas. There are old bottles and stones in a row across the living room window. There is no car, no driveway, no mailbox, no telephone.

As we walk, the house fades behind us into a mixture of nightfall and fog. This is the fourth day of fog on the island, the third day I have been socked in here, and the fifth of a two-day weekend I could not afford to take in the first place. I envy the artist his bought isolation, his having it when he wants it. It is what I came to the island for—a few days away from the Fourth-of-July madness in Bar Harbor—but now that I have been forced into it, it no longer has the feel of luxury. I have tried making the best of it, sitting in VanLoon's shed during the day writing or reading, but my mind seems to wander to my business and, I guess, whether or not I am missed. I envy the artist his being able to set his schedule according to his inclinations. A convenience of wealth. When we talk about it, my friend, who is not rich, shrugs her shoulders and says once you get used to it, the island is really not so bad. But she lives here and has grown used to the solitude. The artist thinks he owns it. I am the only one who would rather be elsewhere.

Everything is gone now, and we are alone on the road. I feel foolish jingling in the dark with my two six-packs of beer, but later they work to get us both mildly drunk, and I am glad about having gone to the trouble. We smoke the last of my small traveling stash in her kitchen, after her parents have gone upstairs, and driving me back to the beach she side-swipes a tree, knocking off the rear bumper. There are no inspectors on the island, no roads to speak of, so it matters little enough that we can laugh about it. We are both too drunk and too stoned to care anyway.

We make very businesslike love in the back seat.

I stand beside the road until the headlights are gone, then turn and walk back to the shed. There are no stars, no moon, no sky. Only fog.

In my island mornings I walk. The island is not very big—one mile by a mile-and-a-half—but there are several paths cutting through the timber lots between the wharf houses and the backside homes. I made my first trips to the island's natural wonders, but they were all hidden in fog, so most of my recent walks have been in and around the shed along the water where the men build their traps and store gear. On a clear day it could be a postcard. In fact, all summer long people stop by in their private sloops just to take pictures, but few ever come ashore. Usually they save their curiosity for Monhegan Island, which is south another 40 or so miles and civilized in the summer by artists, actors and fashionably rich and reclusive New Yorkers. Here, there are just lobstermen.

There is a small diner on the dock that used to serve seafood sandwiches and milkshakes at night to the local kids and would attract an occasional tourist for lunch or dinner, but its windows are broken and boarded up now with failure. Marijuana took the kids away. A few of the local boys scared off the tourists by getting drunk one night and attacking a 40-foot sailing yacht out of Portsmouth with shit-slinging catapults. Now the kids sit outside or in the boathouses and smoke themselves incoherent and terminally listless by 14, while their fathers sit at home or in the boathouses drinking themselves mean trying to ignore the cycle. Sometimes they will beat up their wives or their next-door neighbors, or for special occasions, go upstairs to a bedroom, lock the door, wrap their heads in towels and put shotguns in their mouths. It's happened twice in the last four years.

Sullenness is the way of life here. There is no place to go except fishing, nothing to do but get sad and tired and crazy. The whiskey comes ashore bootlegged from the mainland and is sold out in 20 minutes. A man buys four bottles of Lord Calvert ("Praise the Lord" is the island motto) and holes up on it till the next shipment arrives. The island could have legalized liquor, but its people vote it down every year. Everyone drinks, but legalization means control. It means Ezra would lose about $1,000 a year bringing in the bootleg, which means that he would probably give up the ferry—and, above all, it would mean outsiders. The law means control, and this is a place with none. It is the tradition. So they get drunk on their smuggled whiskey and go up into their bedrooms.

The fishing has been good, as close as an outsider can gather. Everybody has pots down, and everybody seems to be making money. Not everyone works as hard as the next person, but a man willing to go out in a little hard weather every now and then, willing to pull his whole string twice a week, and someone who's a little bit lucky and maybe even a little bit smart about where he puts down his gear...well, a fellow like that can make himself about $30,000 a year. But people don't talk much about fishing. They are too suspicious and careful. Pots get cut off lines or pulled by someone else, locations get changed, stories spread about someone or other keeping "shorts" or putting down pots a long way from home with someone else's marking on them.

Lobsters are life on the island, and life is a game the islanders play at very hard. And that is the most striking thing about the wharf. There are only 26 full-time families on the island. Most of them are related; most of them have known one another their whole lives. Each family keeps a shed on the wharf, and each shed is padlocked and chained. It is hard to imagine who they are so afraid of. One another? I suppose that's the rationale, but it seems too far-fetched—to have to lock one another out after all these years? People whom you see every day, who have as much or as little as you? Perhaps they worry about strangers breaking in, but that, too, is unlikely. There are so few, and the ones who do come are either invited guests or tourists in sleek yawls from Rockport or Seal Harbor who are not very given to cluttering up their boats with stolen lobster traps and someone's painted buoys. There are no wild animals on the island—the deer and rabbits were hunted off and run down by dogs long ago—and yet everybody owns a shotgun. Who is it that they are afraid of, one wonders? Who is it that they hate so much? Themselves, probably, and their lives. They are suspicious that they are wasted and atrophying, boys caught playing out a useless legend of brave men and the sea that they no longer live or believe in. So they lock their sheds, load their guns and drink too much. They grow fat on the fear of themselves.

Turns out there is another refugee like me. He has been staying on the other side of the island camping out on one of the beaches, but he has run out of food and patience. And now he has come down to the wharf to wait. He is a small, pudgy man in his mid-thirties, very delicate and animated. He wears a small, dark-blue Basque beret and gold wire-frame glasses. He teaches Latin at a private school in Boston. He speaks with enthusiasm about his students, about their eagerness to learn and how far they seem to grow in a year towards becoming adults. I tell him that I tried teaching once and gave it up for what I am doing now: writing and climbing and writing about climbing when I can fit the two together. It is closer to a living—the way I wish to live, at any rate—than teaching. I do my usually poor job of explaining this to Phil, and he is not so stranded that he has to listen to such loose-ended apologies. He changes the subject to the island.

Coffee boils on the wood stove in the corner. The door facing the harbor is open to let out the heat, and a few darkly silhouetted masts and shoring timbers loom, just barely visible in the mist. The ocean has disappeared in gray. Boats and houses are only faintly there, occasionally defined by a voice or the sound of an engine. Few of the men have come down today. Commerce sits behind the stillness and locked doors—waiting, like everything else, for a break in the weather. Phil and I turn to books. There is a growing, geometric silence. Coffee is poured without comment, wood goes on the fire when it is needed. We seem to have already exhausted our unfamiliarity to the point where further questions become prying and any more answers are unnecessary.

We are both reading when a mad boy named "100 Percent" pokes his slanted tow-head through the open door. He is ten or 11, with glasses sliding down his flat nose and a broad slash of a mouth always half-open. He was sitting on the dock trying to fly when I arrived four days ago.

"Brrrooommmm!" he says, "I just lay back, she say brrrooommmm! I can't be givin' her no other, she says, heh? Just moves her slow down a twist, say yes, yes, she say brrrooommmm..."

He turns the accelerator of his insane motorcycle, his tongue touching the bottom of his nose with concentration. Phil stands up. "100 Percent" looks oddly at him and moves back a step. He has no way of knowing that all of Gaul was once divided into three parts and that, therein, secretly is the key to maturity. I throw a scrub brush at him.

"Hey!" he squeals, just bobbing out of the way. "Don't throw that at me. Please, hey. Okay, mister? I'm coming now. I'll fix everything, you betcha. Brrrooommmm!"

He drops down out of sight, running off over soft, bent boards and the broken trunks of dead boats, leaving the door vacant and the empty harbor open to us.

Bat Head and Littlejohn come in after breakfast and ask if I'd like to spend the morning hauling traps with them. It is warmer than it has been, and there is no place else left to go, so I agree. It is 8:30 in the morning and they are already mixing the Lord with their coffee as we pull clear of the harbor.

Bat Head is 20. He owns his boat and keeps a string of about 200 pots. That is less than most of the full-time fishermen on the island, but Bat Head is not an intensely motivated person. He has tried life on the mainland twice. The first time he spent about six months in a trade school. The second time he spent three months in a juvenile offenders' reformatory. He has not been back for more than two days in a row since. He never could quite get adjusted to so many people "ragging his ass," as he puts it. He lives with his parents now in a small prefab on the north end overlooking a wide boulder-strewn cove littered copiously with motor blocks, broken bottles, driftwood, pieces of broken boats, refrigerators, garbage, nets, storm wash and traplines. I have been told that Bat Head is the second meanest man on the island.

Littlejohn is 15 but looks less, except around the eyes. There he is 50 and crapped out. He is very drunk before we arrive at the first string. Nothing in the first two pots, three shorts in the last one. Bat Head pulls them and tells Littlejohn to boil them down in the galley. In 15 minutes we are eating illegal lobsters and washing them down with straight shots of the Lord. Bat Head tells me that Crazy John Lasagna, who is the meanest man on the island, once ate a dead mouse. It was at a dance in the schoolhouse, and John was very drunk. He had brought down a can of dog food for VanLoon's retriever, and the dog wouldn't eat it. VanLoon, who was drunk and in a very nasty mood himself, said it was because the dog food was made of dead mice. Voices were raised, dares and threats exchanged, and finally, a bet made. A short while later John produced a dead mouse and ate it in front of everybody. VanLoon then ate the dog food, which was his part of the bet. The story has been told all up and down the coast, Bat Head tells me. In Rockland even the police know who John is and think he's crazy.

Halfway through the second string, Littlejohn, who is too high on the Lord now to even stand up, decides it is time he drove the boat while Bat Head pulls traps and stuffs the bait bags in them. The bags are small mesh nets filled with putrefied fish that attracts the lobsters. The smell of it is overpowering and Littlejohn, who has not been too accurate in his work, has the stink of it all over him. Pieces of fish dangle in his hair and on the front of his shorts. Staggering from the back of the boat, he lunges at the wheel, pushing Bat Head back against the hot exhaust pipe, and steers us directly over the trap line. None of this is a good idea. Bat Head, who is not very tolerant where his boat or being burned are concerned, strikes Littlejohn directly in the middle of the forehead with the Lord Calvert bottle, and he collapses like a card house. We pull the rest of the string and then turn back towards the harbor, which is around the other side of the island now. We have 62 keepers at $2.08 a pound. We have been working three hours, pulled approximately 80 traps, finished a bottle-and-a-half of whiskey and eaten about 15 shorts.

Bat Head throws some water in Littlejohn's face as we pull into dock and Littlejohn comes to, crying hard and holding his forehead. We leave him in the bottom of the boat sobbing about wanting to be left alone. After expenses, the take comes to about $75. Bat

Head gives me ten and drops off $20 for Littlejohn with his parents. He tells them what happened and that Littlejohn ought to be up after a little while. They do not seem shocked, only vaguely alarmed and fairly respectful of Bat Head. He is, after all, the second meanest man on the island.

    On an afternoon walk I cut off the main path into the woods and find a small lean-to behind a thicket of blown-down trees. The floor is sandy and dry and there is a bed made from fir boughs and a poncho. In front there is a stone fireplace with a new grill leaning in the ashes and the foil discards of many pans of instant popcorn. At the foot end of the shelter there are three crude shelves. The bottom two hold food cans mostly, and a couple of hash pipes. On the top shelf there is a small hand towel and a pack of lubricated condoms. On my way back I meet Littlejohn's father. He seems almost to materialize out of the fog on the path in front of me. He is looking for Littlejohn, who is not at the boathouse or with any of his friends. There is worry on the edge of his voice. Bat Head, he says, gets a little carried away sometimes.

    Hamburger Sam is at the shed when I get back. He is very drunk, this being Friday night and him trying to keep to some civilized schedule. Sam grew up on the island, but he is only a summer resident now. During the year he teaches literature and creative writing at a small college in the Midwest. The lobstermen usually go about their business Saturdays and Sundays just like any other day, but Sam is keeping to his mainland calendar: During the week he reads and writes, on the weekends he sinks into an alcoholic oblivion. He is building a log cabin near Land's End Point, and his hands are rough from the work. He is showing them to Phil and discussing the work when I come in. They both advise me that the weather is supposed to be better in the morning and that Ezra will be taking the ferry in to Rockland and we will be able to catch a ride in with him if we want. Also, there is going to be a party at the schoolhouse tonight. Praise the Lord. Everyone is going to be there, and there will even be a band. Phil seems genuinely excited about the news. I am keeping my expectations low until I can look back on them from the mainland.

    It is dark now, and Phil has gone over to Sam's for some supper and a few beers before the dance. The room is absolutely quiet. I sit in a rocking chair VanLoon made from trap staves and look through the door out onto the black harbor. The only light comes in a zig-zag from the weigh station, where Mithaud is bringing in his haul late. It is not that I can tell Mithaud's boat from any of the others, it is that Mithaud always brings his haul in late, especially if he knows everyone at the station will be pissing and complaining about not getting home in time to shower before the dance. They could close without him of course—they have before—but catch is rare in a fog like this, and they need his. If they don't weigh him in, he isn't beyond going into Stonington on radar and selling his lobsters there. There is no love lost, I am told, between Mithaud and the rest of the island.

He first came here about 12 years ago, leaving a construction job in Lewiston and moving into a house he bought on the point side of the harbor. A hard worker by nature, he immediately built a dock, put up his own shed, bought a boat, a few traps, got himself a lobster license and settled down to earning a living with his new neighbors. But "new" is not necessarily a respected characteristic along the Maine coast, and Mithaud was wrong for the island at least three other ways, as well. First, the people here have never much liked strangers; two, they especially don't like strangers who come and put down 300 traps; and, three, God's immortal teeth forbid that this stranger should be a French-Canadian Catholic.

So, all these things taken into account, Mithaud's first years on the island were obviously lonely and antagonizing. His pot lines were continually being cut, sand was poured into his engine, his windows were broken, he could not get credit at the island store and, worse, every time he brought in a haul the weighing station would close just as he got there. If he left his catch overnight, it would be gone in the morning. If he took it into Stonington to be weighed, he burned up just about every penny he made in gas. But Mithaud was not the kind of man given to throwing in the towel so easily. In fact, just the opposite happened. He got very angry. He began playing his half of the game. When he found his lines cut, he pulled other people's traps, took their lobsters and cut the gear loose. When he had a window broken, he would break someone else's window; when he found sand in his engine, he would put sand in someone else's engine. It didn't matter whose.

Then, after this had gone on for about a year, one night he broke the picture window overlooking the harbor from the house of Crazy John Lasagna. Mithaud was at the post office on the dock the next morning when John came down. People who were witness to it say that John was never in a more evil temper. Sensing blood, a crowd followed him over from the general store and spread out in a semi-circle around the front of the post office to wait and listen while he went inside. First thing John did coming through the door was to pick up Mithaud by the lapels and hold him against the wanted posters with one hand, pulling the other one back over his hip and into an enormous fist. He was red-faced and wild with anger and about to lay Mithaud's face, once and for all, right through the wall when Mithaud told him to go ahead, hit me, you Wop son of a bitch, but goddamit you better kill me, because if you don't, soon as I get out of the fucking hospital first thing I'm going to do is to get a gun and blow the shit out between your *diego* eyes. Now Crazy John might well have been the meanest man on the island, but he was no fool, and it occurred to him that he was way out of his class with Mithaud, who was hanging there by his throat showing no fear at all and probably wishing for nothing other than that John would go ahead and hit him so he could sneak up behind him some night and shove a shotgun up his ass. John dropped him, made a few idle threats and walked out. Mithaud tucked in his shirt, picked up his mail and has not had a line cut since. He is still pulling other people's pots and bringing in about twice the haul of anyone else on the island, but nowadays the weigh station stays open for him even if it means being late to the big dance. People grumble about it, but grumbling is the tone of island life, and hardly anyone hears anymore. Except for Mithaud, who is probably enjoying himself.

"Dennis, you whore's ass, how the hell are you?"

Crazy John claps his free hand on VanLoon's shoulder and spins him halfway around until they are face to face. With his other hand he holds a bottle of gin to VanLoon's mouth. The gin splatters down the front of his shirt and over the pants of several of the men standing around holding beer cans and laughing. VanLoon gags and pushes himself free.

"John," he whines, "you motherless, pasta-sucking bastard." He coughs and pats at the wet stains on his shirt. John laughs again and takes a long drink off the bottle. The others laugh with him like a choir.

My friend is standing near the piano with her younger brother who is already drunk and looking at the floor like a man trying to find a soft place on the hard ground to lie down. I walk over to her and we try being heard above the noise of the band, which is playing top volume twang on cheap guitars and fuzzy amplifiers. They are four boys in about their mid-teens, all with long hair and pale, lightly frightened faces.

"Did you hear about the weather?" my friend shouts. "I heard the fog is already breaking on Deer Isle. That means it ought to be clear here by morning."

Jubilation. A moment of silent prayer to the Lord. I take her hand and turn to watch the crowd. This is farewell, and I am suddenly feeling almost affectionate toward these people. There are about 50 of them in the room, maybe a third of them children under 18. The men and the women, except the ones dancing, seem to keep to opposite sides of the floor. The men all hold beer cans, or paper cups filled with Lord Calvert and ice. Only a few women are drinking. They are hard-looking ladies, most of them mainlanders who met their men and followed them to the island. Nothing connects them other than their universal wish to be somewhere else. Surprisingly, there is very little infidelity on the island, my friend informs me. Everyone knows too well what is going on next door. When the men want to get laid, they make up some excuse for going into Rockland and take care of it there.

A lot of flirting seems to go on. It is probably innocent enough, but it is very aggressive. The three or four women who are drinking are the drunkest in the room. One of them, a lady in her early forties, comes over and asks to be introduced. She is wearing a black pantsuit, the top of which is a very tight jersey with a deep V-neck. Her cleavage is remarkable, formed by a suspension bra whose design must be roughly the equivalent of, say, the George Washington Bridge, for serving up two melon-like breasts to just below several soft steps of chin. The woman's makeup is very heavy, and her breath smells smoky and stale. My friend mentions our names, and the woman takes my hand, coming in close until our knuckles are brushing each others' stomachs. Hers is Spandex elastic. Fascinating, she says, fascinating. She has never met a climber before. At least not a rock climber. She giggles about that, and her hand drops lower. Yes, very aggressive flirting. Later in the evening her husband will attack one of the younger men over just this sort of thing. He will have his shirt ripped in her honor and his son will get in a couple of good licks with his engineer boots before bystanders pull everyone apart.

By 10:30 or so the party has gained its full momentum. Bat Head, who it turns out plays a hot C & W guitar and has a voice like Roy Orbison, has taken over from the band with VanLoon on drums. The musicians are standing off in one corner of the room with terror in their young eyes, wondering what is ever going to happen to them and their fancy equipment and whether any of them will ever see the mainland again. Crazy John is standing with them, wearing the drummer's sunglasses and telling them about dog food. Out on the floor the fat women are up and doing the jitterbug now with wiry husbands too drunk to grab hands as they spin past one another in staggering pirouettes. Singularly, and sometimes as couples, they will tilt too far too fast and fall. A lot of people are beginning to make contact with furniture. Mr. Arbris, who owns the general store, misses a fast side-step in a solo dance of his own creation and falls into a row of chairs, taking most of the buffet table with him. When they bring him out he is covered with punch and chicken parts and bleeding from the nose, but his feet are still moving to the rhythm of Bat Head's borrowed Gibson. Hamburger Sam and Phil arrive about this time. Sam seems sober and composed, but Phil has the glow of the Lord upon him. Amazing grace, as they say. He is flushed, smiling dumbly and clapping his hands quietly on his buns in time to the music. My friend observes this and shakes her head. "Sam's got another one," she laughs, but it will be some time before I pick up on the full meaning of this. I am too busy at this point doing my own odd bob and weave. On this miserable, fog-bound rock, for all its unhappy souls, plundered lobsters and sad hostility, I am dancing for the first time in years. "Brrrooommmmm!" I shout, the Lord rising in my throat to touch every corner of the room. "Brrrooommmmm!"

Unfortunately, getting drunk always makes me forget that I am not much of a drinker. I have never had very much elan at picking the right time and place to vomit. In this case, I hit VanLoon with the first wave, catching him across both legs, then spreading the rest of my stolen lobster over the floor of the coat room for everyone to see. Apparently, no one notices that they are shorts, as I feel several people patting me on the back, impressed with my work and shouting encouragement. The second wave gets most of them. Outside, staggering, I step on something soft between the porch and the ground that moans and twists wildly underneath my foot, but I am heading for the swing-set now and blind to everything else. My friend is kind enough to hold my head, but refuses to drive over me with her car. I tell her that I do not want to live like this, but she refuses to be swayed. Inside the schoolhouse there are screams and thunderous applause. John, she says, is eating a tube from one of the amplifiers.

Delirium and fringe behavior pass into unconsciousness. When I awake I find myself sitting curled in one corner of the front seat of my friend's car. We are not moving; she is talking to someone. Littlejohn's father. He is bending over, his head very near the window and the left half of his face colored by the green lights of the dashboard. His eyes are swollen and his fingers clutch at the suspender buckles of his overalls. "I can't find my boy," he explains. He has lost his hat, and there are tears streaming down his cheeks. "I've asked everybody. No one's seen him. Not since Bat Head beat him this morning. Hell, he didn't have to go and hit him. You know Littlejohn. Shit, he ain't no harm when he's like that." His next words are lost in a gasp. He stands, turning away from the car for a moment trying to compose himself, then turns back and bends over. "If you hear anything..." My friend cuts him off with consolation. He's probably just wandered over somewhere to sleep it off, she tells him. You'll see. He nods and says thank you. I watch him out the back window as we pull away, standing there with one hand in his pocket, slouched over and wiping at his face with the back of his other arm. Dust stirs around his feet and mixes with the remainder of the fog to swallow him, another lost soul on this island of misplaced persons.

The morning comes up clear. I pull on my pack, swallow some coffee and a few aspirins and spring down to the wharf. The ferry is in and loading supplies. I pay my $4, throw my pack on and crawl up some piling to snap a few photos while the other passengers straggle in. 100 Percent comes down on his mystic motorcycle, finishing the last lap of some never-never race he will always win, pulling back at his thin-air hand brakes, skidding in tennis shoes to a stop at the edge of the pier. My friend comes down with sandwiches, and we talk about seeing each other back on the mainland. Phil arrives behind the band to say goodbye. He is going to be staying with Sam for a few days. My friend grins. Behind me the dock hands push up the ramp, and I have to leap five feet for my ride home. The decks already feel like freedom. I turn and wave goodbye, and Phil and my friend wave until we are clear of the harbor.

I take a seat in the stern beside a lady in a cashmere shawl and watch the island fade behind us until it is just a lumpy, dark disk on the horizon.

"Do you live on the island?" she asks.

The question comes completely by surprise.

"No," I tell her, after a moment. "No one does."

—MOUNTAIN GAZETTE, #45

ALICE BROWN

Confessions of a Sauna Junkie

BY
JACK ALEY

I am a self-confessed sauna junkie. Two or three (sometimes even more) times a week, I simply must fire my small cedar-lined room with the double-glazed window facing west (for natural evening light, of course) and endure the heat of hell. If I miss just one sauna night, I get restless. If, for some (ungodly) reason, I miss two or three in a row, it's cold turkey for the kid. I get cantankerous and more obnoxious than usual. My joints begin to freeze up; my eyeballs' humour starts to congeal and my head turns to cardboard. Life ebbs out of me. I start barking at my Constant Companion and start kicking the kids and the cats, whichever get in my way. The only cure is a couple of hours in the box...undergoing that mysterious, sweat-induced metamorphosis of soma and psyche. When I emerge from the hellish rite, all squeaky clean and beatified, I'm fit to live with again. The kids and the cats can relax for a couple days until the saunatropic urges begin building in me again. The sauna is my pot, my boob tube and the monkey on my back. I wouldn't, couldn't, have it any other way.

The history of my addiction to saunas starts inauspiciously enough. The YMCA in the Illinois town where I grew up had a steam room. It was a nice place to go after a handball game. I remember how good I felt after I got out. Nothing heavy. I did discover that the longer I endured the heat, the more miserable I felt in the steam room, the better—(and dare I say it) almost newborn—I felt afterward. I stored that little experiential tidbit away somewhere.

I languished through college and graduate school without access to (or even direct knowledge of) saunas. That's probably why, applying hindsight, I have so few fond memories of college and graduate school. I think how much more bearable those hideous exams in organic chemistry and that Othello seminar would have been had I a sauna to look forward to afterward. Today, I might even be a doctor or an English professor rather than growing vegetables, cutting wood and writing confessions for a living.

Like many restless Americans of my generation, I lived in Colorado for a while after my futile schooling and took to the mountains whenever I could. It was there, in the quasi-Bavarian movie set of Vail (of all places!) where I first ventured into a sauna.

It wasn't real, of course. Nothing in Vail (especially the orange Saab police cars and the mustachioed Marlboro men who went with them) seemed real. My first sauna was in one of those candy-assed jobs tucked into the frail, sheetrocked recesses of a "Texas townhouse." (I don't know where they ever got the name "Texas townhouse" except that they were somehow related to all the Dallas money and accents floating around the place. In general, names didn't make sense at Vail. Sinclair Lewis could have written a dandy Rocky Mountain *Babbitt* at Vail. Such a wealth of material.)

The Vail sauna was electric, the empowering box labeled Am-Finn or some such ridiculous hybrid. In my innocence, however, I accepted it as the real McCoy and rediscovered what I had learned several years back in the Illinois steam room. The more miserable I felt in the sauna, the longer I endured, the more (dare I say it again) newborn I felt afterward. (I hate to keep using newborn—it sounds so rapturous—but it was like that.)

I even took to jumping in the snow after my more miserable stints in the redwood closet. My leaps took me from a Texas townhouse balcony into snow drifts but a few steps away from the center of Vail. My behavior appeared to cause my usually tolerant hosts some embarrassment. They suggested trying a shot of five-star cognac after a bout with 210 degrees. That was very nice too. But in the long run, I've found the snow cheaper than the Martell's.

That winter, I gleaned a glimmer of understanding about the nature of the process I was learning to endure so happily in the sauna. I went so far as to reason the process had some relationship to literature. I was teaching a high school English course at the time in which I used the nature of chemical bonding as a tool for interpreting drama. The spiel went something like this: The more energy released in a chemical reaction—that is, the more explosive it is—the firmer the resulting bond of elements; ergo, the more intense the literary experience, the more powerful and sublime the resolution of it. Sophomoric, perhaps. But the kids seemed to dig it, and I figured I had to do something, someday, with a double-major in pre-med and English. God, I even reduced *King Lear* to a case study in inorganic chemistry. "See kids, Lear loses both his clothes and his sanity on the raging heath (the crucible, of course) before Shakespeare permits him a serene death. In contrast, Old Prufrock is unredeemed and unrewarded because Eliot makes of 'Do I dare disturb the universe' a rhetorical question." I understood this stuff intellectually, I think. But it was not until the initial saunas that the paradoxical curves of extreme experiences compelled me emotionally. All this suffering for redemption crap had something to it. At one sublime extreme, there was Lear on his heath. At the ridiculous other extreme, there was Aley in the sauna. Important things were becoming clear…if not particularly reassuring.

 The restlessness that carried me to the Rockies (I didn't spend all my time at Vail) also swept me away…around our untidy little world. There weren't many saunas along the way. I missed them, not consciously as I remember, but enough so that when I stumbled into Istanbul, I snooped around for a Turkish bath. The one I found was located in a fetid alley somewhere near St. Sophia. I entered the dank, decaying, mausolean place with every western hang-up firmly in place. I reluctantly divested myself of both my clothes and few valuables, sure I'd never see the latter again. The vaulted steam room was replete with the tiered marble benches I'd read about and seen in a couple of movies. Inside, it was tropical but certainly not as hot as a sauna. I was just beginning to feel the place out when one of the several, small, dark men in the room started making advances. For God's sake, he wanted to give me a rubdown. Shocked (I've always been a little naïve about these things), I retreated to a protected corner of the steam room and did my best to let the slow emissions of steam soothe and relax one very uptight initiate to Turkish baths. The result: the curve of stress and relief in a Turkish bath is nowhere near as pronounced as in a good sauna. Besides, the languid vault seemed somehow tumescent, even degenerate. There was absolutely none of that hyperborean crispness I associated with saunas. I've never taken another Turkish bath. And, miraculously, I didn't lose my clothes, my money or my heterosexual purity.

Sweating through the summer months in Asia was very much like being trapped in a huge, overpopulated and dirty Turkish bath, one with no exit. When I finally did get out, from Kabul over the Hindu Kush to Tashkent and then Moscow, I was 30 pounds lighter and possessed of an involuntary lower intestine that had endured the bacterial wilds of Asia and could only tolerate Mexaform and boiled tea…in moderation. I resembled a tall, badly groomed prune. The bear-like (what else) visa man at the Russian embassy in Kabul laughed derisively when he saw me. It took me two weeks to get a visa.

My few days in Russia, in the stolid hands of Intourist, Russia's only and singular "tourist agency," did me little good. I even had trouble with the mineral water. But on the last day, the day I was to take the train from Moscow to Helsinki, I gambled. I traded in an Intourist chit on a plate of beef stroganoff and some red wine. It was a gamble. I lost. That night, I fouled my bed on the train between Moscow and Helsinki.

I arrived in Helsinki in about as bad shape as I left Kabul. But at least I was in Finland. Finland! At last! Pine trees, reindeer, flawless complexions, edible meats and cheeses, and, of course, saunas. I was in heaven.

My first night in Helsinki, I ferreted out a sauna, ironically an electric one located in a university dorm. The first full day in Finland's capital, I went sightseeing. My journal of those days says the picture I liked best in the gallery I visited "was of a real Finnish sauna being tended by a little girl." After two days of sauna-taking and yogurt-eating in Helsinki, I began to feel better, horny even, a sensation that had all but abandoned me in Asia.

The next two weeks in Finland were to rekindle my love affair with saunas and set the stage for my permanent addiction. I planned to hitchhike through the eastern lake district, then north across the Arctic Circle, stopping at public campgrounds along the way. Finnish campgrounds, it turned out, were not your average campgrounds. They were miniature Valhallas, and they had saunas, real ones, the ones with wood heat, warming rooms and a cold lake just outside the door. I was going, quite literally, on a sauna trip through the homeland of saunas.

I got to the lake district the first night and, as my journal recalls, the euphoria started: "I spent the first night in the veranda of an archetypal sauna, woodburning and a few sylvan feet from a clear, cold spring lake. In late evening (still light at 10 p.m...moving north now) with two Germans and three Finns (the sauna pros). I scalded and breathed utter satisfaction alternately for about an hour. One Finn just kept it steaming and once drove me out. I was sure my skin was boiling off. But the hurried dips in the lake brought that fantastic invigoration and relaxation that only saunas seem to develop."

The second night, at a lovely campground in Kattka, my journal indicates what the sauna was doing to my head: "The sauna was a beauty. I was the only non-Finn, and it was still rosy at 10 p.m. after my sauna, so I took a walk in the incessant twilight looking for animals along the dirt road. It was so quiet, the light so peculiar and steady, the situation so strange and haunting. A certain timelessness and vague contented communion with things just not disturbed very much...I thought such winters here, such winters...like looking at an old man, tough and wrinkled and lying in the sun and me wondering about his winter and respecting his mere being and held a little bit in awe by it."

Christ, lyricism almost. That Illinois steam room never triggered anything like that in me. It seemed these saunas not only had the power to cleanse me and make me feel somehow virtuous, but they could also turn me on as well; me, the rational guy who always fell fast asleep at the first whiff of hashish or pot.

The climax to my sauna trip in Finland came in the Arctic Circle city of Rovaniemi, an outpost devastated by the Germans and rebuilt with the kind of severe, economic good taste often germane to northern peoples. A public sauna was included in the reconstruction (it probably was the first thing they built). Upon hitching into the city, I of course hit the sauna first thing. My journal takes dutiful note: "Bless the Finnish for their saunas. The one I had tonight at the sauna and swimming hall in Rovaniemi was fantastic...It was the biggest sauna I've ever seen. Had indoor and outdoor pool. Swimming on the Arctic Circle! Again the sauna is creating its great effects. Rovaniemi is the highlight of the sauna circuit."

I was completely hooked after the trip through Finland. The sauna was a part of my life. I still dream about that Rovaniemi sauna and the others that helped restore my health after the exhausting tour of Asia. I was only a day or two out of Finland, transfixed by the never-setting sun in the Norwegian fishing village of Tromso, when I wrote my parents a letter. I recounted my trip through Finland and vowed I would build my own sauna someday.

In the interim, I returned to the States, suffered routine (but powerful) culture shock and stumbled into a job as a wire service reporter in Maine. Those were pretty crummy years for I was discovering I could scarcely tolerate office work and, basically, was unemployable. To forget my troubles, I ran a lot and became a sauna whore. I started sneaking into motel saunas and college saunas and ski area saunas, and I held pass number 576 at the Portland Jewish Community Center, a pass that was good for nine saunas during the year. These were all pseudo saunas, of course, smelly and electric and sorely underheated. They were a far cry from those on the Finnish sauna circuit, those wonderfully fragrant, woodheated cedarlined rooms with windows facing west (for natural evening light of course). I suffered being a sauna whore, but I had to have the heat.

After some time in Maine (a place with uncanny similarities to Finland, by the way), I did discover the existence of some colonies of Finns in the state. One such colony was in Kingfield, near the Sugarloaf ski area. A Finnish couple, the Pillmans, opened their wood-heated sauna to the (largely skiing) public on weekends. For two winters, I hardly ever missed. When I wasn't in the Pillman's sauna, I managed to get in a little skiing. As a sauna whore, I kept my priorities straight.

After three years of getting my fix of heat in crapola saunas, I'd had it with being a sauna whore in Maine. Enough was enough. I'd learned conclusively that Americans had done to saunas what they'd done to a lot of imported stuff...to chocolate and beer and yogurt and yoga and downhill skiing and the list goes on. We'd sexed it up, put it on a production line, marketed the hell out of it and in the process ruined it. An American sauna (actually a contradiction in terms) was a Hershey Bar.

So I did what I had to do to rehabilitate myself. I needed a worthy object of my passion for saunas. I bought some land, a fairly modest parcel in southeastern Maine with a tumbledown old Cape. Like many of the so-called "back-to-the-landers," I mouthed the platitudes about going back to basics, the owner-built house, the organic garden and the airtight wood stoves. The purists in that elite subculture could never have understood that the real reason I bought the land was so I would have a place for my sauna. I didn't give a particular shit about independence; I wanted heat and ritual. Whenever I started talking to my few Maine friends about saunas, their eyes went blank with boredom or incomprehension.

I spent the first few months on my newly purchased land trying to learn the arcane art of carpentry through rebuilding the old Cape. I wanted the hands-on experience behind me. The house itself wasn't so important. But by no means did I want to botch the building of my sauna.

I actually started getting the sauna together months before construction started. In June, I got hold of 700 feet of white cedar logs from northern Maine and had them milled locally. I spent part of one afternoon lovingly stacking the cedar to air dry.

Also in June I found a Finn in Owls Head (a coastal town near Rockland) who built sauna stoves in his garage. Demand at the time was low...a couple per year. I got one of them. The night I drove my old pick-up the 70-some miles to Richard Ilvennon's was auspicious...soft and clear. Ilvennon greeted me with warmth and a fish dinner. I was a stranger but I was understood. The fact that I wanted a sauna stove at all constituted a bond. Ilvennon took saunas once a week at his parents' home nearby. So he had a pretty good idea of why I'd driven an old truck so far to pick up the crude but substantial stove he'd welded together for me.

I drove carefully. I might as well have had 200 pounds of eggs in the back. Many nights during that summer, I'd go out into the shed and sit and stare at that squat source of my future ecstasy. Some nights I fondled it.

By early that fall, I'd picked up enough about carpentry to forget reconstructing the house for a while and turn to the top priority on the place...getting that sauna together. It was with some awe and trepidation that I began to frame the 12-foot-long by eight-foot-wide by seven-foot-high spaces in the far corner of my shed. Never had I wanted something to be so perfect.

I work best under deadlines, so I gave myself one for the completion of the sauna room—October 12, a Saturday. I figured if I planned a sauna-warming party for that night, I'd have to have a sauna. I invited the poet, the proctologist and the artist down the road: everybody in Maine I knew, all 12 of them. I also invited (or rather summoned) Zee, an old ski-bumming buddy who, despite his penchant for wearing Harvard athletic department sweatshirts, knew his way around tools. Zee made the mistake of arriving from Vermont a couple of days before the party. The sauna was only about half done. I was beginning to panic, so we worked through the nights. I drove Zee unmercifully. Two hours before party time, we were hanging the sauna door and lighting the stove at the same time. I breathed a sigh of relief, and Zee and I took off for a celebratory run. I really wanted to be primed for this one. I wanted the inauguration of my sauna to knock my socks off.

If a perfect sauna results in achieving oblivion (one of my definitions), then my first sauna in my very own sauna was perfect. I passed out just as the first guests were arriving. The combination of little sleep, the long run and the sauna itself, topped by some red wine, did me in. That was okay because I really didn't feel like sharing the moment with too many people anyway.

Few, if any of the subsequent saunas at my place have been as Dionysian as the first (I later heard everybody had a good, drunken time). In fact, sauna nights have evolved into rather quiet rites, one I and my Constant Companion have come to guard quite jealously. (She is a very physical woman who immediately came to share my raving affinity for saunas.)

We've found that the often rough passage to the calm-after-the-sauna is best made in absolute quiet. None but the two other sauna addicts we know seem to understand the importance of shutting up in the sauna. Talk blows the trance, ruffles the concentration needed to give body and mind over to the alchemy of intense heat. Up to a couple of years ago, we'd often have people over on sauna nights. But inevitably, in their innocence, they'd start getting social and kill it. Now, on sauna nights, we turn out the lights and don't answer the phone. We are very selfish about our saunas, we suffer no fools.

Oh, one or two people are sometimes welcome because they understand how to behave in church. Noel, a card-carrying Bohemian, craftsman and ace apple-picker who lives down the road, is one of the cognoscenti we accept from time to time. But even Noel has transgressed grievously. One night, after he'd been mainlining raw garlic for a couple of days, Noel came over for a sauna. What intense heat can do to a molecule of digested garlic being flushed through the human skin is fantastic. The result is nauseating. Poor Noel was banished from the sauna until he'd flushed his system of the offending bulb. Another time, Noel brought into a sauna a towel he'd been storing in mothballs. Within a few seconds, the sauna smelled like the inside of a hot truck. We didn't ask him back for two months after that. Noel understands better now. He behaves well in my sauna except for the times he becomes talkative. But I've found a cure for that. When he opens his mouth, I make sure he's on the top bench and then douse the rocks with a pail of water. The resulting wave of fiery wet heat that engulfs the room is enough to shut anybody up, and the quiet rite may resume.

Not much interferes with our sauna night. Only a phone call from more than 1,000 miles away is sufficient to get me out of the sauna and then I make the caller wait. We did, after a year or so of intense use, blow the top off of Richard Ilvennon's stove, which by that time, had settled into permanent warp. A heavy steel plate welded over the top fixed it, and we were back on line within a week. But it was a bad week for me...and for the kids and the cats. One sauna evening last summer, I noticed the sauna stove wasn't drawing very well. I went outside and discovered that the corroded chimney pipe had finally disintegrated just above the elbow. It was hot and dry outside and the fire's smoke, plus a few sparks, was being disgorged directly into the shed's cedar siding. Rather than extinguish the fire and forget the sauna, I wet down the side of the shed and (albeit a little nervously) settled in for my sweat. I guess junkies do dumb things.

While every sauna we take is subtly different, they basically seem to divide into summer and winter saunas, the difference having something to do with the metaphysics of sweat.

The nature of a winter sauna, if you are familiar with saunas at all, is what you might expect: the contrivance of extreme difference in temperature (nature contrives the cold and you contrive the heat) and commuting between extremes, the better the sauna. One winter's day a few years back, I skied (took one or two runs actually) at Sugarloaf where the wind-chilled temperature was 100 degrees below zero. That same night, I was in the Pillmans 200-degree sauna. Three hundred degrees is an extreme variance. Such winter saunas are about polarities and stretching the mind and body between them...stretching each as far as possible. Fifteen minutes in the heat, hit the snow, rest; 20 minutes in the sauna, hit the snow, the ice water, rest; permit yourself a couple swallows of beer and into the heat again.

Three times is good, four is better, but five, even for the hardened sauna taker, is to begin flirting with fainting. I've had my winter sauna as high as 255 degrees, but at that temperature the wood is almost impossible to touch and you want to touch the wood because rough-cut white cedar makes a dandy back scratcher. These days, we'll settle for 225 to 235 degrees in the winter sauna. Any lower than that and my hard-assed Constant Companion refuses to go in more than once or twice. "It's not worth it," she says with disdain. "You can't fly on 210 degrees." My god, two sauna junkies in one family and the kids are coming on strong. It's almost too much.

The summer sauna is oxymoronic. You know, cold heat, dark light and all that. As a paradox of the senses, a summer sauna is hot cold. It doesn't make sense but it doesn't have to; it is. A good one goes something like this: It is 95 degrees outside at 4 p.m. on a July afternoon. You've been cultivating potatoes or (if you're really flipped out) cutting down trees, and there is sweat-saturated earth clogging every pore of your body. Your soma is suffocating; your psyche is wilting. You are so hot you want to die. The surefire relief is not the icy tub, cold drink and shade tree; they come later. First you fire up the sauna and on a hot day it only takes 30 to 45 minutes to get the room up to 200 degrees, which is all the heat you want. (We are not now dealing with the extreme winter polarities...just the matter of 100 degrees or so.) About 5 p.m., cocktail hour for addicts of another ilk, you strip off your steaming and stinking work clothes and jump into the sauna. God, you thought it was hot outside. This is terrible...hell upon hell. But you endure a few minutes and the flow starts from your core. You didn't think you had any sweat left but you do, plenty. It erupts through your clogged pores like a million hot springs. Hang in there for five or ten minutes, then retire (or crawl) to the shade tree and rest; it is still 90 degrees outside but they are oh-so-much-cooler degrees.

Repeat this deceptive art two or three times and, oxmoronically, you have beat the heat. It's not everybody's idea of doing it. But it makes perfect sense to a sauna junkie.

I'd feel less comfortable telling you about my addiction to the sauna were it not for the fact that it is a perfectly practical dependency. Unlike most things people cannot do without, the sauna is a big money saver. Because of the sauna, we never go out. We just go out to the sauna. We haven't been to a movie or a party or a restaurant in months. Our sauna nights begin in late afternoon with the lighting of the fire and end in late evening with the champagne dessert to our root crop casserole (also cheap and good). The very few times in the past couple of years we have yielded to the ever-weakening desire to go out and be social, we've spent most of the night faulting the food, the flick or the friends (sometimes all three) and fervently wishing we were home in the sauna.

Since the sauna is (almost incidentally) the family bath, I long ago did away with the ancient, current-gulping hot-water heater I inherited with the old house. The absence of the heater alone saves us almost $200 a year in electricity bills, which, because of our wood-heated existence, run under $10 a month anyway. We also use the sauna to dry clothes, make yogurt and heat the adjoining warming room I just finished building.

Since Sauna I (the first 20 or so I recorded like Super Bowls), I calculate I've fired up my sauna 320 times. The 321st is heating up at this very moment. I've been down with a mild case of the flu and, with the help of the sauna, am going to try to blast it out of me. If it works, I'll be well. If it doesn't, I'll probably be sicker than hell and can get more sympathy and breakfast in bed. You see, you just can't lose in a sauna.

—*MOUNTAIN GAZETTE, #70*

GREG WRIGHT

Gone Fishin'

BY
JOHN NICHOLS

On Saturday morning Toby woke up early because he was going fishing with Bobby Salazar. He had crashed at six a.m., setting the alarm for ten-twenty. When it started beeping the cats began skittishly prancing around, confused and disturbed. Toby cursed, swung his feet out of bed and sat on the edge of his mattress, dumbfounded, groggy, despairing, hurting all over—his neck, his hip, his right leg. Sometimes he got cramps in bed that made him yowl just as he awoke, but not this morning, thank God. His lips were so chapped and swollen they felt about to explode and drop off his face. His throat was as dry as if somebody had poured sand into it all night. The ache in his sinuses was massive, brutal, thundering. The awakenings he experienced nowadays were like Olympian hangovers, earthquake survivors, the aftermath of slow and ponderous car wrecks. It was great to be sixty years old, qué no?

He sat there rubbing his eye sockets, too blown out to reach over and quash the alarm's irritating beep. Then he finally roused himself out of terminal lethargy, pushed the button to Off, removed the fat blue pillow from on top of his answering machine, and stared at the blinking red light, trying to make sense of its bap-a-dap-dap sequence. Evidently it was saying, "Three messages, buttmonkey, I dare you to push me now when your brain is still in warp eight."

Later, gator. Toby shuffled into the bathroom, almost tripping over Carlos. He pee'd in the sink, splashed some water around, brushed his teeth. Cookie came in and started eating her Science Diet lamb and rice bits. Why was the cat dish in the bathroom? Because when it had been in the kitchen skunks had come through the kitty door to have at it, no kidding.

From the medicine cabinet mirror a six-hundred-year-old man, a sort of alcoholic, homeless Albert Einstein with grubby hair cartoonishly fluffed in many directions, grimaced at Toby, gap-toothed just like our hero, and said, "You look marvelous."

Toby said, "Thank you, screw you, up yours."

From the top bureau drawer he fetched his lanoxin and an aspirin, and started toward the kitchen barefoot. But there was so much grit on the eight-thousand-year-old carpet that he doubled back, grabbed a pair of socks from the pile beside his bed, sat down and tugged them on. In the kitchen he poured a glass of apple juice and drank his pills, then parted the filthy lace curtains so he could better look out at another hideously sunny day in their winter without beginning. "What's the UV count today," Toby muttered, bitterly sardonic, "eight billion, four-hundred eighty-seven million, five hundred and forty-six thousand?"

He and Bobby Salazar wouldn't need sunblocker on the Río Grande today. A mud plaster over lycra Aztec ski masks was more like it. Toby still had one of those sleep-created hardons that he observed for a second rather bemusedly, wishing there was someplace pink to stick it. Carlos rubbed against one shin, so Toby bent over, scratched behind his ears, then rubbed his backbone by the tail hard for ten seconds. He grabbed a Kleenex out of the box on the toilet tank and blew his nose, big gobs of snot.

Back on the bed, groaning as every bone in every joint in his body screeched loudly, Toby grabbed up a couple of Jobst varicose vein pressure stockings, gathered his strength for two minutes, then tugged them on, whimpering from the pain. It required a superhuman effort. If he lived three more years how would he ever have the strength to keep pulling on the stretch hose?

His heart blurped in and out of atrial fib. Toby put on a pair of filthy dungarees, an old flannel shirt, then he trudged outside to the tool shed, unlocked the door, and paused for a moment intimidated by the disarray. A bicycle, garden hoses, tool boxes, paint cans, nail bags, nuts and bolts in mayonnaise jars, a dust-clogged Casio keyboard, all of Toby's digging and chopping and hoeing tools and scads of other hardware crap cluttered the six-by-ten space. Toby had to be careful not to straighten up in there otherwise dozens of nailpoints in the ceiling would perforate his skull. Another fabulous Toby Scott Floyd construction job. Left to his own devices with a hammer, a Skil saw, and a T-square, Toby could out-bumble the Three Stooges.

His fishing equipment lay on an old foldout table with both leaves down behind the bicycle in a mess of detritus guarded by two dozen black widow spiders flashing their Day-glo red hourglasses at him. Behind the pile of refuse that included his fishing vests and numerous busted reels stood a half-dozen rod cases in various stages of collapse. Toby gingerly retrieved his vest, a couple of reels, and his one good aluminum rod case, and humped out of there before the black widows started spitting cobra venom in his direction. The vest he shook out spiritedly for thirty seconds, then he plunged back into the raging tool shed inferno to retrieve a burlap sack for carting his catch.

Long ago, in another century, Toby had kept up his equipment. In those days he'd been a fanatic for the Río Grande, fishing several days a week throughout the late summer and autumn, scampering up and down steep rattlesnake-infested bajadas like a professional athlete, him and Bobby Salazar and Bubba Baxter, and all the women that Toby had loved: Mona Perot and Brenda the Cop, Jo Ellen and Beverly Sinclair, Sherri

Franzetti, even Rebecca, even Penny Sullivan a couple of times, Loretta Larson, too, and Iris Candelaria. Those girls had loved to fish or tag along. It was always the same and always wonderful. Weather warm or icy, the magnificent towering canyon walls, hot winds at gale force stinging their faces, big basalt boulders everywhere, green water roaring and splashing, redolent sagebrush on their fingertips, blossoming sulfur-yellow rabbitbrush, the canyon wrens and owls calling and bats fluttering at dark when they started out, and then languid sessions on the Dodge tailgate or in the Impala, exhausted, happy drinking beer and eating sandwiches beneath the stars and various stages of moon or cloudy, growling skies. He and the bimbos fucked in the truck, they fucked in the Impala, they fucked on pine needles among ponderosa pines on the trail down, on the trail out, they fucked in grasses beside the river, against those enormous gray boulders, they fucked whenever the urge or opportunity presented itself, crusty with sweat and fish stink, their moans lost in the river roar, their happiness uncompromised by earthly preoccupations.

At night the Milky Way traveled almost exactly overhead, north and south. They made love totally wasted and then drove home thirty miles down from Big Arsenic or Bear Crossing or El Aguaje, or from Cedar Springs or Miner's Trail, or from Francisco Antonio on the west side of the gorge, completely happy and satiated, shamelessly drinking beer, tired, almost drunk, ecstatic, the heaters whirring—what a wonderful life.

Toby arranged four bottles of St. Pauli Girl in his red and white cooler, added ice and a churchkey, then made sandwiches of baloney, lettuce, Swiss cheese, and tons of mayonnaise on Potato Bread, wrapped them in tinfoil, dropped 'em in a sack. He added a couple of apples for good measure, then cranked up the D 150 and headed north on Valverde to the Salazar residence, where Bobby was waiting impatiently, frowning at his wristwatch: "What the hell took you so long, mofo?"

"Shuttup," Toby growled, "I got three hours sleep and I'm not in the mood." He was exaggerating by only ninety minutes.

"Time's double when you're goin' fishin'," Bobby grumped, by rote, for the eight millionth time, flinging his stuff in the back of the truck, then popping a Diet Pepsi after he'd swung up into the cab.

Toby hit the next speed hump so hard half the equipment almost bounced out of the bed. "Jesus Christ," Bobby wailed, "why don't you freakin' step on it?"

"Times double when you're goin' fishing," Toby said.

They were OFF, and as excited as little boys. Just because you're sixty don't mean you can't have fun. "Snow on the roof," Mona Perot always liked to say, "don't mean there ain't a fire down below."

Toby drove north on Placitas, hung a Louie on the highway, passed the blinking light four miles later, outdistanced civilization, and, for a minute before Arroya Hondo, they had mesa on both sides, wide and lavender gray, with that electric blue sky and brilliant sunshine cascading in crazy droves all around them, The Nightmare Non-Winter From Hell...but the river water was low and pretty clear, and for sure the fish would be biting.

When Bobby stuck a cigar in his mouth, Toby screamed. "Not in my truck, you asshole. Don't you dare strike a match!"

"Toma lo suave," Bobby said. "What do you think I am, a neanderthal? Hey, ese, gimme some credit, okay?"

"You light a match, I'll roll this truck over deliberately."

Bobby lit a match, saying, "Go ahead."

Toby punched his shoulder. Bobby whacked him back. Toby said, "Did you bring me any candy?" Bobby said, "Up yours and your candy, whattayou think I am, an idiot? Buy your own candy. Plaque up your own stinkin' veins with your own money. I refuse to participate."

"OH MY GOD WE'RE FREE!" Toby hollered.

Bobby puffed on his unlit cigar. They were behind a state cop going exactly fifty-five so Toby had to rein it in, cursing the marrano who must've known they were headed for trout.

"Did you bring some beer?" Bobby asked.

"Does the Pope swim upstream?" Toby replied.

"You know, you oughtta grow up one of these days," Bobby said. "That childish bent is reflected in everything you write."

"I never said I was Dostoevski," Toby said.

"Dostoevski? Spare me, dweeb. In your sueños. Shame on you. Vonnegut's more like it. On a good day. On a lucky day."

"I'm gonna catch a twenty-inch rainbow," Toby said.

"Oh yeah? You and whose ejercito?" Bobby answered.

"Five dollars," Toby said. "Biggest fish."

"Don't bore me with your peasant gaming," Bobby scoffed. "I'm out to contemplate the natural beauty of La Naturaleza, and I wouldn't cheapen the experience by half-assed wagering for all the oro en el mundo."

Toby said, "Look at the sky, look at the sierra, look at San Antonio Mountain over there, look at the sagebrush, look at the empty highway, ain't it great to be alive?"

"Enjoy it while you can," Bobby said. "It's a quick trip for all of us."

The lads smiled in unison. Then they laughed...

But neither of them caught a single fish. Ouch. Bobby had six hits and missed them all; Toby had seven hits, same deal. Anymore, he had the reflexes of a sloth. They started climbing up from the river at dusk cursing the gods and goddesses of the goddam Rio Grande. "That river beat me up again," Bobby moaned. "Me comio los huevos. It's like a cheap hooker from Enseñada, tetas the size of Vesuvius, but with her legs chained together."

"You're so gross," Toby muttered. "I'm ashamed, you chauvinist pig. Clean up your act, bro'."

"We shoulda brought dynamite," Bobby said. "Grenadas de mano. Sulfuric acid."

"Stop bitching," Toby said, huffing, puffing, gasping, grunting as they climbed the steep trail leading out of Big Arsenic toward the rim two miles away. "If it was easy, we'd hate it."

"I hate it like this," Bobby complained. "I lead a hard life, I want a little positive reinforcement en mi vida loca. I'm sixty-one years viejito and I spent sixty-one hours this semana peering through a loup repairing a bunch of stupid digital pedazos de mierda I wouldn't even wipe my ass on let alone latch on my wrist. One in fifty relojes anymore that come across my mesa were made when craft quería decir algo. I spend my whole chingada vida fixing drek for pendejos who wouldn't know quality if it came in their oreja. I deserved a trout today, fuck you and your estúpido gabacho locuras. Puto. Me cago en las botas de tu madre."

Toby sang, "Allá en el Río Grande, allá donde vivían …Vivían truchas grande, que nuestras plumas no querían—"

Bobby tried to punch him, miscalculated, and fell down, banging his right hand on a prickly pear cactus. Half the stars in heaven cringed when he screamed.

Neither of them had a flashlight. "Get away from me, o seguro te mato," Bobby snarled, seated on a rock, fumbling to pinch out the hideous hooked needles aggravating his palm. "You made me do it, you cabeza de basura. A pox on your casa."

Toby whistled through the gap in his teeth, contemplating the miraculous clarity of the constellations. "Hey, there's Cassiopia," he said merrily.

"Tú madre."

"And Ursa Major, I do believe."

"Tú padre."

"And Orion's Belt—wow! Qué cielo máravilloso!"

"Tú hermana, tres veces, en el culo."

"Ain't that big dipper glorious?" Toby exulted.

"Tú abuelita. En la boca."

"La tuya," Toby chortled. "Man, I love this river."

"Screw this river," Bobby grouched. Getting up again, gingerly flexing his aching fingers. "Let's lárganos from este infierno. You got any water left? My bottle is empty."

"Yup. But I need it all. If I go into A-fib it's instant dehydration, I'll die, you know that."

"Gimme that agua," Bobby said, "or I'll slit your gringo garanta."

Toby gave him the water. He watched Bobby guzzle, then took a gulp for himself and had never felt more divine in his life than right at this moment. The river, the gorge, the disasters did that to you. It was all bloody wonderful. And what a sky overhead, sparkling with flaming hydrogen, black holes, exploding nebulae. Toby could smell icy space, mingled with pine pitch, adobe dust, and sagebrush.

An hour later they reached the truck, groaning, moaning, cursing, aching, sweating, panting, coughing. "Hijo madre," Bobby said, "who taught us to suffer like this? Me I understand, I'm chicano, lo entiendo, it's in the sangre, somos todos jodidos. But you're a gringo, you're supposed to know better, it's not genetic with you, so what's the catch?"

It was too cold to sit on the tailgate. Toby cranked over the engine, hit the fan button, pushed the heating levers all the way to the right, and let the decrepit old Dodge idle while they ate their sandwiches and sucked pensively on the icy beer. Toby had always been a glutton for this ecstasy of exhaustion, he adored the anguish, the ache, the physical collapse so sweet as he mingled into it the sandwiches and cerveza, warm air blowing against his fatigue like angel breath. And there they were, two old guys, sixty and sixty-one respectively, side by side in the cab of an idling dinosaur with 194,866 miles on the odometer, wondering if they'd ever grow up.

Sixty seemed awfully old to Toby, yet he still felt infantile, innocent, inept and childish most of the time, amazed that anybody took him seriously. How had he earned a living for thirty-eight years, raised children, pretended to be wise? "It doesn't get any better than this," he said, staring through the windshield at darkness, at the dimly visible shapes of piñon and juniper trees.

"Tonight is kinda special," Bobby replied. "We only go around una vez in life."

"We've sure had some times together," Toby said.

"Ain't that la verdad."

Toby finished his second sandwich about the same second Bobby licked a last mayonnaised crumb off his stubby fingertip. They clinked beer bottles again: "L'chaim," "La Vida," "Salud, amor, dinero, y mucho tiempo pa' gastarlas." Then Toby switched on the headlights and put it in gear, popped the clutch, headed out. Between Arsenic and Cerro they saw a couple of small does and maybe the twitch of a disappearing coyote. On such a deserted road, with the moon shining brightly, Toby could cut his headlights. They cruised for a way in darkness, feeling awed, peaceful, happy. Adios, Río Bravo. Hasta la próxima.

"We're gettin' old," Bobby said. "We shoulda killed that river today."

"Shoulda, woulda, coulda," Toby laughed.

Actually, on reflection, those funky old boys had killed the river today. And they would do it again, and again, and again…and again…and again…and again.

—MOUNTAIN GAZETTE, #80

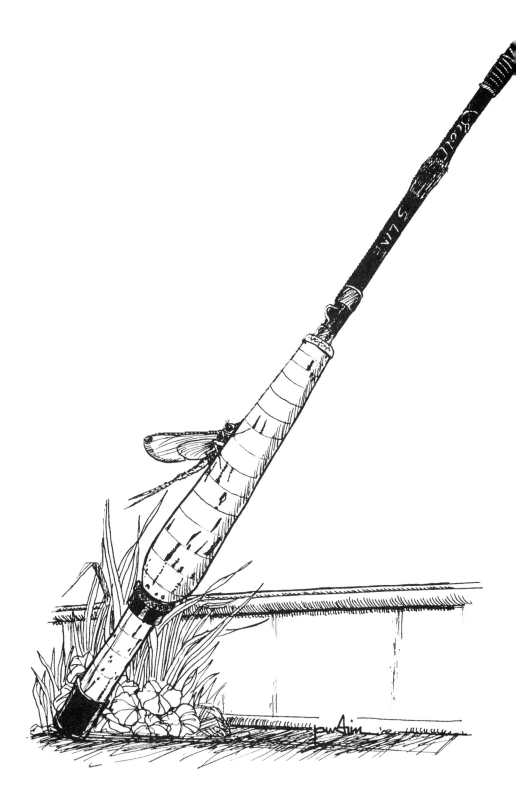

Obituary
The Deceased: Scott Fly Rod
Born: Many fish ago
Died: Many fish later
Cause of death: broken heart

BY
MICHAEL HOLZMEISTER

I broke my favorite fly rod last year on a float down the Gunnison River, and I still haven't gotten the thing fixed. It's a Scott made when the company was still making fly rods in Telluride. Scott rods carry a lifetime warranty against anything that might befall a fly rod, like mashing it between a rod case and an oar frame. All I have to do is send it in, and Scott will send it back good as new, yet I keep it cased and drag it around with me unrepaired.

The rod is one of the older Scotts, before they started making stiff rods with little feel or flex. It casts easily, letting you know when it loads up, ready to spring 40 feet of fly line across the water and settle it down with barely a ripple. The nine-foot graphite wand makes my fishing feel complete. I can cast well and catch fish with my back-up rod, a stiff, matte-black Loomis, but the grace has left my casting ever since I broke the pewter-gray Scott.

When I heard the graphite crunch as I tightened the strap, it seemed perfectly appropriate that I should break my favorite fly rod on this particular trip down the Gunnison. After all, this was a fishing funeral.

I had never heard of a fishing funeral before Elizabeth brought it up when her doctor told her that her cancer was terminal. She died a couple of months later, a day after the Super Bowl and just as Black Magic was winning the America's Cup. I watched the yacht races in her house every night as she got sicker and the tumors multiplied and grew. No part of her body was spared.

A lump sprouted up on her tongue, and I swabbed it with lidocaine. Lumps like golf balls grew on her back. Her ankles and knees swelled as the rest of her shrank. The tumors took all of her body's energy, and she could only lie in bed stoned and constipated from all the morphine she swallowed to relieve the pain and feed the addiction she developed. When you're about to die, doctors seem to be pretty generous with the dangerous drugs.

On the Sunday before she died, a pained moan accompanied every single labored breath. And when she could close her eyes and sleep, her troubled breathing snapped her awake in a panic and the fear was clear in her eyes. The next day she slept and breathed easily but not deeply, and that night, she died.

Her sister and I drank the bourbon Elizabeth saved for special occasions and stayed up all night.

The wake was a couple of days later. There were bagpipers and people from years past who I had all but forgotten about. Elizabeth's bar was packed with mourners celebrating with stout and whiskey, but for me, drinking didn't feel right, and I watched tears roll down the faces of people buzzed enough not to care. I mourned, and still do, privately, and I woke the next morning with no regrets and a clear head.

Elizabeth was the first person I loved who died. The finality of it and the memory of her last moments lying in her bed won't let me get to sleep sometimes. I start to doze off, and the picture I keep in my head of her room on that winter night startles me awake as if it were a memory along the lines of leaving the stove burning or the headlights on. I want another chance to talk to her, and when I'm half asleep, I'm afraid I'm missing it.

Her instructions for the fishing funeral were quite clear. Me and Tom and James were to take Elizabeth and her family fishing on the Gunnison River in early July so we might hit the Green Drake hatch. Elizabeth loved that time of the year when the big mayflies fluttered off the water as we floated along. She would catch fish, giggle and generally have a good time. There's a grassy spot near an old quarry where the river splits and widens and boulders make all sorts of eddies where the trout can hide, and we used to stop there for lunch, and the Drakes would come off the river, and the fish would go nuts eating them. I would fish, but Elizabeth would lie on the grass and take a nap in the sun. Sometimes, we would float down in the evenings, stopping at the grassy bank for happy hour and even have dinner there. That is where she wanted to be.

So, last year, me and Tom and James loaded everything and everyone in boats and floated down to the grassy bank, fishing along the way. Other than a lack of Green Drakes, it was a beautiful Gunnison Country summer day. All the cliches apply. We were bathed in warmth. There was a light breeze. The sun shone brightly. Anyone in the mountains should know what I'm talking about.

We got to the grassy spot and unloaded. Someone produced a bottle and we had a toast. Then we unscrewed the box that held Elizabeth. Her ashes were in a clear plastic bag, and I grabbed a handful of her, and flung her into the bushes. Tom took a handful and poured her into the river. James did the same. Her sister, Mary, flung the ashes into the water. I grabbed another handful and threw Elizabeth into the bushes again. Her bits of bones tinkled through the leaves, and the fine white dust settled on everything. Elizabeth was stuck to everything and she was everywhere. We had another toast before we left Elizabeth and the grassy spot, and as we were getting back in the boats, I broke my Scott about three inches above the cork handle where the graphite is the strongest.

The bottom half of the rod is still in one piece, and if you were to look at it leaning in the corner, you'd think there was nothing wrong with it. You can put the rod together, and it looks whole and perfectly usable. But with the slightest motion, the butt section collapses, and the rod's fatal flaw becomes apparent.

The graphite is shredded and splintered all the way around the rod, but everything is still attached lengthwise so that when there is no pressure on it, the bottom half of the rod maintains its tubular integrity. Once flexed, it goes flat, the strands of resin and graphite spreading apart and releasing a little dust. My Scott has lost its backbone.

I haven't sent it in for a year, and I miss it terribly when I'm trout fishing. As the time passes, I feel a little more foolish for holding onto a worthless, broken fly rod when all I have to do to get a new one is call UPS and request a pick-up.

I wish every loss were so easy to replace.

—MOUNTAIN GAZETTE, #83

Mountain
Gazette 52

75 CENTS

GALEN ROWELL

Coyote Song

BY
DICK DORWORTH

You may say that I'm not free,
But it don't worry me.
—Keith Carradine

The highway between Wilson and Jackson crosses the Snake River about a mile outside Wilson, over a concrete-asphalt-metal bridge of uninspired, though functional, design. Past the river, the road continues for a half-mile before entering a long right turn leading to a quarter-mile straightaway and then turns left into another straightaway. That is the only section of the Wilson-Jackson highway we are concerned with here.

The road itself is not special. Just a ten-mile stretch of classic two-lane black-top connecting two western American towns. The only thing unusual and unique about this particular slice of highway is the contradictory unusualness and uniqueness common to *any* piece of the road we are all traveling. That is the fact observable to the patient and interested, that he who pursues the road, no matter how sporadically, will, like every gypsy who ever used unspeakable cruelty to teach a bear to dance, someday find himself once again on the same stretch of road during one or another of his swings away from his own everchanging, unvarying nature.

Wilson is little more than a road stop at the bottom of the eastern side of Teton Pass, and that's the way locals like it. Wilson is the site of the Stagecoach Bar, the one saloon in the Jackson area that is common ground for all of the diverse social elements living there—cowboys, ski bums, hippies, climbers, tourists, musicians, horny housewives, college students on vacation or leave, construction workers, restaurant workers, fat cats, lodge owners, condominium salesmen, fishing guides and anyone else in the vicinity hankerin' for a sandwich, some company, a bunch of beers, a pool game, good music, and, maybe, a lay. On Sunday afternoon the Stagecoach jumps. Jumps, hops, skips, rocks, rolls, howls, runs, back-flips and spread eagles. All good local musicians and any passing through gather there to jam. Sunday afternoon in Wilson can get pretty raucous; but because of the local laws, inspired by quasi-religious sentiment, the bars close at 8 p.m. on Sunday. Around 7:30 there is a run on six-packs at the Stagecoach, and by

8:30 there are empty beer cans all over the parking lot, the highway and alongside every road leading out of town.

That's Wilson.

Jackson has its charms, but all in all it's about the worst tourist trap in western America. During summer, Jackson is wall to wall people, bumper to bumper traffic, asshole to eye-lid hustle, junk stores, mosquitoes and all the lost energy of displaced Americans desperately seeking their own misspent history and heritage in the noon and 5 p.m. fake gunfight held daily in the town square. The entrance to each of the four corners of the square is through an enormous arch made from the antlers of elk, a large noble animal indigenous to the area. Indeed, the elk is indispensable to the local economy, which thrives on the trade of the great white hunter in the autumn in much the same manner as it survives on the dreaded white tourist in the summer. Most conscientious wanderers pausing in Jackson overnight or a little longer will somehow drift into the Million-Dollar Cowboy Bar. At one time only the bold, the blind, the unwise or the saintly long-hair would have dared venture into the then aptly named saloon. But times change, and, in one of the ironic moves of the karmic wheel, the cowboys lost *their* territory for a change. Not lost but came to share. And what better way to work out all the old bullshit than by sharing—both the bullshit and the bar.

That's Jackson.

The Snake River drains out of the mountains of Wyoming into Idaho and Oregon and on to Washington where it joins the mighty Columbia, which eventually flows home to the ocean. Some people speak of an ocean of love from which life comes and to which it must return. And because of all this idle talk down the years, it often crosses my mind as I cross over, bathe in, look upon and drink from the fine Snake River that, if that's how it works, then that which begins in love must, inevitably, end in love. And it is simple to make the next step of seeing the true beginnings of things in how they end. That's called hindsight, but I don't hear so much about the importance of beginnings. It is the state of mind that comes before the aim that comes before the arrow is launched toward the target. The river of peace; the ocean of love; and there was even a man who is said to have walked on the water. Who knows? He may have walked on the Snake.

Concrete, asphalt and metal are materials used by the human animal to subjugate, dominate and violate the nature that gave him birth and so far continues to sustain him. The human critter can be exceedingly ungrateful.

The bridge across the Snake is a tool of convenience. From one aspect it's a piece of shit, but it serves a function by allowing people and their vehicles to shuttle back and forth across the river without getting wet. Some people and most vehicles do not take well to getting wet; though coyote shuns the bridge. In a 100 years the bridge won't be there, but the Snake will. There may be another bridge over the same river and different men to cross it; but I cannot repress my curiosity about the state of those men's minds, a 100 years from now.

Uninspired is the state of life of the coward who would rather live with an unacceptable comfortable situation than throw it all over for a chance at joy.

Functional to an engineer or a soldier or a politician or an insurance salesman may mean something very different from what it means to, for instance, a coyote. What is functional to each person says more about the person than about function, and it is an interesting word to throw into a conversation with someone you wish to check out. The bridge does serve a function in the material world.

Construction. Well, shit, boys and girls, we still haven't figured out *how* the Pyramids were built, much less *why*. If modern technology can't answer that one, it puts, at the least, what man calls "construction" in a perspective that cannot help but make the honest scientific mind...pause.

Once past the bridge, the road goes straight toward a turn. Just before the turn, a small farmhouse on a hill can be observed out the left window. Right ahead is a field where the farmer grows hay, and the road bends around a field. It is, perhaps, half a mile long and a quarter mile wide; and every time I've seen the field it has been as groomed and well kept as those beautiful women in international airports who melt your heart and fry your brain, and, when you're graced, sustain your spirit during those long, alone trips around the planet...trips which find you trapped in strange cities between flights to other, even stranger, places where you know you will not tarry long, just as you know it is part of the weaving of eternal tapestry that you must visit there from time to time. And that's why there is a turn at the end of the straight section.

If you had been in the Stagecoach for ten hours, playing pool and drinking beer without eating sandwiches or getting laid; and if you had ingested ten reds and, possibly, snorted holes in your septum with the magic anesthetic white dust; and maybe if there were some other lethal frustration in your life...like ten years (or ten minutes) living with a mate no longer wanted; or a job so boring that it turns the honey of the spirit to carbolic acid, or, at the very, very minimum, a good old-fashioned scrotum-to-brain burn by the all-time honest-to-God, truer-'n-shit wonderful unbelievable down to the center of the earth higher than the cosmos perfect love of your life...then, with such a frustration or physical or psychic handicap bubbling away in your brain and being, clouding judgment with visions of devils and demons and never-ending red lights in the rear view mirror, you might miss the turn and go blazing across the good farmer's field. If you did that, and if your vehicle and everything in it survived, which is not impossible, and you kept going with a slight lean to the left and did not hit any hay bales or coyotes or holes, you would cross the field and run through some willows on the other side from which you would emerge to crash through a hand water pump and continue up a driveway to a small cabin nestled right up against a small forest of aspens.

I once spent the better part of a summer in that cabin.

To reach the cabin by staying on the road, it is necessary to negotiate the right turn, continue up the straightaway, hang on through the left turn, continue a 100 yards, and turn back left onto a dirt road just off the highway. The hoop gate on a barbed wire fence must be opened before driving through and closed after; and there are three such gates before the cabin is reached, each to be opened and closed, both coming and going. The road goes along the edge of the shimmering, murmuring aspens, mostly within the shade of the fine summer leaves; and the road must be driven with as much care as is cared for the vehicle driven. Very often hawks, ground squirrels and coyotes are seen along this road.

The one-story cabin is a beauty for people who do not mind a 100-foot walk to the pump for water, or, in the other direction, to the two-hole shitter; or cooking over a wonderful old cast iron woodburning stove; and cutting wood for that stove; and doing without electricity. It was built of wood by some less than mediocre craftsmen and has a large rock fireplace in the middle of its one room. That summer there was one wooden table and four matching chairs and a dresser and two double beds, which we never used, preferring to sleep outside under clear Wyoming skies or in the bus with all the doors open, listening to the nightly coyote serenade.

I was cruising for a time with a peroxided lady and a child who were both close and distant. On clear days I climbed the variable rock of the Tetons. Stormy days were spent writing at the cabin or in the peaceful Jackson library where there were not only free coffee and comfortable chairs and a big table to write upon, but the quiet of all the sad, lost souls seeking freedom from both sides of every page of every book of every shelf on every aisle of all the libraries man has ever built and burnt and sanctified and censored throughout a history he but dimly remembers…for if he remembered and understood he would not be condemned to the prison of repetition, and the seeking of a freedom that stands, like naked, beautiful, beckoning innocence across the ocean of love, the river of peace, the stream of understanding and the trickle of attempt.

A few days were spent in the front yard with heads full of acid, watching our neighbor tend his fields. One particular day sticks in memory. We were sitting on the ground with our friend the German woman, of fine intelligence and heart. She talked too much and pushed too hard and was never sure about living in unending sorrow over some unacceptable personal tragedy that was talked around but never about, and thus could not be plowed under to fertilize happiness; and the tears she shed inside flooded the world, drowning all not contained within the ark of her mind.

The two interweaving cross currents of our energies revolved around reading *Ecclesiastes* aloud to each other and watching the good farmer work his fields the entire day in the sun. The two were, of course, the whole; and holding them together in our minds was, at the same time, the most serious endeavor; the most hilarious pastime; the most arduous undertaking; the easiest frivolity; grinding work; and the most fun any of us had ever had. The high awareness that it is "all emptiness and chasing the wind" laid us out in hysterical laughter, clapping each other on the thighs and backs and repeating over and over, "all emptiness and chasing the wind." And out of that day and line we were finally able to name a route we had climbed on Mt. Mitchell in the Wind River Mountains a few weeks before. It was a hard, beautiful route on perfect rock that we started right after breakfast and which saw us return to camp at midnight. It is one of my favorite climbs. We named it *Ecclesiastes*, in honor of the joy of the empty chase.

The farmer worked his field in a circular manner, starting from the perimeter and advancing inward, in just the opposite direction of harvesting crops of karma. He was cutting hay that day, sitting beneath the sunshade atop his roaring machine, and a circuit of the field took about 15 minutes. He was a big man wearing a blue Levi shirt and straw hat. I never spoke with him, but for perhaps 20 seconds of each tour of the field we could hear him, above the roar of the machine, singing at the top of his lungs. There was, in the strength of persistence of his voice, a daylight counterpoint to the nighttime coyote song. His deep baritone was filled with joy and revelry which came, we could only assume, from his work. He sang Italian opera; and, though we only picked up on his serenade for a few seconds of each cycle, it was consistent and it is fair to assume he sang the whole day long. And we were there from tea-and-capsule breakfast until sundown.

Or maybe the man was putting on a show for us...the neighbors who never, ever communicated or worked or did *anything* that he could see...and it is possible that he only sang during the part of his cycle which came within our realm. But that is a cynicism I recognize and cannot accept. I never felt he cared a politician's word of honor whether we watched him or not, but I was *aware* he *knew* we were watching; and in a sense that cannot be written about because I wasn't on his side of the page, he was as much a spectator as we...watching a boy and a longhair beard and a blonde and a shapely brunette sitting in front of the cabin across the way...apparently *doing absolutely nothing* the entire day long. He worked his fields with a thoroughness we could not envy because envy gets you hard every time; but we did not refrain from admiring and wondering about it. While I will never *know* what was going on in the farmer's mind, I still would not like to live in a world without wonder; and there was no emptiness in his barn. If there is a wind to chase, the farmer made an inward circular pattern out of his pursuit.

> *If you witness in some province the oppression of the poor and the denial of right and justice, do not be surprised at what goes on, for every official has a higher one set over him, and the highest keeps watch over them all. The best thing for a country is a king whose own lands are well tilled.*

We read those thoughtful words while watching a careful, conscientious farmer at work upon his land; and our particular vision allowed us to see that there are many kinds of fields to till, and we were learning how much work, and fun, it is. The sun will rise and set again and the earth will abide; but whether or not human life on earth survives, there's no excuse for making the living of it cruel, harsh or unreasonable. Probably we made a mistake not to invite our industrious neighbor to join us.

But the only thing unforgivable about mistakes lies in the ones that are continued and in the song repetition blares forth about the inability or refusal of its singer to learn, for once we truly learn we move on and that's called evolution; and then the circle is not endless but only functional. Sounds in the form of words flowed from the blonde, the brunette, the bearded and the boy as easily as water in a mountain stream, though there were droughts that must have their place in nature but certainly put you through your paces and don't help at all in dealing with the lurking paranoia that must be fought at every step; and, as the killer of trust, is the most vicious of enemies, more dangerous than a shark or polar bear or cobra that can kill only your body since they carry no malice. The dry spells usually happened while the farmer was at the apogee of his orbit of contact with us, for the sound of his singing voice brought us laughter from his pleasure, faith in the feeling that someone in the neighborhood had their shit together; and then there would come the sound of our own voices talking about the farmer and ourselves and what we all might possibly be doing, should be doing, could be doing and damn well will be doing, and, actually were doing. It was fun to hear him singing.

Who is wise enough for all this? Who knows the meaning of anything? Wisdom lights up a man's face, but grim looks make a man hated. Do as the King commands you, and if you have to swear by God, do not be precipitate.

I remember the sadness, humor, terror and beauty of assurance striking home; assurance that the farmer would keep on working his fields in the pattern he had chosen beneath the sun that would continue to rise and set upon the…if you can believe *Ecclesiastes*…eternal earth; assurance that we would accept our destinies and take what we would from them according to how hard we enforced our own will and fought for what we wanted; assurance that the particular pattern by which each of us expressed the love within was not so important as the intensity of that love; assurance that there is not understanding without mystery; and assurance that no matter how much intelligence we use and how hard we try, there is an element outside ourselves that the irreligious call "luck" that will cover mistakes or destroy creations according to laws we don't comprehend except that finished work on one particular pattern moves us into a different standard that is only another segment of a much larger pattern seen only through the eyes of the Buddha nature in its entirety, unless we drop a stitch along the way and have to do the whole thing over again, which brings on the assurance that all is contained within the mind and that both everything and nothing is ours. It's a strange, wonderful…ah, balanced, universe, for even if it is all emptiness, there is fullness in the chase; and if that's all we got we might as well make fun out of it instead of some of the other things we might make.

> I know that there is nothing good for man except to be happy and live the best life he can while he is alive. Moreover, that a man should eat and drink and enjoy himself, in return for all his labours, is a gift of God I know that whatever God does lasts forever; to add to it or subtract from it is impossible. And he had done it all in such a way that man must feel awe in his presence. Whatever is has been already, and whatever is to come has been already, and God summons each event back in its turn. Moreover I saw here under the sun that, where justice ought to be, there was wickedness, and where righteousness ought to be, there was wickedness. I said to myself, "God will judge the just man and wicked equally; every activity and every purpose has its proper time." I said to myself, "In dealing with men it is God's purpose to test them and to see what they truly are. For man is a creature of chance and the beasts are creatures of chance, and one mischance awaits them all: death comes to both alike. They all draw the same breath. Men have no advantage over beasts; for everything is emptiness. All go to the same place: all came from the dust, and to the dust all return. Who knows whether the spirit of man goes upward or whether the spirit of the beast goes downward to the earth?" So I saw that there is nothing better than that a man should enjoy his work, since that is his lot. For who can bring him through to see what will happen next?

Accordingly, before bedding down that night under summer sky, we made ourselves a feast worthy of kings and queens and princes and laborers; and we washed it down with a couple of bottles of good wine, though not so much as we had and would again consume in the evenings of less hard-working days when unstoned heads drifted into more illusory perspectives of reality that the slight to gross wine OD makes real, or, at least, bearable.

That night and every other we ever slept at the cabin the coyotes serenaded us with their wondrous song from the center of the universe. I love coyote's song. I miss it when my life takes me away from coyote life, when coyote sings me to sleep on the bed of Mother Earth. Coyote, as every Indian and all spiritual gypsies of the cosmos know, is hunter, trickster, teacher, fool, creator, protector and wife stealer; or, as poet Barry Gifford (*Coyote Tantras*) writes, "Coyote drifts in and out, a searcher, a wastrel, supersensitive vagabond of the universe; never settled; always moving; dropping in here and there along the way. Coyote is no idealist; but he never gives up. What is most important is that he is alive; and whatever shred of nobility he wears rests in his awareness of that life. Never aimless, always grinning; forever looking, always lost; ever lonely, never making excuses; Coyote speaks for none but himself." Coyote sings for himself in the night, but he sings for us too; and in the bus or on the ground in the warm down bags that would not be zipped together too much longer past that long ago Wyoming summer, we listened—carefully—to his songs of cold, lonely space travel and the distances between galaxies and the warmth and humor and wisdom of the chase, the hunt, the song itself and of the teachings you can pick up from coyote or the songs of the humpbacked whale or the flight and swoop of the hawk or the shy grace of the deer or the brute wild strength of the moose that tell you way down there in the *central* nerves of the solar plexus to be very, very careful of men who only understand nature through such manmade abstractions as politics, religion, war and power, and have not spent enough time in relationship to the true, eternal nature that, in functional fact, sustains and gives life to them and their abstractions and to the coyotes and trees and bears and birds and bees and elk and wolves and marmots and flowers and fish and rivers and oceans and all the other interacting forms of life on planet earth that men like that are so unconscious of.

One early morning I woke from the restless sleep that is the lot of the wanderer who has been too long in the same place, but isn't moving on just yet. We were sleeping in the bus with the back open, and the sun had just hit the farmer's field. It was early morning chilly, but a hot day was coming. Something nagged at my sleep-filled consciousness. And then it came again—a solitary, soulful, painful and sick coyote call from very close by. I came instantly awake, for something was deeply and terribly wrong with that call. It was not a howl of the proud loneliness and joy and interstellar communication found in the normal coyote song. It was a yell of such pathos and pain and nearness that I became both afraid and angry in the same rush of clear feeling; afraid for the animal itself and afraid, since he undoubtedly was one of the coyotes who had serenaded us in the night for several weeks and who we had seen on many occasions, for a friend. And also afraid of what a pain-crazed critter might do; and angry because I could only think of two things that could put a coyote in that sort of pain—poison and traps—both from the murderous hand of man, and, as a man, angry at that cruel, uncaring potential within myself.

Motherfucker, I said to myself. Motherfuckers. Sonsabitches. Bastards Killers. What's wrong with that poor fucker? The woman and the boy, masters of more sedentary souls than mine, were deeply asleep. I crawled out of the bag, quickly dressed, picked up the axe we used for splitting wood, and cautiously went down to the willows at the edge of the farmer's field. I hunkered down and crept through the willows until I could see the field, full, by that stage of the growing cycle, of hundreds of bales of hay waiting to be picked up. There I saw the damndest thing.

Dragging himself up the field from the south was the most pitiful, wretched coyote ever seen on planet earth. He was pulling himself along mostly with the power of his forepaws. His ass-end sort of clawed and dragged itself along behind; and the two halves of his body seemed to be disjointed, as if his back were broken or some carbolic poison and pain were wrenching the poor creature's innards in indescribable agony. He passed maybe 50 feet in front of me, too intent on his own destiny to notice me, which, of course, is the fool aspect of coyote. Every so often he would crawl upon a bale of hay, raise his muzzle to the sky, and give out that terrible, caricatured howl that had awakened me. I watched, fascinated by the scenario and by some inner resource operating in that sad beast who, I could not forget, was coyote, pre-historic animal of myth and fable and story, and, to the Indian who knows this land better than the white late-comers, creation Coyote, the trickster Coyote, Panama Red of the most ancient hipster. Just as this coyote was finishing his call of affliction from atop a bale directly in front of me, the farmer's dogs, two big hounds of indiscriminate heritage, went berserk with awareness of their cousin's plight. I could see them running in circles, jumping in the air and raising dust in the farmer's front yard. Their barks were ecstatic and out of control, but it was evident they weren't leaving their master's front yard.

    Coyote flopped off the bale and continued his wearisome journey north through the field. I had decided by then it must be poison because I could see he hadn't been hurt in a trap, and his back looked intact. My curiosity wouldn't allow me to quit my seat at this show. But I was pissed. There are certain sorts of shitheads (I use that word literally) on earth who set poison out for coyote, not caring about coyote, rabbit, fox, mouse, hawk, ground squirrel, groundhog, bear, eagle, porcupine, skunk and even domestic dog who, thereby, leave this life in agony and bewilderment, wondering what evil unnatural fate has come over them. Cocksuckers. May they eat some of their own poison and see how it feels, if they got any feeling left. No! No! No! Richard, that's not the way either. You can't answer for another man's actions, intentions or karma. You got your own to take care of. But you can, by rights and necessity and duty and fun, say what you think and express what you feel; and setting poison out for coyote and his friends is not the way and will buy the man who does it some unholy dues; but that's not the point somehow, surely not to the animal with a gut full of crippling pain and a spirit full of a cruel gift from brother man. I felt terrible about that coyote; and not hate but disgust for the pitiful excuse for a human being who had done it to him. Teacher/trickster coyote dying so ignominiously was patently unacceptable; for how could he teach or trick or find nobility in his own awareness of life with a belly full of pain?

    A few yards up the field he dragged himself again atop a bale and repeated his cry of agony, muzzle to the sky. The hounds were in a frenzy. By then the farmer was out in his yard, loading gas and water and tools in his pickup, which prior observation had taught me he would next drive down to the field to begin his day's work, that day involving the loader sitting idly at the southern end of the field. Sometimes the dogs accompanied him, and my feelings were mixed about the possibilities. My attention was divided between watching coyote finish his sad song and nearly fall off the bale before continuing to drag himself up the field, and watching the farmer call his dogs into the back of his truck and drive down to the field.

    Shit, the dogs are going to kill the coyote, I said to myself. I didn't like that. I also didn't like the coyote's suffering. I was stuck upon my own dislikes until, as the pickup approached the loader, I realized what I really disliked was that these dogs would never mess with a healthy coyote. All they were doing was letting out the bully that always grows from the indignity of being a domestic animal. Fucking cowards! Buzzards! Scum! Vocabulary, as usual, falls short of feeling, but *no way* was I going to relinquish my spectator's seat at whatever this play was going to be; besides, I was both spectator and participant, like every man. The farmer stopped next to the loader, and I was struck by his unconcern about the two frenetic, howling hounds. The dogs leapt from the truck in a full sprint north. The farmer never even turned to watch.

I, on the contrary, swung my vision to what I was sure was going to be an ugly battle to the coyote's death; and the next few seconds seemed a couple of hours, for everything slowed down as the flow of life tends to do when attention is complete.

There is an evil that I have observed here under the sun, an error for which a ruler is responsible: the fool given high office, but the great and rich in humble posts. I have seen slaves on horseback and men of high rank going on foot like slaves. The man who digs a pit may fall into it, and he who pulls down a wall may be bitten by a snake. The man who quarries stones may strain himself, and the wood-cutter runs a risk of injury. When the axe is blunt and has not first been sharpened, then one must use more force; the wise man has a better chance of success. If a snake bites before it is charmed, the snake-charmer loses his fee.

As I turned my attention north, I was aware of the Grand Teton (the great tit of the great Mother Earth) overlooking all. I saw the coyote increase the rate of its struggles and thrash about between the bales as if seeking shelter among them. The hounds closed the distance as fast as they could run, howling the whole time, the thrill of the kill driving them dog crazy. Suddenly, not 50 feet from the coyote, I saw a second coyote crouched down behind a bale; and even from my perspective I could see the grin upon his face and the life within his eyes. He waited until the hounds were about 70 to 80 feet from his partner before he broke cover. At that instant the crippled coyote, like Lazarus springing from the grave, blossomed into full-stature coyote and turned on the hounds. One of the grand sights of my life was seeing a couple of full-grown mongrel hounds exchanging ass-holes for noses while involved in a full stride known only to the heat of the hunt, and get that stride headed in the opposite direction. One of them tried to back pedal, causing his rear quarters to come underneath, and he wound up skidding on his back; but he came up in a scrambling sprint with the greatest actor I have ever seen right on his ass end with coyote's own magnificent tail laid flat out behind, floating like a flag of coyote wildness in the wind of the newly directioned chase. The other hound just put on the brakes. He tumbled end over end in a couple of good head-first rolls before he, too, could get back up with his powerful legs moving in the other direction, the hidden coyote of patience right on *his* ass. Those coyotes chased the two hounds around that field at full speed and the farmer went about his work without paying the slightest attention to the whole spectacle, as if he had seen it 1,000 times before; and I laughed aloud with the show and at my new knowledge and at the pattern of education; and I watched the coyotes chase the dogs without catching them around the field and around the field and around the field and around and around and around and around.

—*MOUNTAIN GAZETTE, #52*

ROBERT CHAMBERLAIN

Climbing the Walls in Berkeley

BY
KAREN RECKNAGEL (CHAMBERLAIN)

The first thing was the weather. The California drought. Even a sun-worshipper realizes that each place has its own normal climate and seasonal patterns, and for me the relativity of n-dimensional coordinates, windowpane days of heavy, healthy, nonstop soaking winter rains imparted a vaguely let-down feeling. Something like buying a ticket to your first XXX-rated movie and walking into the theatre during some redeeming social continuity scene, where the only heavy breathing turns out to be your own anticipation.

On the other hand, it seems like spring's been hanging around all winter, groping about in the vacuum ordinarily filled by all that rain. Patient, puzzled sunshine. One bud, one leaf at a time. January fragrant with Japanese cherry, February sprouting plum and quince, March stiffening into magnolia, April fooling with fall's marigolds. Not that it's all so tiresome, or gives no pleasure, but simply that for a New Englander from the Rockies it requires a certain reordering of the awareness so as not to seem like some sort of Telegraph Avenue handout…(What'd I go *through* to deserve this?) More to the point, where's the big wave? The equinox? The ritual celebration? What if somebody gave a party and it lasted all year?

In the course of this reordering, while reading physics and molecular genetics and listening to a friend threaten me with Infinitesimals—"the ghosts of departed quantities"—I found myself thinking about the whole concept of a continuum. Beginning with Einstein's space/time and wave/particle theories and wandering off to other possible examples like Nature/History, Myth/Cliché, Trail/Freeway, City/Wilderness, to name a few arbitrary mid- or endpoints. Or spring in California…

Consider, for instance, "Myth." If the serpent that beguiled Eve is the dragon keeping Hercules from his golden apples; if Chronos is Saturn (Odin's brother?) who devours his own children; if Deucalion doubles as Noah, et cetera, then through the gathering goop we begin perhaps to see that Thoreau's "Swamp" is the paradisiacal "Garde" is the primeval "Woods" (for the trees…). Or for the lack of them, for that matter, Jocasta returns as Mrs. Robinson. The crystal vision of Man in the Mirror fragments into acres of TV screen reflecting millions of Captain Marvel images per minute. The medium becomes more than the message, and Salvation is safely locked up in the Gawdawful Given Moment...

Oh well. Apocalyptic tedium—yes. But I remember trying to make all the green lights on University Avenue one night and wishing, O for a little spray vial of Rocky Mountain Cloud #9, Mr. Denver's Mace, instant immunity against the notion that, jellyfishlike, we might finally be evolving into a true social organism, sections of which cannot help but oscillate at the same frequency. Cloned consciousness, to match our Jungian "collective unconscious," with a little Soma greasing the rails to ensure smooth rapid transit.

Granted the jet-lag in human perspective, what remained every-day-relevant was not the relativity of n-dimensional coordinates, nor the physico-chemical implications of "myth" as a form of molecular memory, nor all those myriad particles bumbling around in one bottle with Maxwell's demon at the door. It's still scaled more to getting on and off the freeway. Or, as another friend put it, cocking one eye at a Bombay gin and the other at your deliberations on rest-mass and the speed of light: "Yeah, but you gotta watch out for the wind factor..."

And so it seemed that the first lesson in urban "survival" was to go one step further in this scaling process, consciously adopting the viewpoint evolved by generations of city-dwellers, Orientals, sailors and others who live with a minimum of stress in crowded or limited environments: a sort of miniaturizing and internalizing process whereby one allows a whole universe to occur in one's backyard. Or flower-box, even. The Emily Dickinson terrarium effect, so to speak, a kind of regressive corollary to childhood when everything was bigger than life...

Walking across campus to classes soon led to discoveries of secret little places, which, each for its own reason, held a certain charm; places that in certain moods, I could even imbue with qualities of mystery, in the way that children excite themselves with fear of the dark. The ancient enormous ginkgo tree stretching out its great branches near the Life Science Building became a favorite place to rest and read. Or again, often I would go out of my way a bit to follow the asphalt path through a small grove of flowering plum and towering young cedars, over Strawberry Creek on a stone-walled footbridge with little stone steps curving the path up the opposite bank. Breezes met here to exchange the fragrance of eucalyptus and resonated humus, the deep shadow dappled with medallions of blossom and sunlight, and almost always people passing who would say hello.

One tree, one blossom at a time...Of course, the Women's Lib Auxiliary of Fear, Inc., caught wind of such wild places and decided to gift potential rapists with a wider perspective of the world by hacking back all the undergrowth and shrubbery. And I further miniaturized and internalized my way over to another area of the campus, figuring there'd be no disappointment from lack of supply if I learned to indulge myself in people-watching.

Do we sometimes fantasize awareness itself ...?

Soft, warm sunshine. Unobtrusive, insistent. Saying, look, I made it all this way, through decaying ozone, smog, noise and all that other *traffic* out there, in less than eight seconds, and somebody's damn well going to feel good about it... Lying on a grassy knoll outside Evans Hall, I look up from a book to watch two young bodies clinging to a 20-foot stone wall terracing a nearby building. One guy, dark-haired and strongly built, has started to climb first, body flattened, arched against its own shadow on the warm ochre sandstones. Up and sideways. He reaches the top, then leans back over the edge, apparently to coach his spidery-but-slower young friend: outstretched fingers tighten carefully; and then, visibly, tension relaxes into balanced poise and the ascent is completed.

On the terrace, a few words and demonstrative gestures are exchanged, whereupon the second climber mounts a skateboard and weaves off down the sidewalk. His dark friend nods as he passes me on the grass and is lost a moment later among the throng waiting at the Humphrey Go-Bart bus stop. Obviously, I muse, here is a man totally unconcerned with how the hell (in the name of Timothy Leary) one crystallizes oneself out of this supersaturated city cyclotron. Probably spends his weekends at Yosemite and his summers in Nepal, or Peru.

Often during this unsprung winter I get a chance to watch while one of a couple of young climbers hang like flies underneath or between the balconies of Evans Hall, or scale that sandstone wall, or use ropes, even, to tie down the side of an apartment building. Part of me wants to say, "Hey, go pick on something Real," while the voice in the other ear whispers, "O go climb a tree—you just wish you had something you could do with a stiff vertical whenever you found one, *Parce qu'il existe...*"

Gradually one realizes that one has left nothing behind, that perhaps survival was not really the question, that perhaps, after all, There Is No Question. Still and only the immense spectrum of awareness of possibilities from which to choose (like courses from the college catalogue), leading again to the easy feeling of dining at the buffet table of the "elite." As if by that label, by such verbal backpatting, we could escape the knowledge that we still only get once to do it all (better, best, highest, fastest, with or without snakes, et cetera); that it's difficult to do *nothing* next, to un-do, to not-become; that freedom's just another word...

Often there is an urge to watch the sunset from the roof of my apartment building, looking toward the Bay, a watery island surrounded by city. Other people wander out on nearby rooftops, singly or in pairs, quiet. The bay haze swells thick with red, while to the south the skyline hills respire softly in stage-set grays beneath the smog, like long-anaesthetized patients. From mid-campus the campanile tower bells chime the hour, then continue with a stately melody, harmonizing with the evening's peach while at the same time creating a peculiarly poignant restlessness. (I suspect the bellringer, an excommunicated music student, I've been told, now resides in the belfry with several half-tame bats. The only evidence I have, however, is that for noontidings on St. Valentine's Day, listeners were treated to a medley that included a perfectly solemn rendition of "O Come, All Ye Faithful." I admit to a secret desire to meet him.)

Anyway, restlessness and tranquillity being not always so compatible, the tendency for cream to float makes a true virtue, not of patience so much as of the ability to concentrate on matters at hand. Not that studying is any sort of a drag; far from it. It just isn't very physical. And walks up Strawberry Canyon or down by the marina after a while take on a stop-gap quality. There is the sensation of waiting. The right wave, the one that begs you to match yourself against your fantasies...

Sure enough, just as the symptoms of cabin fever rise to throat level, comes a call from Bear Valley: "Get on up here, it's a blizzard!" The Real Thing, one's own slice of the Whole Enchilada. We throw two pairs of skis and a couple of soon-to-be-neglected textbooks into the back of the jeep and head eastward under irritatingly clear starry skies. The suspense mounts: Not until well above Angels Camp do we finally hit snow, either on the ground or falling, and in the meantime skepticism runs rampant. Lordy, maybe it's California-Dreamin' forever, and they've slid the Sierra south for a Hollywood spectacular, so the next thing we'll run into will be the on-ramp for Enchanted Interstate #201, with a Texaco totem and the Last Chance Jack-in-the-Box lighting a white-on-green legend: Spokane—840 miles. Or worse yet, that somebody's figured a way to bypass the usual dam, pipeline and river-diversion methods of water-stealing and has hijacked the weather direct. Using our snow to rinse the grime off the plastic palm trees...

But then suddenly snow, thicker and thicker, glistening and swirling in the headlights, whitening the road, deeper and colder and closer, till we've passed the last plowmark, with no tiretracks left to follow, navigating on the narrow assumption that beneath the least amount of snow lies the road. Then an interminable straight stretch where, as on the sea, time and distance collapse into one another, and the speedometer wavering around 25 mph becomes a mechanical credibility gap. Just as the conviction hits that we're not moving at all, suddenly looms up to the left a rearing wooden bear and a rustic sign—and we're there.

Two in the morning, half-smiles squinted at the sheets of cold flakes dancing in the mercury-vapor lamplight as we clamber out of the jeep and poke around in a station wagon (totally camouflaged as an igloo) for snowshoes left by our friend. With these we have a chance of surfacing the distance to his house.

Winter! Better yet, *weather!* A host of city-tensions softly hissed away under the tiny stinging tongues feathering our cheeks. If one snowfall is enough for winter, this is it. Lodgepole pines like sleepy sheeted Klansmen line the trail, and there is no wind, no intermittency or hesitation, just Sierra powder falling, heavily, steadily, silently, encompassing the darkness.

Fresh tracks tomorrow!—and I fall asleep dreaming a warm suit, a gossamer second skin to ski in...

> I made my Song a coat
> Covered with embroideries
> Out of old mythologies
> From heel to throat;
> But the fools caught it,
> Wore it in the world's eyes
> As though they'd wrought it.
> Song, let them take it,
> For there's more enterprise
> In walking naked.
> —Yeats

"I'd holler, but the town's too small."
—Big Bill Broonzy

"People say I'm hollerin'. Man, I feel like hollerin'."
—Charlie Mingus

"Why don't we do it in the road?"
—The Beatles

—MOUNTAIN GAZETTE, #46

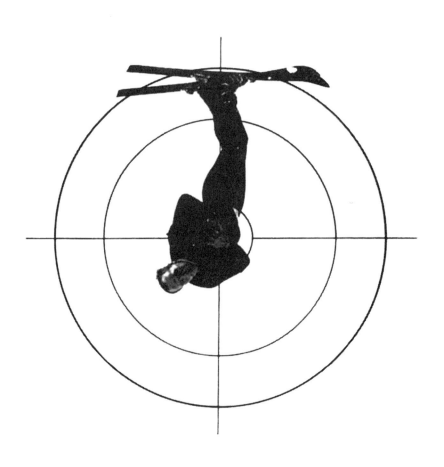

The Mogul Problem

BY
TAD HALL

Violence, violence,
It's the only thing that'll make you see sense
—Mott the Hoople

With a racket of metallic chattering Belsen fled across the freezing mushgullies above the Lodge. Cursing silently he violated a Dogger Precept by wishing desperately for another 30 centimeters of his 110-cm Kneissl *Mini-Magnums*. His heel-thrusts, specifically designed for dealing with the truncated, flat-topped moguls that were specifically created by that technique, failed signally to slow his panic slide across the cheese-grater surface, served, in fact, to accentuate his inability to carve a turn or set an edge on the hard snow. Belsen shivered as his body heat escaped out the armholes of the down vest he wore outside his sweater in classic Dogger fashion.

I didn't see Makro's goons there, Belsen thought. Hands correctly held like thalidomide-induced flippers at nipple height (viz. "Proper Hand Position," *Some Dogger Precepts*: 1974 Fifer Press), Belsen jammed his desperately chattering skis into another scratchy slide. *Didn't see them lurking there on their 220-cm Atomics.* Belsen sneered. *Vigilantes. Sacred Guardians of Technique.* Belsen shivered again as he thought of the hogleg .357 magnums Makro's Gunners carried; and he reached inside his vest to finger his own snub-nosed .32. *Still, it was a classic kill. The Belsen-Mongoose, I'll call it*: Belsen thought with pleasure of noting the new kill technique in the big Dogger Book at the Worm's Turn, and demonstrating it before thousands of cheering bump-humpers at the big Hot Dog Jamboree in April. *There might be an ethics penalty*, Belsen thought, *contravention of the Mammoth Convention Articles pursuant to Non-combatant Skiers. Hell*, he shrugged, *she was on 200s.* Belsen experienced a spasm of intense pleasure as he remembered the way the girl's head had flopped away from his modified outrigger with kick to the jugular, and the way the crimson had spilled over her green nylon breasts.

Warned by the scrabbling clatter of Belsen's approach, a bearded X-er lay on the slope, his absurdly long and narrow skis splayed in a herringbone as he sighted down the barrel of his carbine. *Fucking X-ers,* thought Belsen, *they're so paranoid. Comes from skiing uphill.* There was the sudden drone and crack of a bullet ricocheting past Belsen's ear, and Belsen became aware that the X-er's huge gunmuzzle had moved off him and was pointed up the hill, toward the source of the shot. Belsen managed a tiny mulekick as he jumped the stony ridge to put solid rock between himself and Makro's trigger-happy Gunners. *Doggers got more class,* he grinned. *Doggers rule!*

Belsen almost made it to the warm anonymity of the crowded bar before he saw the cold blank faces of Makro's men. He tried to climb a banister and railing, but a huge red Kastinger descended on his fingers, and he found himself staring into the grim, black-bearded face of Edwin Makro himself.

The Gunners escorted Belsen to a concrete cubicle in the bowels of the lower tram terminal. Above them the great counterweight descended slowly. Makro's men strapped Belsen into a metal chair mounted in the concrete floor.

"Belsen," said Makro in a hollow voice, "you have been tried and found guilty of one count of ecocrimes: to wit, conspiracy to fabricate unskiable moguls; and one count of felonious malesthetics: to wit, promulgation and execution of incorrect techniques."

Belsen became perfectly rigid, from his Aspen Lid to his Scott boots. What about the girl? "Who defended me?" he cried.

"There was no defense possible," Makro continued in the same cadaverous voice. "Your crimes are manifest and irreconcilable with the society of skiers."

Two of the Gunners fixed upon Belsen's feet and legs a heavy metal boot, from which thick cables ran to a box in the wall. The sweat started out on Belsen's face, and he cried out desperately, "Why can't we co-exist? There can be something for everybody!"

Makro shook his great dark head. "No," he said, "we can't co-exist. The way cattle can't graze where a herd of sheep has grazed before them. After you and your kind have been over the snow, Belsen, it is rendered unskiable to any save the shortest skis."

Belsen whimpered in terror. He looked around him at the cold grim faces and the racks of skis taller than a tall man and whined: "You fuckers ski on 220s at the very shortest! No wonder you have trouble with the moguls! Couldn't you compromise a little?"

Makro shook his head again and raked his hand through his beard like a mogul-cutter. "You can't compromise with gangrene, Belsen," he said harshly. "You can't just trim the nails." He thrust his face suddenly into Belsen's, and the Dogger cringed away from the burning fanatical eyes. "We are fighting for our way of life, Belsen," he hissed, "for our *Technique*. We tried all that stuff. We tried creating phallic shame in the tram line. We tried writing scholarly articles in mountain journals, to demonstrate to you the error of your ways and to give you ecological guilt. We tried to scare you off by mocking you from the chairlifts and skiing across the tips of your midget skis while you wiggled your way through your grotesque moguls. Nothing worked, Belsen. You claimed you were having fun! *Fun!* And the sport became ever more corrupt: for every dogger we frightened or shamed into a correct aesthetic, the big ski shops hustled 1,000 non-athletes onto the slopes on short skis. And you were their inspiration, Belsen, you and your kind. *They copied you and Technique was debased.*"

Makro stepped to the box in the wall and pulled a lever. "We want you to understand," he said quietly, "what we go through skiing your moguls. We want you to understand."

Belsen experienced a rolling sensation in his encased feet and legs, as if he were starting slowly across a field of moguls. Instinctively as he rose to the top of these simulated moguls he attempted to suck it up, pivot and slide down the far side, but he felt with terror his skis begin to track away, railroading him off-balance over the next bump and picking up speed. Suddenly the moguls were no longer friendly, helpful platforms, but huge malevolent fists that slammed up under his feet until his teeth rattled in his head and he fought for breath. He tried to cry out, but the intense jolting compressed his ribcage, and his aching knees balked at the harsh asymmetrical rhythms as he struggled for balance. With an inner cry of despair Belsen felt his ankles shatter and turn to mush.

One of the younger Gunners shuddered as Belsen's final cry echoed around the concrete chamber. Makro turned to the young man, stern yet fatherly, and put a hand on his shoulder.

"How can we teach an aesthetic?" he asked rhetorically. Makro's voice was weary, his shoulders bent with the weight of promulgating a great creed. "How can we legislate an ethic? All we can do is kill them, or die out ourselves."

The young man nodded slowly, and then in a sudden spasm he knelt and spat into the face of the dead hotdogger.

The paranoid X-er caressed the muzzle of his military issue 30.06 and smiled slowly, listening to the gunfire from the alpine slopes above. He sat on the deck of the Pine Tar and licked a stick of raspberry-flavored klister, his carbine leaning against his knickerbockered thigh. *We'll let them kill each other off,* he thought. *Then we'll tear down the lifts and it will be ours.*

—MOUNTAIN GAZETTE, #20

MARK FOX

Bitches in Heats

BY
CINDY KLEH

The last group left in the bar asks for its tab, and it's totalled and on their table in less than 30 seconds. Cool. My butt may be able to skate out of here early for a change. Just then, the door squeaks open and slams shut. Three guys sit down at the bar and ask if there are any specials on draft beer.

Special deals are just during happy hour...and this is last call.

Last call? They look at me incredulously. It's only midnight!

Checking my watch, I realize that the health insurance policy I purchased a few days ago has just become effective. For the first time in eight months, I have, at least, catastrophic health insurance, a $2,500 deductible with no fancy extras like dental, vision or pregnancy. If I end up requiring a few stitches I'm out of luck, but if I become one with a tree or fall from a chairlift, it'll kick in. Purchasing health insurance does not mean that I'm finally becoming more mature and making rational, responsible decisions. I just can't win snowboard races without it. It gives me the freedom to take more chances, to push the envelope a little and worry less about the consequences.

I'd have insurance year-round if I could afford it, but there always seem to be those frivolous expenses like gas, rent and food that come first. I mean, wouldn't it be bad for your health to be without food and a place to live? Sacrifices have to be made for common riff-raff like me to live among the trophy homes of mountain resorts. Often, one of the sacrifices is health insurance. Sure, there are nine-to-fivers available with full bennies, but if you have to be somewhere between nine and five almost every day, why live in the mountains?

Hey, I plead to my thirsty patrons, I've got a boardercross race tomorrow, and I have to get up at six in the morning. To my eternal astonishment, the honesty ploy works. They order a round and are gone in 20 minutes. Must have been my please-for-your-own-safety-don't-screw-with-me-right-now look. Even with an early closing, I'm a zombie walking around in the dark the next morning, dressing from a pile of clothes thrown on the floor the night before: socks...long underwear...knee brace...elbow brace...hockey pads...snowboard pants...I'm out the door within a half hour.

I drive up Loveland Pass cradling a huge travel mug of coffee between my legs. The only signs of life are the lights from the grooming machines plowing up and down the slopes of Arapahoe Basin in the dark. An hour later, a faint sunrise begins to show toward Denver as I reach the summit of Berthoud Pass. I pull over to pee, and squat at the top of the Continental Divide, as always, wondering which watershed, the Atlantic or Pacific, the Platte River drainage, or the Colorado, I'm liquifying. Looking down the northwest-facing steeps, my eyes take in the half-foot of snow that has fallen overnight. Those velvety slopes look tempting, and I wish there was time to poach a quick pow run, but there's not enough traffic on Highway 40 this time of the day to thumb back uphill if I want to get to Winter Park in time to register for the race.

I smile, because I used to never let anything get between me and a backcountry pow day. I used to make fun of people who would spend a whole day trying to go as fast as they could on one in-bounds, groomed, ski-area run. I was into pleasure, not speed. And here I am, driving all this way to spend my day doing just what I used to make fun of: racing as fast as I can on one in-bounds, groomed ski-area run. The truth is, I always hated competition and the way my nerves destroyed me even pondering the concept of competition. I used to feel like I was going to yak before gymnastic meets in high school.

It was my first boardercross race three years ago that changed my perspective toward competition, or, more accurately, my perspective toward me participating in competition. I realized early on in my sports reporting career that it was better to compete and cover an event than to sit on the sidelines freezing my butt off trying to match bib numbers zipping by to names. It was easier to understand and write about action that I had experienced first hand, and I got to know many of the competitors as friends. I found that I liked the racing scene…and the parties afterward even more.

Sending four women simultaneously down a course of banked turns and jumps, with incidental contact allowed, intrigued me enough to sign up for a boardercross race. I liked the part that luck plays in this sport. You don't have to be the best rider to win a race. Anything can happen, and it usually does. But it was the adrenaline rush that lasted long after the race was over that ultimately hooked me. Sliding into the starting gates was much scarier than hucking myself off a cliff or diving down a steep couloir. That first race was the beginning of an addiction that changed my outlook on snowboarding, and life in general, for that matter. Forcing myself to show up at the starting gates, even though the whole idea scared me silly, was the beginning of a journey. I faced my fear of competing, and it was victory enough just to go through with the first race, even though I blew my knee out. That victory gave me the balls to begin confronting other fears that are not so easy to address.

The first race that I actually *won* was the beginning of a entirely different addiction.

I spot Janet's old truck in front row parking and pull my car in next to it. It's easy to recognize, because much of the rust is covered with bumper stickers with pithy, intellectual statements like "Powder Slut" and "I'd rather be pissing off skiers." Finding a premier parking space this early in the day at Winter Park is easy, much easier than getting my boots on in the front seat of a Subaru. I throw some energy bars into a parka pocket, grab my board, goggles and helmet and head for the lifts.

Janet is always "already there," and she takes a run while I go in the lodge to register and hit the bathroom. The race director pulls out my file before I even get over to the registration desk. It's obvious that Lauren's mind is focused on running the race, and mine is focused on winning the race, so we hug and promise to catch up on each other after the awards.

Janet is waiting for me in the corral, and we bonk helmets.

"How ya been, bitch?" she asks, in her most lady-like voice.

"Great! Guess what? I just got health insurance, effective midnight last night, and I'm gonna kick your ass!" I respond in my charm-school best, pinky extended.

"Yeah, maybe you will...you're always behind it!"

With those formalities out of the way, we board a chair and get down to the important stuff...the stuff I'm dying to know. How many times have you ridden the course? Are there any gap jumps? How much snow did Winter Park get last night? How many women are signed up for our age group?

There are no gap jumps, but there are four whoop-de-doos in the flat section near the end. You could gap any of the whoops to make up time, but if you fall, you're in the flats with no speed for the last tabletop jump. If you have too much speed, that baby will send you.

There are eight women on the starting list. We'll see how many actually show up.

Janet unzips her parka and shows me her brand-new padded BMX sweater. 250 bucks. It has built-in wrist guards and a reptilian back. You could fall from 20 feet on your back and walk away without a scratch.

I ooh and ahh over it. We don't say it, but we both know this extra protection gives her a huge psychological edge. Suddenly, I feel very naked. I've trained like an obsessive maniac all spring, summer and fall to be as buff and quick as possible. Staying in shape took on a new meaning after I got hooked on trying to win these stupid boardercross races. Hikes in the woods became trail runs. Trees became gates and rocks became jumps to olly. I ran backward and sideways and propelled off banks to get my ankles and knees ready for anything boardercross could throw at them.

Now it's going to cost me an additional $250 to be competitive, because suddenly, I must have one of those jackets. Without one, I feel like a Harley going up against a Mack truck!

We talk nonstop to the top, but my mind is already turning over the possibilities. Hmm, eight women. That means there will be at least two heats. It's nice to get in more races for the entry fee, but my stomach tightens as I wonder about the competition.

(I may be dealing positively with my competition-phobia but I'm not even close to totally conquering it.)

We take a few practice runs on the course and run into Gwen in the liftline. Her face is white and she looks like crap. Usually she is a force to be reckoned with, even though she barely weighs 100 pounds. She is known to wear a scuba weight belt under her parka and use every dirty trick in the book. I've heard her talk trash at the start and seen her elbow other riders out of the race. I've even seen her grab the tail of a board from behind and yank a rider off her feet!

But today she is obviously very hungover and useless. As Gwen pukes off the chairlift, Janet and I exchange smiles. She won't be anyone to worry about this fine day. I offer Gwen one of my energy bars, but she has a box of Apple Jacks and a carton of milk in her backpack. Janet and I take practice runs until the starters won't let us on the course anymore. I settle down near the start to do yoga while Janet listens to Widespread Panic on her CD player. Gwen goes off in the woods to puke again.

I meet Shelley while stretching. It's her first race, and she is taking deep drags on a cigarette.

"I am so nervous I feel weak," she says.

"Everyone feels like that on her first race," I respond. "I was a wreck! I thought I was going to have a heart attack, I was so scared. You'll feel much better when it's over."

Thus I establish the experience pecking order. I also know that I am the oldest competitor here, but that holds no honors. I could choose to do only races with a 40–49 age category, but I would be running them alone until I got to the national level. Racing against the youngsters all winter gets me ready for anyone I might find at nationals who happens to be my age and rips on a snowboard.

"Konichi-wa, Kitako! Good to see you." We shake hands as I remember racing against her last season. (Pushover. Too polite to win.)

Kitako is a college student from Japan studying for a pharmaceutical degree at University of Denver. We check out the starting lists for our category and realize we are in the same first heat.

We don't have to race Janet until the finals. High-five.

I meet A.C. and Marla, who are also in my heat. Marla has pearly pink lipstick, matching nail polish and eyelashes thick with mascara.

A.C. has hard boots, plate bindings and a GS board, and she looks like she's built pretty solid. She borrows some wax from me, but we don't have much time to talk because the starters have just called for women's masters, first heat. The boardercross bitches grab their boards and head for the starting gates.

Strapping in, I try to breathe deep and calm myself down with my mantra: It's only snowboarding. Stay low. Stay on your feet.

Still, my heart is racing as we slide into the gates. I can't wait for the starter to yell

"Racers ready!" But there's an injury on course, and we have to wait. Nobody wants to think about injuries in the starting gates. Minutes feel like hours. We tell some dirty jokes to ease the tension and give the starters crap ("Come on...while we're young! Oh, too late.") Finally, we are ready to start.

Three, two, one...the gate drops and we push off. We are all in a pack, and I know that someone will have to give in on the first bank turn. Nobody slows down, and Marla, A.C. and I collide.

I hate collisions. I never did contact sports before boardercross racing. I grew up immersed in toeshoes and tutus, satin and sequins. I was a girly-girl...a child of the '60s. I wasn't even allowed to wear pants to school until sixth grade. I put wedding dresses and cheerleader outfits on my Barbies, and was taught that pushing or being physically aggressive was just not acceptable behavior for a nice girl.

Who would've imagined that one day I would be charging into a banked turn elbow-to-elbow with three other large butts on boards, shoving each other for high-speed position?

Despite the collision I somehow manage to stay on my feet and chase Kitako down the steep section. I pass her on the next bank turn, but my contact lens blows out of my eye. I can see it stuck to the inside of my goggles, and keep navigating gates with one eye winking. This works for the next three gates, then I misjudge a turn and skid out enough for Kitako to pass me. She stomps the landing on the table top jump and crosses the finish line ahead of me, but since the top two advance, we will both make it to the finals.

While I'm sticking the lens back in my eye, Marla crosses the finish line. I punch Kitako in the arm and laugh.

"You've gotten better since last year."

She admits that she has more races under her belt since the last time we met.

A few minutes later, A.C. comes down. She tells me my board gave her a concussion. She doesn't seem mad, so I offer to buy her a beer. As we relax in the sun watching the guys race, I find out she's a patroller at Berthoud Pass in the winter and a bush pilot and river raft guide in Alaska in the summer. We hit it off like a house on fire, and she invites me to move up to Alaska with her in April. I can't give her an answer (that's the kind of decision that can only be made drunk, and it's at least two hours too early in the day to start hitting the tequila), but we agree to meet for some bump runs after the race.

I chow down two hard boiled eggs, some Saltines and loads of water (snowboarder's breakfast: $1.60 plus tax). Now that the first race is over, my stomach is settling down.

We watch the men's races from the start while we wait for our final run. Janet, Shelley, Kitako and I hang out. Guys never chit-chat at the start of a race. They may nod or shake hands, but they are competitors, warriors, gladiators (at least in their own minds). Women couldn't care less about macho posturing. They like to talk a lot to ease the tension.

Shelley looks much calmer. She took second in her heat despite her nerves, and now she has a different look in her eyes. A focused gleam that says "watch out, bitches" to the rest of us. We line up, still adjusting bindings and shimmying boards back and forth in the starting gate. We grab the hand posts and look down the course.

The gate drops, and Janet and Shelley move ahead of the pack. Suddenly, Shelley tucks and passes the mighty Janet.

Kitako is ahead of me, but I'm so close behind she can hear my board scraping turns. I know I have to pass her, but I keep waiting for the right moment. I almost pass her from above on a bank turn. I have the speed, but my back is toward her, and since I can't see her, I back off.

It's getting near the bottom and I have to do something, but I don't want a collision. That crash that knocked me out cold last year...going down in a sled, bruises, sore muscles, living in the fog and headache of a concussion for a week. Ouch. Can't think about that; can't *not* think about that. Stay focused on the course. Stay low. Stay on your feet.

We pump through the whoops side by side, watching each other out of the corners of our eyes. Kitako launches off the tabletop a second ahead of me and bails on the landing. I have two choices: to land on her or slide out and take my chances. Our boards crash, but I'm able to get up first and inchworm across the finish line.

Janet takes first. (Again.) (She's won enough Palmer boards to start making them into patio furniture.) Shelley had the race won, but was disqualified when she missed a gate near the bottom. Oh great—another woman in the 26-and-older age category to worry about.

The four of us find A.C. in the bar and spend the rest of the afternoon together in search of powder bumps. Turns out, A.C. is an incredible rider, and I was fortunate to take her out in the first heat. Great, yet *another* woman to worry about.

It starts to snow as the awards party gets under way. I look out the window while sucking down a cold one, and think about tomorrow's powder day brewing outside.

"Hey, can I buy you another beer?" A.C. asks me.

"Ehh, better not. I gotta drive home in a snowstorm."

"Crash on my couch tonight. We'll get first tracks in Parsenn's Bowl in the morning. I can get you a free ticket."

"Twist my arm, bitch."

—MOUNTAIN GAZETTE, #79

R. CRUMB

My Friend Ed

BY
DOUG PEACOCK

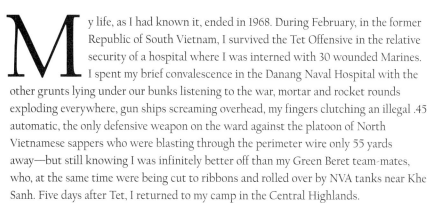

My life, as I had known it, ended in 1968. During February, in the former Republic of South Vietnam, I survived the Tet Offensive in the relative security of a hospital where I was interned with 30 wounded Marines. I spent my brief convalescence in the Danang Naval Hospital with the other grunts lying under our bunks listening to the war, mortar and rocket rounds exploding everywhere, gun ships screaming overhead, my fingers clutching an illegal .45 automatic, the only defensive weapon on the ward against the platoon of North Vietnamese sappers who were blasting through the perimeter wire only 55 yards away—but still knowing I was infinitely better off than my Green Beret team-mates, who, at the same time were being cut to ribbons and rolled over by NVA tanks near Khe Sanh. Five days after Tet, I returned to my camp in the Central Highlands.

There, as senior medic on an A-team, I pieced together Montagnard children who had been caught in the cross-fire for over a week until I began to lose my mind. The day I packed my bags for home, March 16, 1968, American soldiers ruthlessly murdered 347 Vietnamese civilians 40 miles to the north in a place called My Lai.

Back in the world, it was early spring. I bought a Jeep and drove west, hoping to look up my two oldest friends from the Midwest. One worked as a teacher on the Navajo Reservation. We had been together at the University of Michigan, where I had been associated with civil rights politics and the beginnings of the New Left. I had once brought Martin Luther King, Jr. to the campus to speak. Two weeks after leaving Vietnam and a day after joining my friend in Fort Defiance, Arizona, Martin Luther King was assassinated. My response to the murder bothered my old friend a great deal: I was not surprised that someone had finally shot him down. I had known that King had powerful enemies and—in 1968—this was what I expected of the world. By the time summer rolled around and Bobby Kennedy was killed, I was beyond the pale of politics; the world had gone quite mad and I had to deal with it accordingly.

The other friend lived in Colorado. This ex-roommate was an intellectual, a city-dweller who had never camped out in his life. But, by late 1968, that had all changed. My friend was now a big Sierra Club guy, a dedicated environmentalist, serious backpacker and mountain climber. The reason for these changes was a man named Edward Abbey. Earlier that year, he had published a book called *Desert Solitaire*. This book changed lives.

That was the first I had heard of Ed Abbey and I still hadn't got around to reading his books when Bill Eastlake called me the next winter in Tucson and told me to come on over. In 1969, William Eastlake was the grand old man of Southwestern letters, and so remains despite the poisonous seasonality of New York publishing. I hopped on my motorcycle and drove up and down the desert roads at the foot of the Santa Catalina Mountains until I located Eastlake's house that, back then, was beyond the edge of the giant cow-plop from the sky that has become Tucson.

Some people were there, possibly writer-types, but I didn't know them. The winter air had chilled me to the point my hands shook as I pulled out a baggy of Bugler tobacco and rolled a joint-like cigarette. The cold palsied my fingers and I had trouble striking a match. The man sitting next to me gave me a light. He was a tall rangy man with a short dark beard. We talked about mountain lions, a subject about which he had just written for *Life* magazine. He worked as a seasonal ranger at Organ Pipe National Monument in southwestern Arizona and his name was Ed Abbey. He invited me to visit.

A week later, I threw my sleeping bag in my Jeep and drove down to Organ Pipe bearing gifts: a six-pack of beer and a bottle of whiskey. That was how you visited people in those days. (Abbey had first met William Eastlake in Cuba, New Mexico, nervously knocking at Bill's door with bottle of gin in hand.) I met up with Ed after work in the government World War II Quonset hut he shared with Bill Hoy and other rangers south of Ajo. From a nearby room, someone nostalgically played Grofe's "Grand Canyon Suite," and Ed noted he was taking a lookout job on the North Rim, one of the four corners of his desert world, staking out the territory the big gulch held in his heart. We drank and talked until early in the morning, late for a working man. I was new to that part of the country, and Ed directed me to Dripping Springs, one of two natural, permanent water sources in Organ Pipe. The late winter morning was warm, and yellow blossoms of brittlebush foretold the approaching spring. I hiked a short trail to a small cave. Within the grotto lay a milky pool, pale and oracular, draining into a brushy draw swarming with honey bees.

I retreated along the well-used foot trail, hard-packed by the boots of hundreds of visitors. Abbey had told me Organ Pipe, though lush and lovely in desert terms, was relatively tame compared to the great expanse of land adjacent to the west and northwest—the wild and empty valleys and ranges of the Cabeza Prieta.

"You can't rebuild a life that had no structure to begin with. In 1969, I was still reeling from a year-and-a-half of Vietnam. I had no desire to re-enter society nor any talent for reform. The precise problem seemed to be that I wanted a life but not the world they said I'd have to live it in. At 26 years of age, my slightly cynical suspicions about the human race had been verified by experience. Though estranged from my own time and without a clue as how to live a life, I was not lost; I knew that my real homeland, the one I would fight the authentic war over, the one I would die for, was still out there in the wilderness deserts and mountains of America. This man Abbey had a distant but deep passion, a humorous fanaticism about defending the wild I found attractive.

The next summer, derangement surfaced again as I entertained the notion of going back to Southeast Asia as a photojournalist. I scored a graduate fellowship in Intensive Vietnamese, a language I already knew, and traveled to University of Hawaii to snorkel and watch fish for three months. While I was there, I heard Ed Abbey's wife died of leukemia so I wrote him a note. Some time went by and he wrote back wondering if I wanted to take a trip into the canyon country.

I picked Ed Abbey up in Kanab. We drove east along the Vermillion Cliffs, turned north on a dirt road tracing a bench above the Paria and on into Cottonwood Canyon. We bounced on toward Kodachrome Basin with the dark spine of the Kaiparowits Plateau lying in the gray distance. I had seen Southern Utah before but not much and never anything like this: feminine landscapes of bentonite, soft clay hills with lenses of blue and green, badlands disappearing under the coarse angular red scree from the cliffs above—the massive faces of wind-blown Navajo and Wingate sandstone, their dark patina stained and decorated by seeps and runoff.

South of Kodachrome Basin, amid a wasteland of rabbit bush and juniper, we pulled off the rutted road to an old drill-rig site. The rig was gone and the hole was capped, but part of the frame remained along with sections of pipe and used bits; the site was deserted but not abandoned. The drilling outfit was coming back, probably to look for coal deposits to fuel a proposed coal-burning power plant at Kaiparowits 40 miles to the north. Ed found a spanner and fit the big wrench around the cap and removed it. We dropped in a rock, then pieces of pipe and chain to see how deep the hole was; nothing, only wind-sounds whistling down the casing. Ed found some more junk lying around and I located a pile of used-up diamond drill bits. All this went down the hole.

"Should take them a while to drill through all that junk," said Abbey. Then he grinned. "Someone has to do it."

We passed through the xenophobic, nasty little town of Escalante, then turned back south on the sandstone rim rock. We paused at Dance Rock, where Mormon pioneers stopped their wagons to hold a dance in 1879. These men, women, children and babies, under orders from Brigham Young, left their homes in South-Central Utah to establish a new settlement at Bluff, near Four Corners. The settlers had to lower their wagons and livestock down into Glen Canyon, then up the other side, and on to 1500-foot-high Comb's Ridge, finally blasting through with hammers and drills. Those old Mormons were tough, noted Abbey, even-handed in his credits.

We crept southward over the rock toward the heads of Hurricane, Coyote and

Davis gulches. This 500,000-acre de facto wilderness was not as wild as one might think. We passed a total of four vehicles parked at the heads of the various canyons. They were the rigs of backpackers, we guessed, by the looks of the bumperstickers: "Save Black Mesa," "Think Hopi" and "Save the Whales." A Volvo station wagon with Sierra Club stickers and two occupants stopped us to talk; one of them, the man, apparently recognized Ed Abbey. I readied my pack for our hike.

Cinching up my cheap backpack, I caught the tall Sierra Clubber with knobby knees and hairy legs staring at me as I zipped my .357 Ruger magnum into a particularly accessible side pocket. He looked back quickly to Ed and continued his diatribe.

"If you hadn't written that book this place wouldn't be so crowded."

Ed was being much too generous and forgiving of this spit-dribbler, so I strode over—back in those days I couldn't help myself—and stuck my finger in his face.

"If we want any more shit out of you I'll squeeze your fucking head," I said.

"You don't have to be so violent," screeched the woman.

"True. All true," said Abbey.

Soon, we parked our rig, scouted the rim rock for routes, then shouldered our packs and edged down a wash that quickly grew into a canyon. Logs of agatized fossil wood, washed down from the Kaiparowits, littered the upper wash, and some of the logs were a couple of feet across, the annular rings now silicified into lovely red and yellow agate used by the Anasazi to chip arrowheads. The canyon sunk into the slickrock, a deep, narrow slot with a ribbon of sky above. Off to the right lay an alcove under the stain of waterfall that coursed in from the rim during the rains. On one side, a horned anthropomorph was carved into the rock between two smaller figures with wide shoulders and tapered bodies. One of the petroglyphs was vandalized by Mormon cowboys who had carved initials over the ancient etchings. We were broke for food, and Ed heated up a can of chili by propping the can between two suitable rocks then igniting twigs, one at a time, under the can. I was impressed with his skill and this stark efficiency in contrast to my own style of wilderness bonfires.

Late in the afternoon, we turned up a short box canyon. A creek trickled out, and in the damp sand were tracks of deer and coyotes. We moved upstream above the range of cattle, passing three small beaver dams of willow. Just as the last sunlight filtered down into the canyon, we came to the dead-end. The sheer cliffs closed in on three sides and at the bottom of the box canyon was a plunge pool of striking beauty—clear water surrounded by cottonwood and canyon ash—an oriental scene of dappled light and stark dendritic shadows cast against the red rock cliffs.

The next morning we hiked down the main canyon until the creek turned sluggish then disappeared under the rising waters of Lake Powell, the man-made outrage visible from space that had drowned the loveliest of all canyons: Glen Canyon.

With a floating willow stick, I scratched an obscenity against the Bureau of

Reclamation in the mud. Abbey sat silently on a rock. As we turned to go, Ed asked me if I knew anything about explosives. I replied that I had been cross-trained in demolitions as a Green Beret and had filled in for our wounded Demolition Sergeant in Quang Ngai Province during the summer of 1967.

Two days later, we hiked out and sat out a thunderstorm in the International Scout I had borrowed from a girlfriend in Tucson. After four dry days of backpacking fare, we wanted drink and real food. I found some beer and a chocolate bar; Ed came up with some cheese and a bottle of bourbon. After several permutations, we discovered that beer, cheese and crackers were the best compliment, followed by an after-dinner drink of whiskey and snack of chocolate. I had a cheap little tape player and I slipped in a late quartet of Beethoven, the transcendent Fifteenth. Thunder crashed and sheets of rain ran down the windshield; beyond, through the mist, lay a ghostly wasteland of dull-red and golden slickrock, fossilized hummocks of sand dune, dotted here and there with a bush of cliff rose or a juniper tree.

Later, I would look back to this time and wonder what the hell it was we saw in one another. Ed was 15 years older than I, so there was a paternalistic edge to our friendship. He seemed more sullen and grouchy than charismatic to me. And, 20 years ago, I was hardly a prize myself; I only smiled when drinking beer and the slightest sudden movement, noise, or trauma would bring out what I called "the cornered-ferret aspect" of my otherwise charming personality. Ed was actually tolerant of this erratic conduct, believing it a fated and necessary part of the determined crazy category of whacko Vietnam Vet behavior.

On the other hand, I had read *Desert Solitaire* and could see that a big chunk of the modern conservation movement had its origins right here and understand why a land ethic grew out of this Southern Utah slickrock and why its protectors tended to be so militant. *Desert Solitaire* was something larger than just a book about the desert. It was about the power of the land, of human connections to the earth, an idea of freedom. Ed's book was a call to arms.

A month later, Ed was on the eve of a broken heart and about to enter his "Black Sun" summer as a fire lookout on the North Rim of Grand Canyon. I returned to Tucson, moved into a tiny shack on the edge of the desert, and landed the only real job I ever had after Vietnam: substitute hippie mailman. Five months later, Ed Abbey showed up at my doorstep alone. His girlfriend had left him. Ed stayed at my place for a month, sleeping outside in the back of his VW station wagon until he found a stone house to rent at the foot of the Santa Catalina mountains.

By the time the fall rains arrived, Ed was still crushed and forlorn. In his desperation, Ed had sent the ex-girlfriend pictures of his new paramour—a long-legged Las Vegas show girl, in full dancing regalia—in a reckless attempt to win back the woman he was still secretly hooked on.

One day a letter from the girl arrived at the stone house telling Ed all about her new boyfriend, and how Ed ought to find a woman his own age. The tactic had backfired. Ed was now inconsolable.

How could she be so cruel? he wondered. I muttered something lame about how maybe he shouldn't have sent the provocative pictures, but he just wasn't listening. I tried cheering him up with some music, putting my favorite Mozart on the turntable, the "Sinfonia Concertante," a surpassingly serene piece. Shortly into the slow second movement, Ed stopped the music.

"Too sad," he said. "I can't stand it anymore."

We took a walk up Esperero Canyon. The torrential summer rains had scoured the bed of the dry wash and the bench above the bedrock gorge was covered with green grass. Where the bench terraced out against a low cliff of volcanic breccia, sculptured giant puffballs lay in recesses like misplaced soccer balls. The mushrooms were fresh and I was thinking about cooking up some for supper. But Ed wasn't interested. We continued up the wash startling a bobcat—an unusual daytime sighting. Ed took out the girlfriend's letter and crumpled it on the bedrock. He struck a match to it. We watched the pages flare and curl with some ceremony. It was still flaming as I turned and headed back down the canyon, already gathering mushrooms for dinner.

The Seventies were a time of flux. I quit my mailman job and chased a girl to Cape Cod. Ed lived in the stone house and worked as a writer. I came back and moved into the stone house with Abbey, then flitted off again for Northern California. Ed took up with my old friend, the woman who had lent me the International Scout on our Escalante trip; she moved in with Ed for a bit. The three of us went camping in the Cabeza Prieta.

Ed and I took backpacking trips into the Superstition Mountains, the Dripping Springs Mountains, the Gilas and the Galiuros. We did several more truck-camping trips into the Cabeza Prieta. Abbey got the two of us a job working for The Defenders of Wildlife as "custodians" of a large, private wildlife refuge in Aravaipa Canyon. We split the job since both of us wanted at least six months to travel. Actually, it was more a non-job since there was nothing to do but live there. We explored the country and, in November of 1972, I caught a glimpse of the last Arizona lobo, a wolf with a considerable reputation as expert ham-stringer of calves. In the end, however, we both found the country tame. I lasted a few months; Ed hung on for nearly a year. We were restless.

That winter, we started taking out billboards and bulldozers, and plotting against strip mines, copper smelters and logging operations. One night in the desert along I-10, we worked on a giant billboard. It showed four old white guys dressed in golf outfits laughing over cocktails at the club. The sign said: "Time to Relax" and "Green Valley Retirement Community." Amid the yucca and bursage Ed Abbey, John DePuy—a landscape artist and Ed's oldest friend—and myself labored against the wooden legs of the billboard, two with a bucksaw, me with an axe. A broken chainsaw lay on the ground. DePuy murmured, "drink" and we broke for a quick beer. John whispered, "Time to work" and everybody went back to the sawing and chopping. We dropped our tools and hit the dirt as vehicle lights zoomed by. Parked off the frontage road with the lights off was a 1970 Lincoln Continental. Sitting in Bill Eastlake's darkened Continental was lovely, long-legged Janet, Ed's new girlfriend, lusted after by all three friends.

The "Monkey Wrench Days" had begun, though I was unaware of any larger context for our mischief. As a team, we were careless and ineffective, almost recreational in our sloth, abstract in our political theory. It was simply something to do; a fist of anger raised against the blind greed of technology, an anodyne to impotence. The driving force in our misbehavior was probably an idea Ed had for a book he'd just started to write.

Ed Abbey gave me a great gift that year—advice about getting a job. Though essentially unemployable, I too had to earn a living of sorts. Ed advised applying for seasonal work with the National Park Service. "They give you a quitting date to look forward to," he said.

I got out the atlas and put in applications for all the places I might want to visit during summer months, including a new national park in Washington State called North Cascades.

During December of 1974, Abbey and I drove my pickup into the Cabeza Prieta, the last trip we would take out there together for a number of years. At that time, Ed and I were unattached and without families. We had spent a sniffling, lonely Christmas Eve at a topless bar in Tucson drinking whiskey. Thinking we could improve on that one, we packed up and drove a 150 miles west over Charlie Bell Pass into the Cabeza Prieta. We sipped beer all the way from Three Forks and were a tad plastered by the time we hit Charlie Bell Pass. We got my '66 Ford truck stuck several times creeping down the dark treacherous road to the well, hanging up the ass end of the truck, jacking it up in the dark, rocking it free, and then dropping down into the Growler Valley. We continued on for one more six-pack around the north end of the Granite Mountains, where we got stuck again, finally crawling into our sleeping bags shortly after midnight. Ed and I drove through Montrose well west into the Mohawk Valley. At the low pass, we found bighorn sheep tracks, always a big deal since desert bighorn were rare. Later, on New Year's Eve at Eagle Tank, it sleeted and snowed on us—an unusual occurrence.

We sat out the rain for two days under a tarp, stoking an ironwood fire. We charted all the desert islands we wanted to visit in the Sea of Cortez. Ed scribbled notes on the new book he was writing—*The Monkey Wrench Gang*. Ed was especially concerned with the technical credibility of this book (at that time, I didn't know it was a comic novel of environmental saboteurs).

"I want it to read accurately to a bulldozer operator hard rock miner," Ed said.

I said I could help a bit, but some of this technical stuff was over my head. In the back of my pickup I had a little field library of military manuals and "Confidential" materials passed out at the Center for Special Warfare when I was stationed at Fort Bragg. The manual on improvised weapons and demolitions was the most interesting; a little home cookbook for zip-guns, napalm and the making of thermite using common household or hardware metals and oxides. A good place to start.

Back in Tucson, Ed bought a copy of *The Blasters Handbook* and went to work. The manuscript of *The Monkey Wrench Gang* was completed and Ed asked me to do the "technical" editing." I camped on the rim west of Horsehead Rock, looking out into space towards Rustler Canyon in the northern Needle District of Canyonlands National Park, sitting on the rim reading the manuscript for two days with my loyal collie dog at my side. Ed's quick study of heavy equipment and incendiary materials had eclipsed my own grasp of the subject. I found only small items to correct.

During fall and winter of 1975, Ed was living in Moab with his new wife. *The Monkey Wrench Gang* had been published earlier that year. I had finished my last season as a backcountry ranger for the National Park Service; anarchists tend to make lousy law-enforcement agents, and I was no exception. In three years as a seasonal ranger for North Cascades National Park, I had only managed to write a single ticket on a poorly parked Winnebago.

Earlier that spring, I had begun a project filming the remaining grizzlies in the lower Forty-Eight. After an autumn of shooting film of bears in Glacier Park and my usual forays in Yellowstone, I returned to southeastern Utah.

Ed asked me if I wanted to take a hike up Mill Creek. We headed up the creek, which dumped right into Moab. Soon the jeep trail petered out and beaver dams slowed the flow of the creek—sizeable for this country. The creek grew into a large gulch, then deepened into a canyon. We climbed up the slickrock to examine a panel of petroglyphs using ancient steps pecked into the rocks. Images of sheep, deer and hunters were chiselled into the patina separated by symbols, perhaps of mountains, lightning or sunrises. We had no idea what it was about, though it was clear the scene was epic and the place special. Below, the little canyon forked into two prongs. At the junction was a boulder covered with rock carvings of humans with rake headdresses, perhaps feathers, and more feather-like appendages hanging from their outstretched arms as if they were about to take flight. On top of the big boulder were carved meandering lines that represented an accurate map of the area.

The truth is, this had become a rough time in our friendship.

My sometimes turbulent behavior was indeed a cross to bear, and Ed's new wife didn't like me. And it would be less than honest to say the publication of *The Monkey Wrench Gang* did not strain our friendship. Even a famous dolt is still a dolt. The only thing much worse than reading your own press was becoming someone else's fiction. Fortunately, I had my own full-time work with bears, and the question of Hayduke was of more interest to others than to Ed and myself. In one of our lowest moments, the legal staff of Lippincott had insisted Ed Abbey write me a very embarrassing letter about how I should only consider the good characteristics of George Washington Hayduke a reflection of Peacock and not the bad parts. That sort of thing.

I had the letter with me. On top of the boulder carved with the ancient map of the area, at the junction of two great canyons, in the heart of the country the opening sentence of *Desert Solitaire* describes as the most beautiful place on "earth," we struck a match and—as we had burned another years before—ritually vaporized the letter. We watched the wind carry off charred fragments, gathered by the laughing waters of Mill Creek. Neither of us ever mentioned this matter nor the origins of George Washington Hayduke to the other again.

Our petty quarreling was quickly patched up when a common enemy reared its massive ugly head—a coal-burning power plan at Kaiparowits. We resurrected the old Black Mesa Defense Fund and closed ranks. Also, that winter I took my first solo hike all the way across the Cabeza Prieta. Abbey and I had talked about backpacking—as opposed to the usual jeep trip—the 150 miles or so of desert, but thought it might be too hard or even impossible until two women we both knew did it first.

An incredulous Greyhound bus driver dropped me on the off-ramp at Welton, at the northwest corner of the Cabeza Prieta. I shouldered my old backpack and stumbled southward toward Mexico.

Ten days later, I walked out, without having seen a single human being, to Quitobaquito in Organ Pipe. The next winter I did the walk again and Abbey went on one of his own. By 1989, I had made seven solo tracks across this great desert place.

In the meantime, I received another dose of Edwardian occupational advice: "If you like to read and write and your own company, a fire lookout is a perfect job," said Ed. Thus, I started a new career as a fire lookout in Glacier National Park. I had to go up there anyway because that's where the most grizzlies lived in the lower Forty-Eight.

Just before leaving Tucson for a summer up in Montana, I stopped by Ed's house. I had wanted to say good-bye, but he wasn't answering his phone, so I just walked in on him. He was with a woman I hadn't met before, a blond, and I was staring directly at her bare bottom not three feet away through the window. Ed waved from behind her back and she shrieked. I said "hi" and backed off a couple of steps. Later, we all went out to the El Dorado, a Mexican restaurant with an erratic record for good food. I introduced Ed to my new friend Lisa and I met the blond woman, who is still a friend, a fine, warm-hearted woman with a great big ass.

The Monkey Wrench Gang received the usual mixed responses; that is, the Eastern Literary Establishment totally ignored it and, somehow, with absolutely no advertising or publicity, the book managed to sell 500,000 copies on its own. In order to promote the protection of grizzlies, I had to go public. I did a television program with Arnold

Schwartzeneggar for the *American Sportsman*—the old yuk-it-up in the woods with TV-type celebrity, then blow away the grouse or catch the bass. The producer of the program called me the most difficult person he'd "dealt with in the history of the *American Sportsman*." Though Abbey was forced to agree with that appraisal, he was by nature accepting and forgiving; my earthy spirit guide had a few warts of his own. On the home front, our lives stabilized. We both met women, married and started families; the first for me, a new one for him.

We kept on going out to the Cabeza Prieta desert. It continued to be the most important thing we ever shared. Every winter, we camped a few days in Daniel's Wash. We followed an old Indian trail over to Sheep Tank. At the pass south of the natural tank, we spotted a bighorn sheep, a young ram with a half-curl—our first sighting of desert sheep together, a landmark in our travels.

The following Thanksgiving, I followed the tracks of another desert sheep in the Cabeza Prieta from the Growler Mountains to the center of the valley, where I lost the animal's trail. I had taken the bus to Ajo after a reluctant Greyhound Bus clerk had sold my one-way ticket to the woman I later married.

"Lady," the clerk had said to Lisa, "Nobody buys a one-way ticket to Ajo."

At Ajo, I shouldered my backpack and disappeared over the mine tailings into the empty desert. It was dark again the next night when I reached the bottom of the Growler Valley. Even by the low luminosity of the moon, I could see big pieces of Hohokam pottery and Glycymeris clam shells lying on the desert pavement—the "Lost City" of the Hohokam shell-trekkers. From here in the Growler Valley, an ancient shell trail ran south to Bahi Adair on the Sea of Cortez. I lost the trail of the bighorn because of the darkness and because a rattlesnake nailed me in the calf that night, though the snakebite turned out to be a dry one. The next day, I walked out with a story to tell 20 miles to Papago Well where Clarke and Ed Abbey were waiting for me.

I continued to range seasonally between Arizona and Glacier National Park, where I continued to work as a fire lookout. Clarke and Ed visited Lisa and me in Montana just before we all started having children. I hauled Abbey up to the wild place known as the Grizzly Hilton to catch a glimpse of my favorite animal. But it was not to be. Ed Abbey lived and died without having seen a grizzly; he was the only person I ever took to the Grizzly Hilton who did not see a bear.

Time passed and Ed's health took a downward turn. He went out to the Cabeza Prieta, shouldered his old backpack, and hiked into the Growler Valley. He had to turn back; his weakened condition made it impossible to duplicate his previous hike.

"I was just too weak," he reported. This was a big blow, though he coped well. What he cared most then about was spending time with Clarke and their two young children—taking his young daughter, with his little boy perched on his shoulders, by the hand walking through the desert.

"I want to watch those kids grow up," he said.

It was a relatively quiet time in my life. Both Ed and I were busy raising our families and making a living. We saw each other at family cookouts in the desert or our backyards. On Halloweens, Abbey and I would follow the children and their mothers through strange Tucson neighborhoods trick-or-treating, creeping along in my pickup, drinking beer, loaded pistols tucked under the seats.

We went back out to the Cabeza Prieta in Jack Loeffler's truck. That was the next-to-last time the three of us would be there together. It was an especially hot May following a wet spring. The desert made for hard cross-country walking because exotic foxtails and bursage stuck to your socks, so we ambled up Daniel's Wash, lined with giant desert willow trees in full bloom and yellow paloverde a bit past prime. The cholla cactus were also in bloom, and about a third of the saguaros sprouted white trumpets. Never had I seen so many honey bees. The surge of wind blended harmonically with the increased wing speed of thousands of bees treading air under the blossoming paloverde trees. We stood silently under the trees listening to the billowing drone of bees as we heard countless symphonies, sonatas and concertos. Along with love of the wild, music had always been the great glue in our friendship.

In February 1989, I took a last nervous break from watching Abbey's declining health and visited one of the wild desert islands Ed and I had talked about in the Cabeza Prieta—Tiburon Island off the coast of Sonora. By the time I returned to Tucson in March of 1989, it was late. Ed made one last public appearance, a reading. While I looked after Ed's little daughter Rebecca, an FBI helicopter circled overhead recording our conversation with an undercover agent; the Federal Government had begun its persecution and harassment of the radical environmental group, Earth First!, a movement inspired by Abbey's books. That evening Ed and I quarreled over his .357 magnum pistol: I had appropriated the gun and sent it away and he was justly angry. It was our last quarrel and the only time in our friendship I ever passed judgement and went against his wishes.

But from that moment the paternalism disappeared. I went home, troubled, and wrote Ed a letter about our children, explaining why I couldn't let him have the gun. He never got to read the letter, but it was buried with him.

I took the letter over to Ed's house. Clarke came running out the driveway. She had been calling me; Ed was bleeding to death. We set off for the hospital. I was in the lead car, an old Honda, with Clarke, leading the way down the middle of Grant Road, lights flashing, horn honking. Ed was behind in Dave Foreman's car, with Dave's wife, a nurse, administering. A chubby young blond in a red Corvette tried to cut me off just to be smart. She didn't know this was a life-and-death run, of course. She was in the lane to my right, alongside me. A power pole was coming up on the next curb corner. I swerved into her and tried to drive her into the pole. Just like that, just to get rid of her and get the job done. Her quick reflexes and fast car saved her.

Madness.

If there remained any question about Abbey's life, it was answered by his dying.

"Those who fear death most are those who enjoy life least," Ed had written. "Death is every man's final critic. To die well you must live bravely."

At the hospital, he submitted—for the sake of the people he loved—to the high-tech medicine he hated. There, I tried to persuade him, using all of my talents, into things that might have saved his life. I felt a surge of guilt, then remembered we were in this together and that I was now merely executing his last wishes. Finally, he pulled out all the tubes and announced—with the clearest eyes I have ever seen—that it was time to go. We drove him out into the desert to die. I built a little mesquite fire, and we said good-bye. Clarke got in the sleeping bag with him and we waited. But he got better.

"Sometimes the magic doesn't work," he said.

It was two more days to the end.

The last time Ed Abbey smiled was when I told him where he was going to be buried. Then it was back to the Cabeza Prieta, and I smile too when I think of this small favor, this last simple task friends can do for one another—the rudimentary shovel work, this sweaty labor consummating trust, finally testing the exact confirmation by lying down in the freshly dug grave to check out the view, bronze patina of boulder behind limb of paloverde and turquoise sky beyond branch of torote, then receiving a sign: seven buzzards soaring above joined by three others, all ten banking over the volcanic rubble and riding the thermal up the flank of the mountain, gliding out and over the distant valley.

Three years later, I still grin as I crest the ridge above his grave, the earth falls away and mountain ranges stretch off into the grey distance as far as the eye can see; there is not a human sign or sound, only a faint desert breeze stirring the blossoms of brittle-bush. We should be so lucky.

Ed Abbey taught me that saving the world was just a part-time job. He told me to find a job with a quitting date to look forward to. He said being a writer was all right, but you should have a manual trade too, a real job like shoe repair or screw-worm eradication, as back-up. And I remember his dying, how it affirmed his life.

"If your life has been wasted, then naturally you're going to cling like a drowning man to whatever kind of semi-life medical technology can offer you, and you're going to end up in a hospital with a dozen tubes sticking in your body, machines keeping your organs going. Which is the worst possible way to die. One's death should mean something."

He told us the truth and what values we might hang onto to save ourselves and this earth we want to abandon for space and "yearn for some more perfect world beyond the sky. We are none of us good enough for the world we have." For that, I loved him.

On the eve of March 16th, I journeyed to the edge of this desert place. March 16th is a "Day of the Dead" for me, the anniversary of the My Lai massacre and also the day in 1989 three friends and myself buried Ed Abbey here, illegally, in accordance with his last wishes.

I had traveled out here alone to Ed's grave, bearing little gifts, including a bottle of mescal and a bowl of pozole verde I had made myself. I sat quietly on the black volcanic rocks listening to the desert silence, blasting my annual dose of grief into the darkness, pouring mescal over the grave and down my throat until the moon came up an hour or so before midnight. Suddenly, I heard a commotion to the south, the roar of basaltic scree thundering down the slope opposite me. A large solitary animal was headed my way.

I got the hell out of there.

Two days later, I told my story of the desert bighorn ram I heard but never saw to a poet friend.

Well, Doug, he said, "maybe it was old Ed."

—MOUNTAIN GAZETTE, #81

ROBERT CHAMBERLAIN

APPENDIX A

Letters to the Editor

No Eskimos In Talkeetna

I have just received the January 1974 issue of the *Gazette*. Another superb effort, but there are a couple of comments I feel compelled to make. Galen's article, "Alaska: Journey by Land," is most interesting, but his credibility as an accurate commentator on the Alaska scene is tarnished somewhat when he refers to "uncounted drunken Eskimos" as being among the features of Talkeetna. Two paragraphs later, he describes a "wizened, almost blind Eskimo" as "rattling the doorknob of the closed tavern." The fact of the matter is that there are absolutely no Eskimos in Talkeetna; the people he referred to are Indians, probably Athabascans who live in Alaska's interior. Eskimos, one group of which I have represented as their lawyer for the past five years, take strong exception to being confused with Indians.

The other comment I have relates to Geoffrey Childs' review of Bonatti's and Rebuffat's books. Childs mistakenly states that Bonatti's fine book, *A Mes Montagnes*, has not been translated into English and that no such translation is "in the works." This book was published in 1964 in an English translation by Rupert Hart-Davis (London), entitled *On the Heights*. I was most interested in his comments on the French translation which he describes as "fine and straightforward" and with prose that is "conservative and to the point...small words, tight lines, jamming the pages as if he had something to say." To the contrary, despite Bonatti's brilliance as an alpinist, I found his writing style (at least in the English translation) overblown, highly emotionalized and on the whole, a bit unreal.

James Wickwire
Seattle, Washington

Where's Ms. Mountain?

As a recent subscriber, I notice that members of my sex are even more underrepresented in your pages than we are in the high country. If you are not receiving manuscripts from women, it would be worthwhile for you to solicit some. If I dare make a generalization about something that obviously varies with the individual, the sort of non-competitive, receptive attitude toward wilderness which you seem to be reaching for is more common among mountain women than among the men. In any case, since you are trying to show geopolitical, economic and ecological consciousness, you ought to have a little sexual consciousness as well.

Deborah Frankel
Oakland, California

Too Many People

George Stranahan's article "On Coming Up Short …" concerns the single most important issue facing mankind. Beside it, things like detente, civil rights, honor-in-government, arms limitations, space travel, Middle East tensions, etc., are really trivial.

I've occasionally tried to discuss it in groups, only to be greeted by a cold silence. People are aghast at the mere thought that the human race peaked out a while back and is rapidly devolving and would rather pretend that anyone voicing such ideas must be off his rocker. After a good ten years of this, the Club of Rome made it at least partly acceptable. But it seems that all the modern-day Cassandras dwell on the physical aspects of our coming doom and don't touch on what may be the cruelest blow of all—the total loss of human freedom.

As shortages multiply and people have to compete unmercifully for bare sustenance, governments will have to impose iron rule, far worse than in the most drastic dictatorship ever known.

Eventually, no matter how rich you are born or how lucky you are or how hard you work, you will be allowed either the minimum square yards for bare existence or no land at all, because it is all communal. No couple will be allowed to have more than one child, and it will have to be enforced by sterilizing women after their first baby. Nobody will be allowed to travel. Nobody will be able to change his place of residence. Choice of occupation will be forbidden, because certain jobs will be essential just to keep the whole race alive. Water will be rationed to the point where baths are even counted, etc., etc.

One can even visualize a child getting some sort of punch card issued at birth, entitling him to so many liters of water, so many square meters of cloth, so many kilograms of metal, etc., for life, and he can either ration it out carefully or run out early and be allowed to go without entirely.

Sadly, no leader could get elected, here or elsewhere, on a platform of halving the population and drastically shrinking the GNP. Nor could he hold office long enough to implement these necessary changes, if he springs them as a surprise or saw the light after attaining leadership.

So, sadly, we must watch it all happen, forever disgraced, because while many species have developed, flourished, failed to adapt or cope, and died out, man is the first that was able to see what was happening to him, predict its course, comprehend what had to be done to correct it, yet chose to do nothing about it and let it all end.

For this we developed a brain!

Edward Scott
Ketchum, Idaho
(Is this the same Edward Scott that gave us the Scott ski pole? Yes, it is.)

Rockcraft Revisited

I find few things as pointless as the various polemics which, from time to time, have raged across the pages of mountain magazines; and so I am writing this letter to comment on a recent book review with the greatest reluctance...But it seems to me that the accidental conjunction of last-minute deadline pressures and unseasonably hot weather that caused an entire column of Dick Dworth's review of *Advanced Rockcraft*, by Royal Robbins, to come unstuck and fall off the paste-up sheets—and quite properly to be reprinted in toto in the following issue—has laid altogether too much stress on one man's opinion. I would like to offer a counter-appraisal of the book, and indirectly, of Robbins as teacher, so that Dorworth's highly biased, although perhaps personally legitimate point of view, will not remain the *Gazette* reader's only impression of this intensely worthwhile book and its author.

First, the review itself: Dick indulged in a bit of oblique writing here, transforming the review genre into a discursive mini-essay on the personality of the book's author, and largely ignoring the volume and its contents. Aside from questioning the validity of this approach to writing a book review, I would take issue with Dick's main thrust, which was to criticize Royal for the glaring inconsistency between his statements on climbing ethics and his action in chopping bolts during the second ascent of the Wall of the Early Morning Light. Well, it so happens that this is about the only single instance anyone can think of in which Royal Robbins has ever behaved in an inconsistent or self-contradictory way. And on top of that, Royal, who has an uncanny gift for honest self-appraisal, recognized the absurdity of his actions on that one climb and said publicly (in *Mountain*) that he regretted the whole thing. This surely speaks for itself. Most of us, Dick Dworth and myself included, are not very good at admitting our mistakes, and such ability surely speaks for a very deep consistency or harmony between one's professed beliefs and one's actions.

Second, the book itself: In my opinion, this book, along with its companion volume, *Basic Rockcraft*, is the finest instructional text on rock climbing ever written in English. It is not merely the punditry of one of the country's rock heavies, but the distilled and effective teaching/learning patterns that Royal has worked out with hundreds of students at his summer climbing courses over the years. In addition, of course, many of the advanced techniques described in this second volume were actually worked out (we could almost say invented) by the author. The information in *Advanced Rockcraft* is simply the best there is right now.

But back to teaching: I have known Royal for more than 15 years, but the first time I ever saw him in action in a teaching situation was in Leysin, Switzerland, in 1965. That summer we both worked for John Harlin as climbing instructors/guides at his fledgling International School of Modern Mountaineering; and later, that winter, we taught skiing together at a private school in Leysin (Royal, who was a certified FWSIA instructor, directed the program, while I was only an apprentice instructor). In both these instances I had an opportunity to see a unique teaching personality at work. For Royal is not a charismatic teacher at all; he does not con his students into performing above their presumed level of skill; nor does he try to develop self-confidence in his students through any kind of contagious enthusiasm. It's just not his way. Instead, his way is to analyze exactly, precisely and simply what the student must do to perform the move—and then communicate it in a minimum of words. A logical, analytical teacher, therefore, and a hell of a good one. This particular bent, or approach, to instructing kinesthetic skills in outdoor adventure sports like climbing and skiing, is an ideal style to be transferred to the pages of a book where personal presence of a more emotionally involved teaching style would be lost. A few weeks ago, I was struck once again by the effectiveness of this kind of teaching style, when Royal taught me easily and quickly to roll a kayak, with a few well-chosen, well-thought-out descriptions of what I had to do, and a very simple no-nonsense demonstration. I guess what I'm saying is that Royal is not a natural teacher, but that he is a very good one because he has worked out a way of presenting these skills that maximizes his own strengths in communication—analysis and clarity. These qualities abound in *Advanced Rockcraft*. It's no use saying that one can't learn to climb out of a book; perhaps it's not ideal, but many people must turn to texts to make up for the lack of good teachers in their area, or just for reference. Texts, I feel, can be useful and this book along with *Basic Rockcraft* is the most useful of all the climbing texts that have appeared so far.

A final word on responsibility: Royal, throughout the pages of this book, as in his own life, urges people to take responsibility for themselves and their actions. He accepts his own responsibilities in explaining the mechanics of a potentially hazardous or fatal sport; and he stresses that his readers/students will succeed and survive on the crags to the extent that they can accurately appraise their situation and take responsibility for handling it. The connection between this attitude and the problems of "law enforcement agencies everywhere," which Dorworth attempted to draw at the end of his piece, to me at least, seems tenuous indeed.

Royal asserts that "climbing is an exercise in reality." I would venture to suggest that writing book reviews is also.

Lito Tejada-Flores,
Bear Valley, California

Dorworth Is Crackers

I was going to reply to Dick Dorworth's character review (Number 23), and attempt to answer his many questions, rhetorical though they were. But after considering a bit, I lost interest. For how does one come to grips with an article in which the author, in one (typical) short finishing paragraph, embarrassingly changes from conditional third person to second, and who goes on to suggest that "law enforcement agencies everywhere" would be "lots, lots better off" if I altered my behavior to Dorworth's specifications! Dorworth is crackers.

R. Robbins,
Modesto, California

Unsentimental Realist

Ed Abbey comes on like a hybrid between Henry Thoreau and John Wayne. Unsentimental realism is the best thing about his writing, but too often, it seems to me, he oversteps the line from honest toughness to irritating macho male propaganda. Do the feminist attacks on "sexist" ads really deserve his putdown? ("Letters," Number 28)

All this aside, I really enjoyed his blend of environmental activism, nostalgia and adventure in *Where's Tonto?* in the same issue. I wish I was macho enough to sabotage a few bulldozers myself.

Donald G. Davis
Fairplay, Colorado

Our Tax Dollars At Work

I think the article "There Was A River" (*Mountain Gazette #31*) is excellent; I like Bruce Berger's style and content. However [referring to Berger's interview with Stewart Udall], it's easy, with hindsight, to blame the environmental movement for Glen Canyon's loss. Certainly, Mr. Udall would like someone to share his responsibility with him!

For the record, the Sierra Club was much smaller then—about 20,000 members compared to 140,000 now—and had much less financial and ideological backing in the rest of the country. Even so, the Colorado River Storage Project was brought to a standstill in Congress by the Sierra Club, along with most other conservation organizations, until proponents of the bill agreed to delete the Echo Park Dam and add the since-unhonored commitment to protect Rainbow Bridge. I clearly remember that it was Mr. Udall, and his Department of the Interior, who refused to spend the peanuts it would have taken to adequately protect Rainbow Bridge from "Lake" Powell's rising waters, while they were spending a billion dollars to build Glen Canyon Dam and associated works. The only way to protect Rainbow Bridge (other than with a dam) would have been to maintain the reservoir at a low enough water level. Mr. Udall scoffs at the idea of a protective dam, but he wouldn't keep the water down, either... (and) he still hadn't learned enough to refrain, a few years later, from pushing Bridge Canyon (Hualapai) Dam in the Grand Canyon.

I firmly believe that if the Echo Park Dam (in Dinosaur) hadn't been stopped, we would now have a number of dams in our national parks and monuments. Being familiar with both Dinosaur National Monument and Glen Canyon, I have to say that Glen was the more important scenically, but there was NO WAY, at that time, to get protection for Glen. Saving Dinosaur was a narrow squeak at that. It was one of those close, almost fluke victories, like the one that ended the war of revolution of 1776. It is a mistake...to imply that Glen Canyon did not have its defenders. Their (our) voices were raised, but not heard over the din made by government proponents, and the lobbying of the construction industry.

I have no desire to take away from Mr. Udall's more recent environmental accomplishments, but it should be clear to anyone "who wasn't there" that, in the final analysis, a cabinet officer has far more power to influence the outcome of political decisions than the head of a conservation organization, if he will use that power fearlessly. It is much to Mr. Udall's credit (and environmental redemption) that he drew up, on the eve of his departure from office, and got President Johnson to sign the Presidential Proclamation that established Marble Gorge National Monument, and importantly enlarged others.

It is a sad and ironic postscript to this history to note that the loss of Glen Canyon undoubtedly helped greatly in later saving Marble Gorge and the Grand Canyon from a similar fate—well, in truth, that battle isn't won yet...there is still a powerful lobby lying in wait for a more opportune time to strike. The cry for Echo Park, for that matter, is still heard in Utah. And Arizona still wants Bridge Canyon (Hualapai)—this time, "for the Indians!" "Eternal vigilance" is also the price for environment-wrecking dams. Will the "energy crisis" and unemployment statistics provide the new urgency the damn (sic) builders are looking for?

We aren't through with Big-Dam Foolishness yet—old dams never die, they only fade temporarily, or are replaced with something worse, e.g., the nukes, or the pollution-spreading coal-generating plants whose proliferation in the Colorado Basin will not just wreck a river, but a whole water- and-airshed. Hard behind the coal plants is the commercial tourism that is poisoning the Plateau Province that was mostly wild less than a generation ago. Wildness is rapidly being reduced to the ugly American nightmare of honky-tonk, complete with guaranteed high-standard access—"your tax dollars at work," under the guise of providing jobs, a broader tax base and "recreation."

I'm no doomsayer, but it's become clear to me that when environmentalists work for the environment, they are really working for the survival of man. If our so-called civilization continues on its devastating path through the natural environment, it will kill itself off, and mankind with it. The Earth has survived other cataclysms; it will survive man's worst, and rebuild without him. Man is not essential to Earth's survival, but a healthy Earth is essential to man's survival.

Thanks for the soapbox.

Philip Hyde
Taylorsville, California

Hopeless, to the Editors:

Okay, that did it!

Up to *Mountain Gazette* #43, I was only mildly annoyed, but after being asked to wade through George Sibley's "book," which I'm sure was a long time in search of a publisher until he joined the staff of *Mountain Gazette*, and then be hit by the ripe garbage of Speer Morgan's river-tripping is more than I can hold still for. If he's the sort of creep for whom I've been fighting 20 years to save our wild rivers, I'm sorry more of them aren't gone. Aside from the numerous inexcusable errors in his babble, i.e. Colonel John Wesley Powell, Cocina for Coconino, his insipid writing (at least Sibley's ego trip gave us an occasional flair of solid prose), and total insensitivity to what river running is all about, he treats us to a distorted, sick picture of the

Colorado, Salmon and Owyhee rivers through the eyes of an acid-head freak with all the style and grace of a turd in a punch bowl. Spare me.

Feel free to send pages 10 through 17 to Speer for his rear.

Katie Lee
Jerome, Arizona

Nymphets Wanted

Tom Morris' critique of "candy-ass towns" was a revelation to me. Here we are working day and night planning cutsie boutiques, landscaping the creek, appealing to the over-monied crowd and just plain trying to create the old standardized "pink bubble," and now I find out that we don't even qualify for full-fledged candyassness because we lack that vital spectre of nubile teeny-boppers.

Please, if anybody can supply us with chrome-gloss lipstick nymphets, we can import them on the same bus we use to export our urban problems.

Myles C. Rademan
Director of Utopian Planning
Crested Butte, Colorado

Finding The Key

When I get high, I find I understand and perceive things that I never would when I'm straight.

It just happened that my latest issue of *Mountain Gazette* arrived at the tail-end of a two-day-long party. I sat down and was immediately enthralled with the magazine. I finally understood the subtle humor and message that pervades your magazine.

Keep up the good work.

Alex Could
Derby, Connecticut

On est

With reference to your September 11 article regarding the AHP convention ("Breaking Free from the Human Potential Movement," *MG #38*), I would like to correct some of the inaccuracies that relate to what the writer, Mike Moore, has to say about est.

In the first column of the article, the author says that, "Erhard appears to be turning them away by the hundreds. He says he is. Although for a vendor that is 'temporarily sold out,' ..." est does not turn people away. Anyone wishing to enroll may do so. Although there is sometimes a wait of several months after enrolling, in no case is a person told he will not be able to take the training.

In the beginning of the second column, Mr. Moore goes on to say that he would be told by "Erhard and everybody who has taken the training" that "I have no right to write on the subject of est until I have been through the training." No one is told that he doesn't have the "right" to write about est. It is a policy of the est Public Information Office to answer all requests for information. Such information has been given to non-graduates.

Since data for this article was researched, the number of graduates has increased to more than 55,000.

Mr. Moore talks about a "humiliating process" that he claims is part of the training. The processes in the training are methods by which a person experiences and looks at, without judgment, what is actually so with regard to specific areas in his or her life. The intended result of doing a training process is a release of greater spontaneity, not that the trainee be humiliated.

The author says that the majority of the crowd of 800 people listening to Werner's talk at Estes Park were his "natural enemies," but he gives no evidence to support this accusation.

Regarding Mr. Moore's reference to "high pressure marketing" on Erhard's part, there is no advertising of est, there is simply a follow-up by est with people who have been recommended by est graduates themselves—as people with whom they would like to share the est training experience.

Est is not therapy and is not psychology. That is specifically pointed out to people before they take the training. People are told that if they feel they need therapy or psychological, psychiatric or medical services, they should see a therapist, psychologist, or physician, as appropriate.

Est is guided by a professional Advisory Board that reviews, advises and recommends about the policies of est and other matters that may come before the Board. The Board is comprised of outstanding women and men in the professional and academic fields chosen for their ability to make meaningful contributions to the effectiveness of est. The Advisory Board head is Dr. Philip R. Lee, Professor of Social Medicine and

Director of the Health Policy Program in the school of Medicine, University of California, San Francisco, and former Assistant Secretary of the Department of Health, Education and Welfare.

In January 1974, eleven faculty members (including three department heads) and students of the School of Medicine, University of California, San Francisco, published an open letter to their colleagues at that school. They stated: "About a dozen or so students and faculty have taken a two-weekend training given by a group headed by Werner Erhard. We have benefitted [sic] sufficiently, both on personal and professional levels, so that we have asked the Dean to provide the opportunity of presenting Werner Erhard to you in an evening seminar...

"Those who have participated in est see that est has some unique features. It has distilled, out of the new consciousness movement, surprisingly efficient, non-invasive approaches that we see have worked. The features we like particularly were: that est insists that you participate entirely as a matter of your own choice; that est opposes manipulation of even subtle group pressures; that est is not a movement and Werner Erhard is not a guru.

"It is noteworthy that est is engaged in documenting its effect by objective physiological measurement and psychological testing. Some of it is being done by members of our own faculty, and all of it by professionals who have not had prior exposure to est."

Your writer refers to an article by Mark Brewer in *Psychology Today*. Brewer has found a model that he thinks fits the est training and has applied the est training to that model. The problem is that the model he has selected is neither an appropriate nor an accurate representation of the est training and, therefore, Brewer's analysis and conclusions fail.

Nevertheless, Brewer concludes that "it is difficult to condemn offhand anything that produces as high a degree of satisfaction and as strong a sense of new personal worth as est does."

There are also references to "est philosophy" and "phony mystery," but, in fact, est simply sets out to handle the problem that Werner Erhard says "has people getting stuck acting the way they were instead of being the way they are."

It is Werner Erhard's observation that people tend to operate from their beliefs and automatic behavior patterns rather than from their experience. The est training is designed to give people an opportunity to become more conscious of the degree to which they operate from their beliefs and images of the way things are, rather than from their experience of the way things are. Their expanded consciousness allows them to be more aware of systems of belief and automatic patterns of behavior that may dominate their lives and that often are not productive or counterproductive in their lives.

We sympathize with Mr. Moore because of the frustration he experienced in trying to put together a story, but it is a disservice to your readership for him to portray the search for expanded consciousness as something sinister.

William Houlton
Vice President
Daniel J. Edelman, Inc.
New York, New York

Mike Moore Replies

Public relations men are poetic-license evokers—a humorless, unhappy lot by and large. (Perhaps they grow sullen and gloomy because of the nature of their work?) I wasn't surprised to learn that Werner Hans Erhard has one.

Nor was I surprised that Mr. Houlton didn't like my article. He picks at it from the edges. I should like to pick back.

On the right to write: I was merely quoting the master who said in Estes Park that too much was getting into print about est written by people who hadn't taken the training (though I think his implied meaning was that too much was getting into print by writers who hadn't been "taken in" by the training). I damn well know that I enjoy the Thomas Paine given "right" to write what I please, about est or anything else—a right I intend to cling to even if est becomes the majority political party in 1984 and bellbottomed goons from the "Public Information Office" come to my house to take away my typewriter.

(As an aside, I talked with an est graduate a couple of months ago who refused to answer any of my questions on the training. He said he had received a letter from San Francisco headquarters that stated that all questions from journalists should be referred to that same Public Information Office for answers.)

That I undershot the number of est graduates by 15,000...well, that's a serious factual error. One that really scares the hell out of me.

I was using my poetic license again when I speculated that the majority of the crowd that day were Erhard's "natural enemies." It was an assumption I made, based on the fact that most of the questions were hostile, far more than half the crowd left the room at the first opportunity, and all during that week I talked to many more people that dreaded est than espoused it. But I have no "evidence" to support my "accusation."

I know Werner has never *advertised* est (something that's always bragged about in his "advertising"). He has something much more effective: the tens of thousands of est graduates, proselytes—no zealots—that attack with all the charm and technique of the best Los Angeles used-car salesman. I will still submit that they create some pretty high pressure.

Mr. Houlton would have done well to pass by Mark Brewer's *Psychology Today* piece altogether. Erhard himself said that Brewer had written an accurate enough description of what goes on at an est session. Erhard was only critical of Brewer's subjective interpretation of the facts—his conclusions. And Houlton completely misrepresents Brewer's conclusions in his letter. Brewer did say that est was difficult to condemn offhand, but he went on to condemn it in no uncertain terms. His conclusions were far better represented in this line from the last paragraph of the piece: "I personally distrust any organization that transforms and uplifts thousands through nihilism of a belief system that denies all other belief systems as bullshit."

At one point in his letter Mr. Houlton denies that est is a therapy or a psychology (without saying, however, what it is then). But he also wants to wear that mantle and he slips it back on in his last sentence when he talks of the value of the "search for expanded consciousness."

Myself, I'm all for expanded consciousnesses. If we had a few more of those around, I daresay there would not be so many lonely people getting sucked into cynical "therapies" like est.

Mike Moore

A Hustle

The short exchange in *Mountain Gazette* #40 about Mike Moore's "Breaking Free from the Human Potential Movement" (*MG* #38) has moved me to drag this out a bit more.
The original article was a hustle: "...the 'holistic' approach (being) a kind of joyful anarchy" and "...seeking a resolution to the old mind-body dichotomy..."...Horseshit. That sounds like good material for *Time*. The fact is that the holistic approach is the resolution to the mind-body dichotomy, and for most of the "movement" the issue has been dead since the late '60s. At least that's my experience.
Despite, and partly because of, his announcement of good intentions (so innocent!), his "pretty bad choices" (his choice) make me think that *really* Moore wasn't prepared to do fuck-all to "get into the story" except watch, which is safe, dull and, given the situation, unreal as any unrealness he saw. Wishing someone would lay (love and feel) him (and not doing anything about it), choosing lectures that left his head spinning, sleeping or tripping and hanging out in a bar listening to losers free-disassociate strikes me as being self-abusive and very tame.
And how condescending! Aren't the readers of *Mountain Gazette* "almost exclusively white, educated," if not middle class, at least elite (thank you, George Sibley) and guilty/resentful?
Something else—I know I'm fantasizing here, but given what's already come down...those people in the pictures seem to be enjoying themselves. Was it really all that

unreal? Moore's article and Po Kempner's pictures don't seem to fit together...maybe it was just "bad choices" again. We create our own reality. If you don't put out your hand, you're not going to get felt, so I'd "rather put the blame" back where it belongs.

Personal responsibility has never been a popular trip.

Those people are fooling around with real and powerful emotions and energies, more or less ignorantly, and there's bound to be lines that don't go. There's plenty of phony guides too, and a few good ones. And since it's a hell of a lot easier to be hustled by ourselves than by Werner Erhard, it take guts, commitment and boldness to explore those lines. So my contention is that before you give Columbus up for lost (I don't mean Erhard), at least try a shakedown cruise to find out what he's up to. So what's this "breaking-free" crap?

Finally, although I'm a lover of satire, Moore's half-satirical and moralistic remarks about "sexual opportunities" and so on strike me as ugly, stupid and fanciful.

Ted Davis
Vancouver, B.C.

No Raw Sex

Recently we subscribed to your publication because of advertising material that you sent to us indicating that yours was a publication devoted to backpacking and mountaineering. We were disappointed with the first issue but decided to wait until the second issue arrived before we made any judgement.

Today the second issue arrived! We wanted to read about the mountains and the beauty of nature—not raw sex in a tent and filthy language.

Please cancel our subscription and refund our money to us.

Mr. and Mrs. Jack Lewis
Bakersfield, California

Beer Ain't Great

I thought the "World of Anna Creek," glorifying the Big Beer Sloshing Syndrome, was unhealthy and unnaturally macho—which mentioned the tortures a person dying of beer alcoholism goes through—am currently watching it happen—11 pints of fluid withdrawn from the lungs in the last two weeks—all over body rash—the tortures of the damned from itching and burning—the best and very expensive medical treatment—can't drink that old horse-urine anymore but he sure is paying for every bottle ever sucked down!

Please don't use my name. You can just sign me tired of hearing a lot of complaints (when it was all self-induced). Beer really ain't great!

"Tart"
Pennsylvania

Joy to the World

"Airplane," was wonderful. CoG may be less cruel and indifferent than she's cracked up to be. Such manna! Let hope run again throughout the land! Miracles happen yet! Gansa Hillsbery, whoever you are, you are a fine journalist. If it was all shuck and jive, please, please, never let us know.

John Krakauer
Boulder, Colorado

Sex-Crazed

Your magazine is a vulgar celebration of sex-crazed egomaniacs stimulating their bodies with hallucinogenic abandon to the outdoor elements. Keep up the good work.

David Marsten
St. Helena, California

Garbage Lady

I have seen garbage in print before, many times, but never anything quite so confused, incoherent, hysterical and amateurish as a thing called "Lady with a Baby" in your April, issue. Whatever the man is trying to say, one thing is clear: He'd be a lot more at home back in New York City, in a completely industrial environment. Nor is he, as he seems to claim, a libertarian. Anyone who is against wilderness is against human freedom. I urge him to get a job like Tucker, as another PR con man for Con Edison—or Mobil Oil.

Neil C. Scott
Phoenix, Arizona

Superb Lady

We give our SUPERB for "Lady with a Baby!!!" I read it aloud to Wilk one evening, and we laughed, nodded our heads and related to so much of it.

Thanks for that one.

Tulasi and Wilk Wilkinson
Sandpoint, Idaho

It's Your Fault—A Response to Editor's Introduction

I deny having anything to do with this book for the following reasons:

1) Imagine the number of people who are going to be offended by some or all of this crash-test dummy of a book. While readers may feel free to call you (970-468-1887) at home, I don't want calls from the likes of the Bible Thumpers Good Taste Committee, Ecoweenies for Anarchy or the Intergalactic Sisterhood of Women Without a Sense of Humor.

2) You are the editor of this tree-killer and need to take responsibility for your actions and those of Mr. Moore and Mr. Guenin. I believe the judge said this during your most recent paternity trial, but let me repeat her words: This is your child.

3) There is a long-cherished tradition of underachievement among book publishers. Should some fool actually read this wreck of a book and recommend it to a friend, and should we actually get some reviews and sell some copies—there would be certain of my profession who would accuse me of breaking this code of mediocrity and ban me from the society of publishers.

Bear
Tom's Tavern
February 14, 2002

APPENDIX B

Author Bios

- Edward Abbey ("Where's Tonto?"): Abbey authored some of the most famous books in Western literary history, including *The Monkey Wrench Gang*, *Hayduke Lives!* and *Desert Solitaire*. He passed away in 1989.

- Bruce Berger ("There Was a River"): Berger's books on nature and culture in the desert, include *The Telling Distance*, winner of the 1990 Western States Book Award, and *Almost an Island*. This article was expanded into a book of the same name. He lives in Aspen, Colorado, and La Paz, Baja California Sur.

- Jeremy Bernstein ("The Guardian of Sleep"): Bernstein is both a professional physicist and writer. He was on the staff of *The New Yorker* for 30 years, where he wrote about science, mountains and mountain travel. His book, *The Wildest Dreams of Kew*, originally written for *The New Yorker*, was the first modern profile of Nepal. His biography of Einstein, also first written for *The New Yorker*, was nominated for a National Book Award. He splits his time between Aspen, Colorado, and Manhattan.

- Barb Bomier (Truck Painting for "Wild Red Dharma Pick-up Truck") is an oil painter living in Red Cliff, Colorado. Her work is displayed at the Cogswell Gallery in Vail, Colorado, and the Gallery Oscar in Ketchum, Idaho.

- Charles Bowden ("Flat Mountain"): Bowden, author of (most recently) *Blues for Cannibals*, *Blood Orchid: An Unnatural History of America*, *Blue Desert* and *Juarez: The Laboratory of Our Future*, is a journalist whose work appears regularly in *Harper's*, *Esquire*, *GQ* and other national publications. He lives in Tucson, where he has been torturing himself for six years with a book about a border murder.

- Edgar Boyles (Photo, "Hallucinations") is a photographer based in Colorado's Roaring Fork Valley.

- Tim Cahill (Foreword) is one of best known outdoor/adventure writers in the country, and author of such ground-breaking books as *Jaguars Ripped My Flesh* and *Road Fever*.

He wrote the "Out There" column for *Outside* magazine for years. His work has appeared in *Rolling Stone*, *Men's Journal* and many other luminarious publications. He lives in Livingstone, Montana.

● Karen Recknagel (Chamberlain) ("Climbing the Walls in Berkeley"): A poet since childhood, Ms. Chamberlain has worked as a senior writer and associate producer for the PBS nature series, "*Wild America*," and as a founding director of the Aspen Writers' Foundation. Her award-winning poetry has been widely anthologized, and her prose has appeared in numerous publications. She teaches creative writing workshops and is currently *Mountain Gazette*'s poetry editor. She lives near Glenwood Springs, Colorado.

● Robert Chamberlain (Various Photos): Chamberlain began his photo career more than four decades ago, when one of his images became his first published work: a cover of a *SKI* Magazine annual. A degree in philosophy from the University of Colorado and studies in photography and filmmaking at San Francisco State University deepened his interest in making documentary images in a variety of subject areas—and in turning them into black-and-white prints that continue to be published, exhibited and collected. He has one book out, *The Photography of Bob Chamberlain*. He lives in Colorado's Roaring Fork Valley.

● Geoffrey Childs ("Lobster Fishing in America"): Childs is the author of *Stone Palaces* (Mountaineers Books), a highly acclaimed collection of mountaineering tales. He lives in Mazama, Washington, with his wife Diane and son Toby. He manages a small resort hotel and guides.

● John Cleare (photo of Royal Robbins in "A Dream of White Horses"): Cleare lives somewhere in England.

● Barry Corbet ("Hallucinations"): Corbet is the author of *Options: Spinal Cord Injury and the Future*, now in its ninth printing. He was the editor of *New Mobility* for nine years. He lives outside Golden, Colorado.

● R. Crumb (Illustrations for "My Friend Ed" and George Sibley's review of *The Monkey Wrench Gang*) is probably the best-known counter-culture artist in history. He was the creator of Zap Comix, as well as the characters Mr. Natural and Fritz the Cat. Mr. Crumb's many books include the *R. Crumb Coffee Table Art Book*, *Complete Crumb: Mr. Sixties* and *The Complete Crumb: R. Crumb vs. The Sisterhood*. The illustrations in this book are taken from the R. Crumb edition of *The Monkey Wrench Gang*, which was commissioned by Ken Sanders, owner of Ken Sanders Rare Books in Salt Lake City. Mr. Crumb lives in France.

● Dick Dorworth ("Coyote Song"): Long-time *Mountain Gazette* contributor Dick Dorworth was a coach for the U.S. Ski Team in 1970-71. He taught skiing from 1963 until 1994 in South America, Europe and North America, including Sun Valley, Keystone, Soldier Mountain, Bear Valley, Squaw Valley and Heavenly Valley. He was director of the Aspen Mountain Ski School from 1988 till 1992. He now lives in Ketchum, where he works as a journalist and climbing guide.

● M. John Fayhee ("Crossings"): Fayhee is the author of seven books, including *Up At Altitude: A Celebration of Life in the High Country*, *Along the Colorado Trail* and *A Colorado Winter*. He was a long-time contributing editor to *Backpacker* and has written for *Canoe*, *Adventure Travel*, *Sports Illustrated*, *Sierra*, *Walking*, *Lesbo Nuns in Bondage* and many others. He lives in Summit County, Colorado.

● Mark Fox (Photo, "Bitches in Heats"): Fox was a staff photographer for the *Summit Daily News* for more than 10 years. He currently works as a freelancer in his hometown of Frisco, Colorado.

● B. Frank ("Lucette K. Car Obituary"): Frank is a ne'er-do-(too)-well intentional misfit pursuing character development while collecting a life story amidst mountains, canyons and deserts of inland North America. When he catches his breath, he lives in Hesperus, Colorado.

● Gaylord Guenin ("For the Sport of It?"): Guenin was the editor of *Mountain Gazette* for three years. He was the editor of the *Aspen Illustrated News*, the managing editor of *Ski Racing* and publicity director for Bob Beattie's World Wide Ski Corporation. He is one of the authors of *Aspen, the Quiet Years*. He has been a columnist for the *Aspen Daily News* for eight years. He lives near Woody Creek, Colorado.

● Michael Holzmeister ("Scott Fly Rod Obituary"): Holzmeister is the editor of the *Lamar (Colorado) Daily News*. He has also worked as the managing editor of the *Gunnison (Colorado) Country Times* and Denver's *La Voz*. He would rather fish than attend an orgy.

● Ted Kerasote ("Mountain Towns"): Kerasote has contributed to more than 50 publications, including *Audubon*, *National Geographic Traveler*, *Outside* and *Sports Afield*, where he has written the Environment column since 1987. He is the author of three books—*Navigations*, *Bloodties* and *Heart of Home*. He edits the Pew Wilderness Center's annual anthology, *Return of the Wild*. He lives in Kelly, Wyoming.

● Ed LaChapelle (Photo, "Growing Up High") is one of the foremost avalanche experts in the country. He is the author of *ABC of Avalanche Study* and, most recently, *Secrets of the Snow: Visual Clues to Avalanche and Ski Conditions* (University of Washington Press). He lives in McCarthy, Alaska.

● Randy (David) LaChapelle ("Growing Up High"): La Chapelle's new book is *Navigating the Tides of Change: Stories From Science*. He lives in Juneau, Alaska.

● Cindy Kleh ("Bitches in Heats"): Kleh started snowboarding in 1986 at Arapahoe Basin on a swallow-tail Burton Performer with lace-up Sorel boots. Her first book, *Snowboarding Skills*, is scheduled to be published by Quintet in September 2002. She lives in Keystone, Colorado.

● Katie Lee ("The Ride"): Lee has dedicated her life to preserving wild rivers. She is the author of *All My Rivers Are Gone: A Journey of Discovery Through Glen Canyon*. She lives in central Arizona.

● Catherine Lutz (Photo, "Slouching Toward Simpletopia"): Lutz is a writer, editor and photographer living in Woody Creek, Colorado. She reports full time for the *Snowmass Sun* weekly.

● Fletcher Manley (Photo, "Where the Trees Walk") has been a skiing and outdoor recreation photographer since the early '60s and was a cinematographer for Warren Miller throughout much of the '70s and early '80s. He lives in Lancaster, New Hampshire, and still skis with a camera in his pack.

● Harvey Manning ("Where the Trees Walk"): Manning is a legend in the world of outdoor writing. The author of more than 25 books, including *Backpacking One Step at a Time*, *100 Classic Hikes in Washington* and *Hiking the Great Northwest*, Manning was the chairman of the committee of editors that produced the first edition of *Mountaineering: Freedom of the Hills*. He lives in Bellevue, Washington.

● Mike Moore ("Breaking Free of the Human Potential Movement"): Moore was the visionary behind *Mountain Gazette*, which he edited for five years, before moving on to *Outside* magazine. He lives in Chelsea, Vermont, where he owns and operates Steerforth Press.

● N.E.D. ("N.E.D.") was an early pioneer in experiments with various now-controlled substances known throughout the Rocky Mountain area. The substances were known, not N.E.D. He made a living stealing the paint off houses and selling used postholes ("Dig 'em up, Saw 'em Off" was his motto), until his untimely death a few years ago from congenitally hyperactive Isles of Langerhans. The substances may have been involved.

● John Nichols ("Gone Fishin'"): Nichols is the author of the famous New Mexico Trilogy, which includes *The Milagro Beanfield War*, *The Magic Journey* and *The Nirvana Blues*. His most recent books are *Voice of the Butterfly* (Chronicle) and *An American Child Supreme* (Milkweed Editions). He has lived in Taos, New Mexico, since 1969.

● Doug Peacock ("My Friend Ed"): Peacock is the real-life inspiration for the character G.W. Hayduke in Edward Abbey's *The Monkey Wrench Gang*. He is the author of *Baja* and *Grizzly Years*, which was named by *National Geographic Adventure* magazine as one of the Top-100 Outdoor Books of all time. He lives near Yellowstone National Park, where he works with war veterans and champions wild causes.

● John Peters ("The Impsons, Ed & Ma'am"): Peters is a physician living in Battle Mountain, Nevada. This story was excerpted from a book-in-progress titled *Memoirs of a Country Doc*.

● Rob Pudim ("Confessions of a Butterfly Chaser;" Various Illustrations): Pudim is a full-time illustrator whose work regularly appears in more than 50 newspapers throughout the West. He lives in the mountains west of Boulder, where he hikes, rides horses and stalks wild butterflies.

● Royal Robbins ("A Dream of White Horses"): Robbins is one of the most famous climbers in American history. In 1957, he made the first ascent of the Northwest Face of Half Dome and has made many first ascents of the three great faces of El Capitan in Yosemite Valley. He is the author of *Basic Rockcraft* and *Advanced Rockcraft*, both considered seminal works in the world of climbing. He lives in Modesto, California.

● David Roberts ("Fear" and "Hanging Around"): Roberts has written for *Outside*, *Men's Journal* and *National Geographic Adventure*. He is the author of, among other books, *Once They Moved Like The Wind: Cochise, Geronimo and the Apache Wars* and *In Search of the Old Ones: Exploring the Anasazi World of the Southwest*. He lives in Cambridge, Massachusetts.

● Galen Rowell ("Alaska, Journey By Land" and Various Photos): One of the best-known photographers in the world, Galen Rowell's books include *In the Throne Room of the Mountain Gods* and *Mountain Light*. His images have appeared in many publications, including *National Geographic*, *Life* and *Outdoor Photographer*. He lives in Bishop, California.

● John Skow ("The South Side of the New England Soul"): Skow has written extensively for *Time*, *SKIING*, *Outside*, *Playboy* and other publications. He has lived in New London, New Hampshire, since 1964 and is only two months behind on his phone bill.

● George Sibley ("Slouching Toward Simpletopia" and "The Monkey Wrench Gang: A Review"): Sibley has likely produced more verbiage for *Mountain Gazette* than any other writer. He has lived in Colorado's upper Gunnison River Valley since the winter of 1966-67. He has written for a variety of publications, from *Crested Butte Magazine* to *Harper's*. He is currently an academic odd-jobber at Western State College in Gunnison.

● Judith Lacey Story ("Wild Red Dharma Pick-up Truck") is an acupuncturist whose interests include trucks, firearms, skiing and playing in the hills. She lives in Breckenridge, Colorado.

● Lito Tejada-Flores ("Crooked Road to the Far North"): Tejada-Flores was born at 13,000 feet in the Bolivian Andes and has spent much of his life in high places. He tried to substitute climbing and skiing for growing up and pursuing a real career, but along the way became a film maker, graphic designer and publisher. His latest book is *Breakthrough on the New Skis*. And his website, www.BreakthroughOnSkis.com, is an alternative on-line skiers' journal, featuring stories on ski travel, ski technique and impressionistic and personal ski writing of all sorts. He lives in Crestone, Colorado.

● Steve Wishart ("On the Frontier"): Wishart was a well-known personality in Aspen, Colorado, where he tended bar at the famous Hotel Jerome. He was once a member of the Aspen City Council. He passed away about three years ago.

● Greg Wright (Illustrations, "The Impsons, Ed & Ma'am" and "Gone Fishin'"): Wright lives and works in Breckenridge, Colorado.

Contributors we were unable to track down include Jack Aley, Tad Hall, Michael Charles Tobias, Frank Davidson, I. Herbert Gordon, David Westwood, Marc PoKempner and Alice Brown. If anyone knows the whereabouts of any of these people, please contact us, so we can contact them.

Bob Chamberlain Photo

DISCARD YOUR INHIBITIONS

subscribe to Mountain Gazette

Send money fast to: Mountain Gazette
5355 Montezuma Rd., Montezuma, CO 80435

```
Name _____
Address _____
City _____ State _____ Zip _____
        $25 for 12 issues
    & A FREE BUMPERSTICKER!
```

We will NOT - under any circumstance - sell our subscribers' names and addresses to the Devil or any of his kind, such as telemarketers & mail order companies.